11 - 19 PROGRESSION

endorsed for edexcel

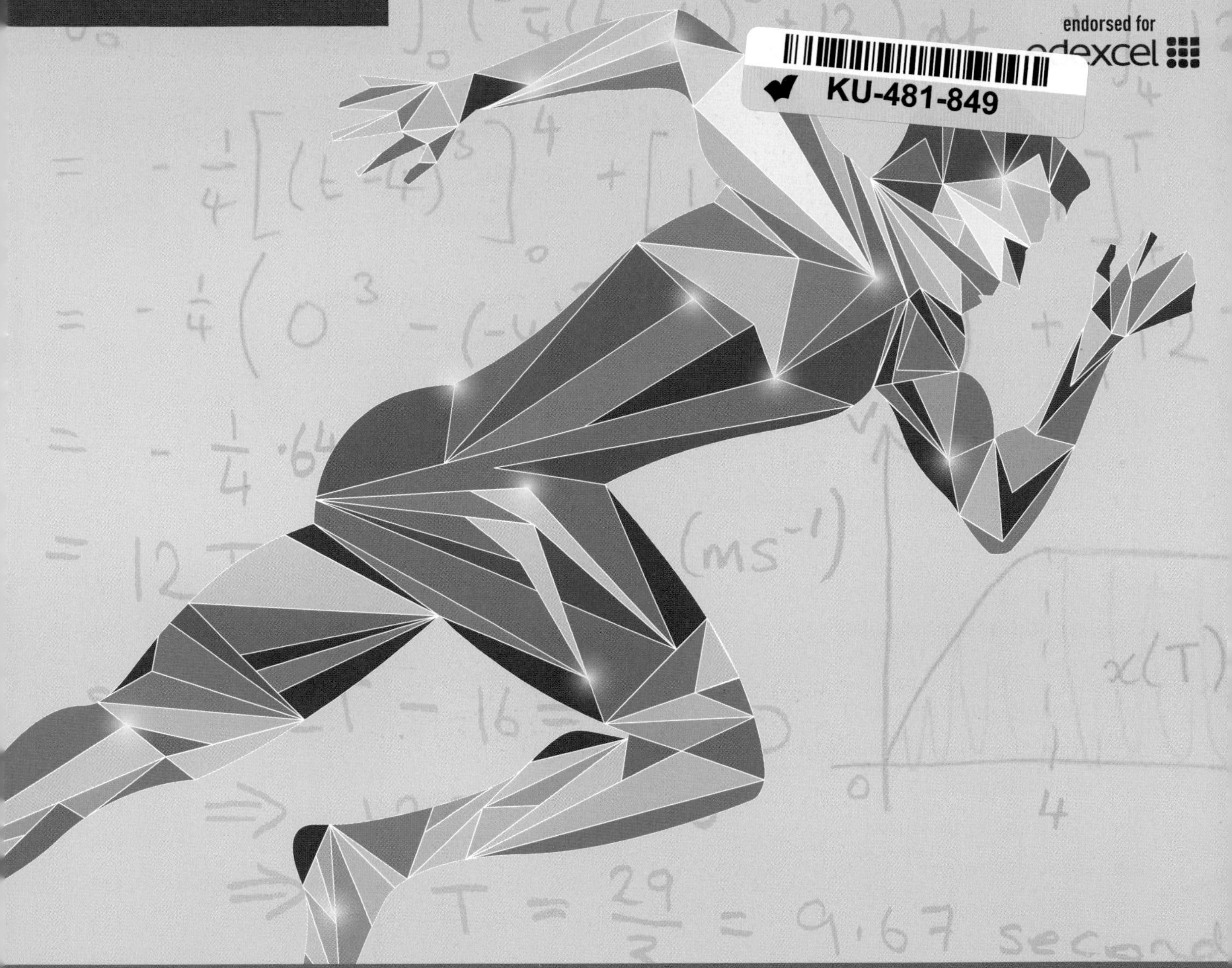

Edexcel A level Mathematics

Statistics and Mechanics

Year 2

Series Editor: Harry Smith

Authors: Greg Attwood, Ian Bettison, Alan Clegg, Gill Dyer, Jane Dyer, Keith Gallick, Susan Hooker, Michael Jennings, Mohammed Ladak, Jean Littlewood, Bronwen Moran, Su Nicholson, Laurence Pateman, Keith Pledger, Harry Smith

Contents

Overarching themes

The following three overarching themes have been fully integrated throughout the Pearson Edexcel AS and A level Mathematics series, so they can be applied alongside your learning and practice.

1. Mathematical argument, language and proof

- Rigorous and consistent approach throughout
- Notation boxes explain key mathematical language and symbols
- Dedicated sections on mathematical proof explain key principles and strategies
- Opportunities to critique arguments and justify methods

2. Mathematical problem solving

- Hundreds of problem-solving questions, fully integrated into the main exercises
- Problem-solving boxes provide tips and strategies
- Structured and unstructured questions to build confidence
- Challenge boxes provide extra stretch

The Mathematical Problem-solving cycle

specify the problem → collect information → process and represent information → interpret results

3. Mathematical modelling

- Dedicated modelling sections in relevant topics provide plenty of practice where you need it
- Examples and exercises include qualitative questions that allow you to interpret answers in the context of the model
- Dedicated chapter in Statistics & Mechanics Year 1/AS explains the principles of modelling in mechanics

Finding your way around the book

Access an online digital edition using the code at the front of the book.

Each chapter starts with a list of objectives

The *Prior knowledge check* helps make sure you are ready to start the chapter

The real world applications of the maths you are about to learn are highlighted at the start of the chapter with links to relevant questions in the chapter

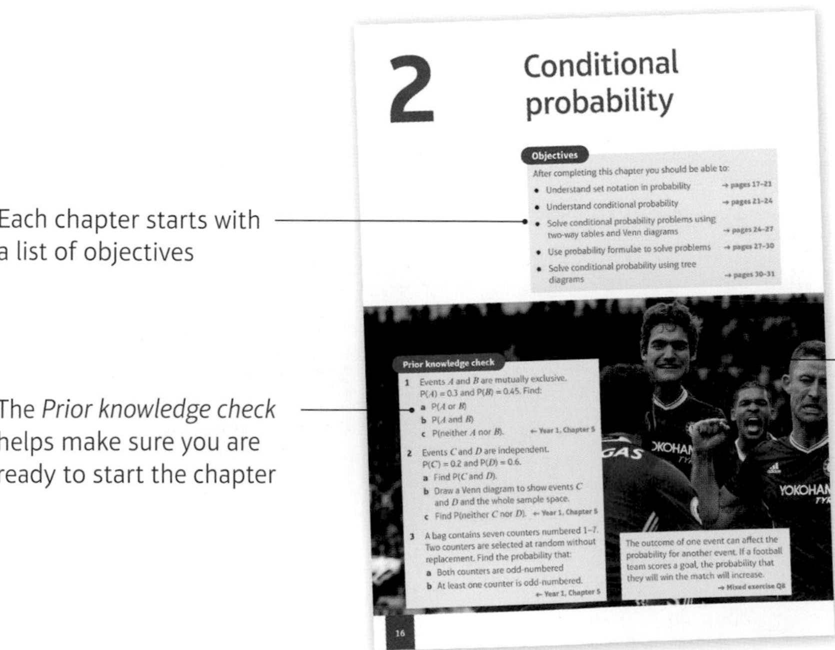

2 Conditional probability

Objectives

After completing this chapter you should be able to:

- Understand set notation in probability → pages 17–21
- Understand conditional probability → pages 21–24
- Solve conditional probability problems using two-way tables and Venn diagrams → pages 24–27
- Use probability formulae to solve problems → pages 27–30
- Solve conditional probability using tree diagrams → pages 30–31

Prior knowledge check

1 Events A and B are mutually exclusive. $P(A) = 0.3$ and $P(B) = 0.45$. Find:
 a $P(A \text{ or } B)$
 b $P(A \text{ and } B)$
 c $P(\text{neither } A \text{ nor } B)$. ← Year 1, Chapter 5

2 Events C and D are independent. $P(C) = 0.2$ and $P(D) = 0.6$.
 a Find $P(C \text{ and } D)$.
 b Draw a Venn diagram to show events C and D and the whole sample space.
 c Find $P(\text{neither } C \text{ nor } D)$. ← Year 1, Chapter 5

3 A bag contains seven counters numbered 1–7. Two counters are selected at random without replacement. Find the probability that:
 a Both counters are odd-numbered
 b At least one counter is odd-numbered. ← Year 1, Chapter 5

The outcome of one event can affect the probability for another event. If a football team scores a goal, the probability that they will win the match will increase. → Mixed exercise Q8

16

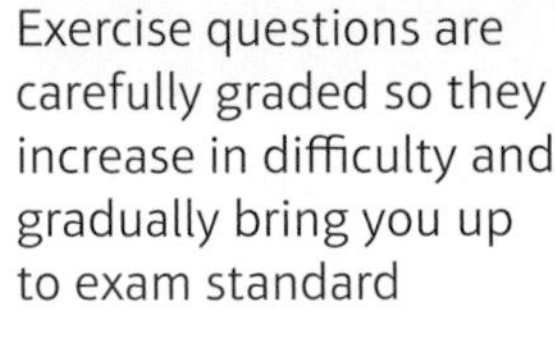

Exercise questions are carefully graded so they increase in difficulty and gradually bring you up to exam standard

Exercises are packed with exam-style questions to ensure you are ready for the exams

Problem-solving boxes provide hints, tips and strategies, and *Watch out* boxes highlight areas where students often lose marks in their exams

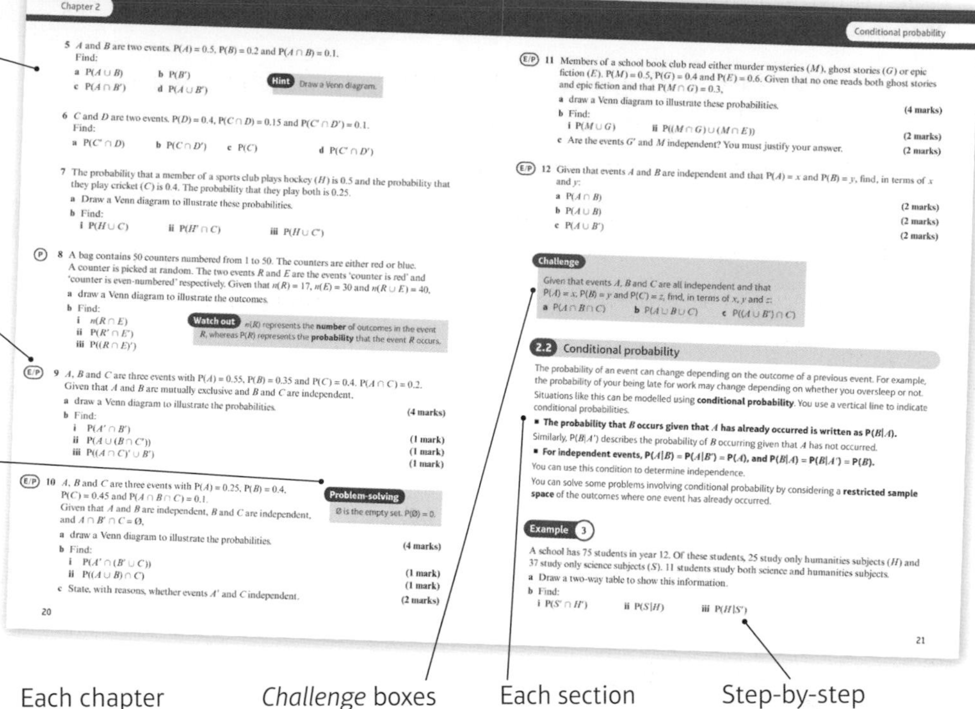

Chapter 2 | Conditional probability

5 A and B are two events. $P(A) = 0.5$, $P(B) = 0.2$ and $P(A \cap B) = 0.1$.
Find:
a $P(A \cup B)$ b $P(B')$
c $P(A \cap B')$ d $P(A \cup B')$

Hint: Draw a Venn diagram.

6 C and D are two events. $P(D) = 0.4$, $P(C \cap D) = 0.15$ and $P(C' \cap D') = 0.1$.
Find:
a $P(C' \cap D)$ b $P(C \cap D')$ c $P(C)$ d $P(C' \cap D')$

7 The probability that a member of a sports club plays hockey (H) is 0.5 and the probability that they play cricket (C) is 0.4. The probability that they play both is 0.25.
a Draw a Venn diagram to illustrate these probabilities.
b Find:
i $P(H \cup C)$ ii $P(H' \cap C)$ iii $P(H \cup C')$

(P) 8 A bag contains 50 counters numbered from 1 to 50. The counters are either red or blue. A counter is picked at random. The two events R and E are the events 'counter is red' and 'counter is even-numbered' respectively. Given that $n(R) = 17$, $n(E) = 30$ and $n(R \cup E) = 40$,
a draw a Venn diagram to illustrate the outcomes.
b Find:
i $n(R \cap E)$
ii $P(R' \cap E')$
iii $P((R \cap E)')$

Watch out: $n(R)$ represents the **number** of outcomes in the event R, whereas $P(R)$ represents the **probability** that the event R occurs.

(E/P) 9 A, B and C are three events with $P(A) = 0.55$, $P(B) = 0.35$ and $P(C) = 0.4$. $P(A \cap C) = 0.2$. Given that A and B are mutually exclusive and B and C are independent,
a draw a Venn diagram to illustrate the probabilities. (4 marks)
b Find:
i $P(A' \cap B')$ (1 mark)
ii $P(A \cup (B \cap C'))$ (1 mark)
iii $P((A \cap C)' \cup B')$ (1 mark)

(E/P) 10 A, B and C are three events with $P(A) = 0.25$, $P(B) = 0.4$, $P(C) = 0.45$ and $P(A \cap B \cap C) = 0.1$.
Given that A and B are independent, B and C are independent, and $A \cap B' \cap C = \varnothing$,
a draw a Venn diagram to illustrate the probabilities. (4 marks)
b Find:
i $P(A' \cap (B' \cup C))$ (1 mark)
ii $P((A \cup B) \cap C)$ (1 mark)
c State, with reasons, whether events A' and C independent. (2 marks)

Problem-solving: $\varnothing$ is the empty set. $P(\varnothing) = 0$.

20

(E/P) 11 Members of a school book club read either murder mysteries (M), ghost stories (G) or epic fiction (E). $P(M) = 0.5$, $P(G) = 0.4$ and $P(E) = 0.6$. Given that no one reads both ghost stories and epic fiction and that $P(M \cap G) = 0.3$,
a draw a Venn diagram to illustrate these probabilities. (4 marks)
b Find:
i $P(M \cup G)$ ii $P((M \cap G) \cup (M \cap E))$ (2 marks)
c Are the events G' and M independent? You must justify your answer. (2 marks)

(E/P) 12 Given that events A and B are independent and that $P(A) = x$ and $P(B) = y$, find, in terms of x and y:
a $P(A \cap B)$ (2 marks)
b $P(A \cup B)$ (2 marks)
c $P(A \cup B')$ (2 marks)

Challenge
Given that events A, B and C are all independent and that $P(A) = x$, $P(B) = y$ and $P(C) = z$, find, in terms of x, y and z:
a $P(A \cap B \cap C)$ b $P(A \cup B \cup C)$ c $P((A \cup B') \cap C)$

2.2 Conditional probability

The probability of an event can change depending on the outcome of a previous event. For example, the probability of your being late for work may change depending on whether you oversleep or not. Situations like this can be modelled using **conditional probability**. You use a vertical line to indicate conditional probabilities.

- **The probability that B occurs given that A has already occurred is written as $P(B|A)$.**

Similarly, $P(B|A')$ describes the probability of B occurring given that A has not occurred.

- **For independent events, $P(A|B) = P(A|B') = P(A)$, and $P(B|A) = P(B|A') = P(B)$.**

You can use this condition to determine independence.

You can solve some problems involving conditional probability by considering a **restricted sample space** of the outcomes where one event has already occurred.

Example 3

A school has 75 students in year 12. Of these students, 25 study only humanities subjects (H) and 37 study only science subjects (S). 11 students study both science and humanities subjects.
a Draw a two-way table to show this information.
b Find:
i $P(S' \cap H')$ ii $P(S|H)$ iii $P(H|S')$

21

Exam-style questions are flagged with (E)

Problem-solving questions are flagged with (P)

Each chapter ends with a *Mixed exercise* and a *Summary of key points*

Challenge boxes give you a chance to tackle some more difficult questions

Each section begins with explanation and key learning points

Step-by-step worked examples focus on the key types of questions you'll need to tackle

Every few chapters a *Review exercise* helps you consolidate your learning with lots of exam-style questions

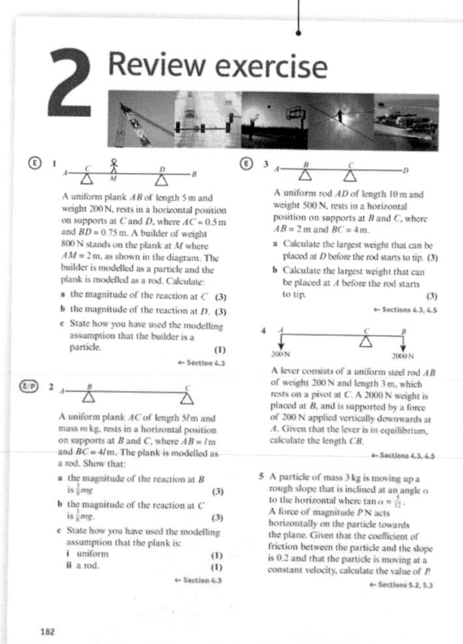

2 Review exercise

(E) 1 A uniform plank AB of length 5 m and weight 200 N, rests in a horizontal position on supports at C and D, where $AC = 0.5$ m and $BD = 0.75$ m. A builder of weight 800 N stands on the plank at M where $AM = 2$ m, as shown in the diagram. The builder is modelled as a particle and the plank is modelled as a rod. Calculate:
a the magnitude of the reaction at C (3)
b the magnitude of the reaction at D. (3)
c State how you have used the modelling assumption that the builder is a particle. (1)
← Section 4.3

(E/P) 2 A uniform plank AC of length $5l$ m and mass m kg, rests in a horizontal position on supports at B and C, where $AB = l$ m and $BC = 4l$ m. The plank is modelled as a rod. Show that:
a the magnitude of the reaction at B is $\frac{5}{8}mg$ (3)
b the magnitude of the reaction at C is $\frac{3}{8}mg$. (3)
c State how you have used the modelling assumption that the plank is:
i uniform (1)
ii a rod. (1)
← Section 4.3

(E) 3 A uniform rod AD of length 10 m and weight 500 N, rests in a horizontal position on supports at B and C, where $AB = 2$ m and $BC = 4$ m.
a Calculate the largest weight that can be placed at D before the rod starts to tip. (3)
b Calculate the largest weight that can be placed at A before the rod starts to tip. (3)
← Sections 4.3, 4.5

4 A lever consists of a uniform steel rod AB of weight 200 N and length 3 m, which rests on a pivot at C. A 2000 N weight is placed at B, and is supported by a force of 200 N applied vertically downwards at A. Given that the lever is in equilibrium, calculate the length CB.
← Sections 4.3, 4.5

5 A particle of mass 3 kg is moving up a rough slope that is inclined at an angle α to the horizontal where $\tan \alpha = \frac{5}{12}$. A force of magnitude P N acts horizontally on the particle towards the plane. Given that the coefficient of friction between the particle and the slope is 0.2 and that the particle is moving at a constant velocity, calculate the value of P.
← Sections 5.2, 5.3

182

Exam-style practice

Mathematics
A Level
Paper 3: Statistics and Mechanics

Time: 2 hours
You must have: Mathematical Formulae and Statistical Tables, Calculator

SECTION A: STATISTICS

1 An electrical engineer makes components for computer systems. She claims that the components last longer than 500 hours on 52% of occasions.
In a random sample of 40 of the components, X last longer than 500 hours.
a Find $P(X > 22)$. (1)
b Write down two conditions under which the normal approximation may be used as an approximation to the binomial distribution. (2)
A random sample of 250 components was taken and 120 lasted longer than 500 hours.
c Assuming the engineer's claim to be correct, use a normal approximation to find the probability that 120 or fewer components last longer than 500 hours. (3)
d Using your answer to part c, comment on the engineer's claim. (1)

2 $P(A) = 0.4$, $P(B) = 0.55$ and $P(C) = 0.26$.
Given that $P(A \cap B) = 0.2$, that events A and C are mutually exclusive and that events B and C are statistically independent,
a Draw a Venn diagram to illustrate events A, B and C. (5)
b Show that events A and B are not statistically independent. (2)
c Find $P(A|B')$. (2)
d Find $P(C|(A \cap B)')$. (2)

3 The daily mean air temperature, t °C, is recorded in Perth for the month of October.

t	$12 \leqslant t < 15$	$15 \leqslant t < 18$	$18 \leqslant t < 20$	$20 \leqslant t < 22$	$22 \leqslant t < 26$
f	2	6	11	7	5

a State, with a reason, whether t is a discrete or continuous variable. (1)
b Use your calculator to find estimates for the mean and standard deviation of the temperatures. (2)
c Give two reasons why a histogram could be used to display this data. (2)
d Use linear interpolation to find the 10th to 90th interpercentile range. (3)

187

Two A level practice papers at the back of the book help you prepare for the real thing

Extra online content

Whenever you see an *Online* box, it means that there is extra online content available to support you.

SolutionBank

SolutionBank provides a full worked solution for every question in the book.

Online Full worked solutions are available in SolutionBank.

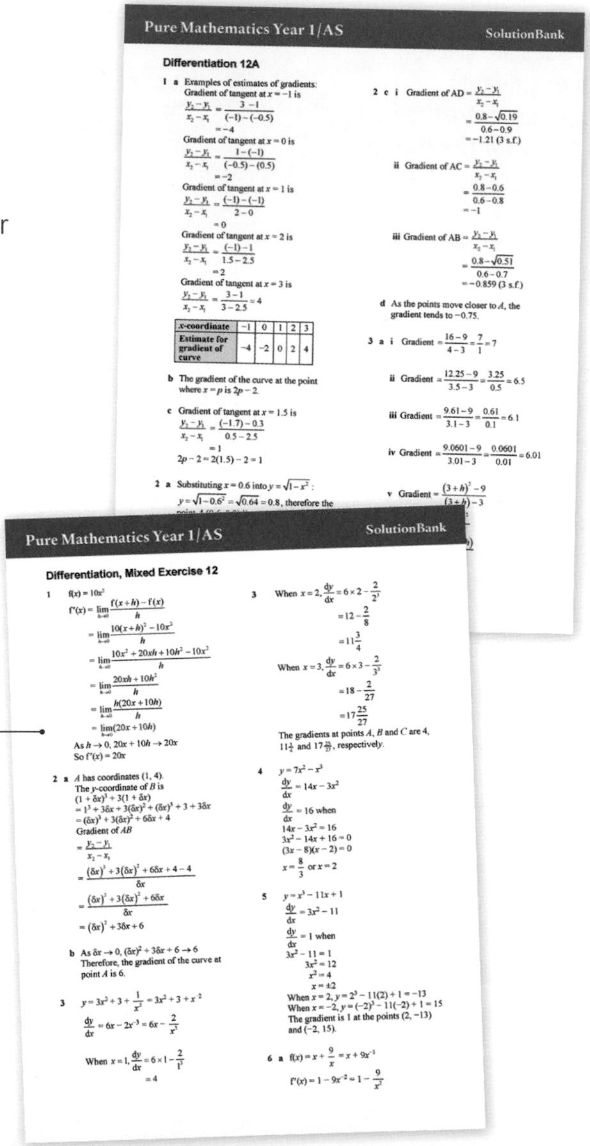

Download all the solutions as a PDF or quickly find the solution you need online

Access all the extra online content for free at:

www.pearsonschools.co.uk/sm2maths

You can also access the extra online content by scanning this QR code:

Use of technology

Explore topics in more detail, visualise problems and consolidate your understanding. Use pre-made GeoGebra activities or Casio resources for a graphic calculator.

Online Find the point of intersection graphically using technology.

GeoGebra

GeoGebra-powered interactives

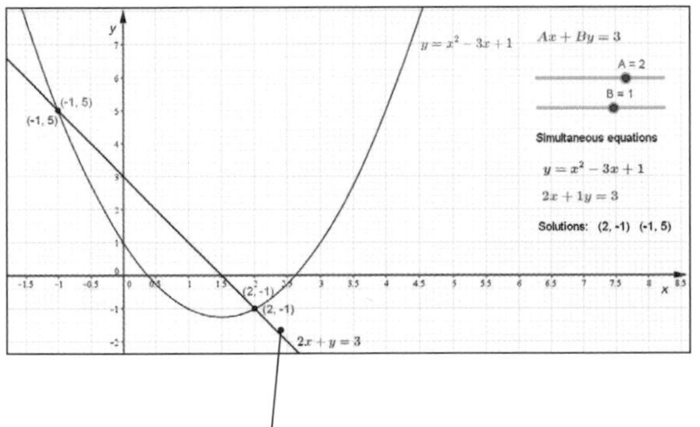

Interact with the maths you are learning using GeoGebra's easy-to-use tools

CASIO

Graphic calculator interactives

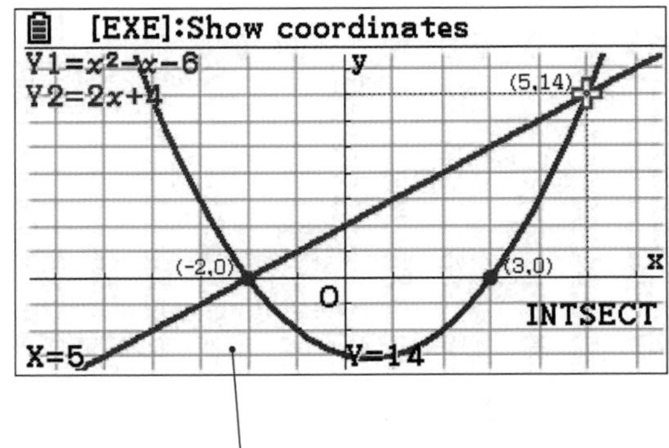

Explore the maths you are learning and gain confidence in using a graphic calculator

Calculator tutorials

Our helpful video tutorials will guide you through how to use your calculator in the exams. They cover both Casio's scientific and colour graphic calculators.

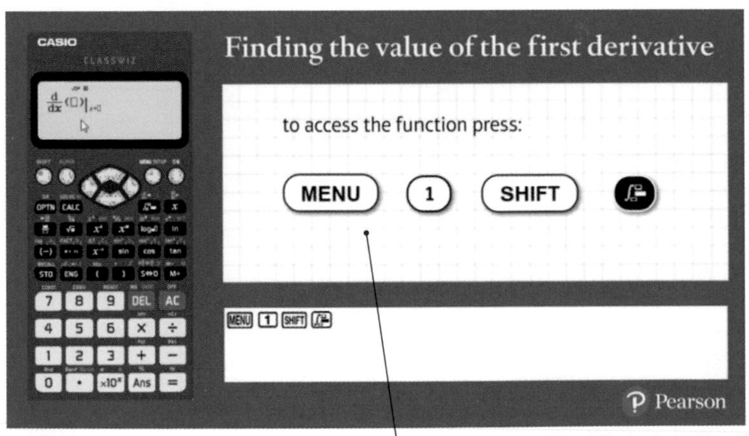

Step-by-step guide with audio instructions on exactly which buttons to press and what should appear on your calculator's screen

Online Work out each coefficient quickly using the ${}^nC_r$ and power functions on your calculator.

Published by Pearson Education Limited, 80 Strand, London WC2R 0RL.

www.pearsonschoolsandfecolleges.co.uk

Copies of official specifications for all Pearson qualifications may be found on the website: qualifications.pearson.com

Edited by Tech-Set Ltd, Gateshead
Typeset by Tech-Set Ltd, Gateshead

First published 2017

20 19 18 17
10 9 8 7 6 5 4

British Library Cataloguing in Publication Data
A catalogue record for this book is available from the British Library

ISBN 978 1 446 94407 3

Printed in the UK by Bell and Bain Ltd, Glasgow

Acknowledgements
The authors and publisher would like to thank the following for their kind permission to reproduce their photographs:

(Key: b-bottom; c-centre; l-left; r-right; t-top)

Alamy: Julia Gavin, 001, 065tl; Reuters, 016, 065tc; **Shutterstock**: Jeremy Richards, 037, 065tr; YuryZap, 070, 182l; Lane V. Erickson, 090, 182cl; Mark Herreid, 107, 182c; mbolina, 128, 182cr; Gary Blakeley, 159, 182

All other images © Pearson Education

Contains public sector information licensed under the Open Government Licence v3.0.

A note from the publisher
In order to ensure that this resource offers high-quality support for the associated Pearson qualification, it has been through a review process by the awarding body. This process confirms that this resource fully covers the teaching and learning content of the specification or part of a specification at which it is aimed. It also confirms that it demonstrates an appropriate balance between the development of subject skills, knowledge and understanding, in addition to preparation for assessment.

Endorsement does not cover any guidance on assessment activities or processes (e.g. practice questions or advice on how to answer assessment questions), included in the resource nor does it prescribe any particular approach to the teaching or delivery of a related course.

While the publishers have made every attempt to ensure that advice on the qualification and its assessment is accurate, the official specification and associated assessment guidance materials are the only authoritative source of information and should always be referred to for definitive guidance.

Pearson examiners have not contributed to any sections in this resource relevant to examination papers for which they have responsibility.

Examiners will not use endorsed resources as a source of material for any assessment set by Pearson.

Endorsement of a resource does not mean that the resource is required to achieve this Pearson qualification, nor does it mean that it is the only suitable material available to support the qualification, and any resource lists produced by the awarding body shall include this and other appropriate resources.

Pearson has robust editorial processes, including answer and fact checks, to ensure the accuracy of the content in this publication, and every effort is made to ensure this publication is free of errors. We are, however, only human, and occasionally errors do occur. Pearson is not liable for any misunderstandings that arise as a result of errors in this publication, but it is our priority to ensure that the content is accurate. If you spot an error, please do contact us at resourcescorrections@pearson.com so we can make sure it is corrected.

Regression, correlation and hypothesis testing 1

Objectives

After completing this chapter you should be able to:

- Understand exponential models in bivariate data → pages 2–5
- Use a change of variable to estimate coefficients in an exponential model → pages 2–5
- Understand and calculate the product moment correlation coefficient → pages 5–8
- Carry out a hypothesis test for zero correlation → pages 8–12

Ice-cream sellers typically find that they sell more ice cream the hotter the day. You can measure the strength of this correlation using the **product moment correlation coefficient**.

→ Mixed exercise, Q10

Prior knowledge check

1 Given that $y = 3 \times 2^x$

  a show that $\log y = A + Bx$, where A and B are constants to be found.

  b The straight-line graph of x against $\log y$ is plotted. Write down the gradient of the line and the intercept on the vertical axis. ← Pure Year 1, Chapter 12

2 The height, h cm, and handspan, s cm, of 20 students are recorded. The regression line of h on s is found to be $h = 22 + 11.3s$. Give an interpretation of the value 11.3 in this model. ← Year 1, Chapter 4

3 A single observation of $x = 32$ is taken from the random variable $X \sim B(40, p)$. Test, at the 1% significance level, $H_0: p = 0.6$ against $H_1: p > 0.6$. ← Year 1, Chapter 7

1.1 Exponential models

Regression lines can be used to model a **linear** relationship between two variables. Sometimes, experimental data does not fit a linear model, but still shows a clear pattern. You can use logarithms and coding to examine trends in non-linear data.

For data that can be modelled by a relationship of the form $y = ax^n$, you need to code the data using $Y = \log y$ and $X = \log x$ to obtain a linear relationship.

- **If $y = ax^n$ for constants a and n then $\log y = \log a + n \log x$**

For data that can be modelled by an **exponential** relationship of the form $y = ab^x$, you need to code the data using $Y = \log y$ and $X = x$ to obtain a linear relationship.

- **If $y = kb^x$ for constants k and b then $\log y = \log k + x \log b$**

Link Take logs of both sides and rearrange to convert the original form into the linear form.
← Pure Year 1, Section 14.6

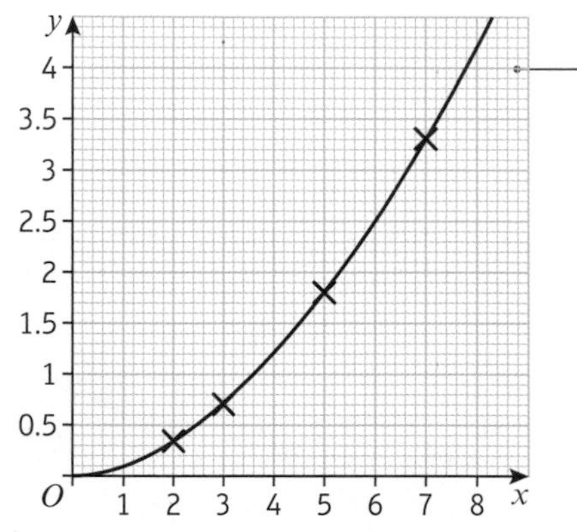

The points on this scatter graph satisfy the relationship $y = 0.1x^{1.8}$. This is in the form $y = ax^n$.

Plotting $\log x$ against $\log y$ gives a straight line.

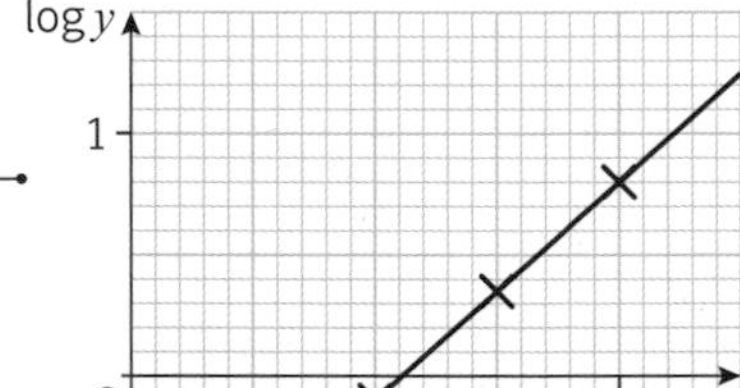

The gradient of the line is 1.8. This corresponds to the value of n in the non-linear relationship.

The y-intercept is at (0, −1). This corresponds to $\log a$ hence $a = 10^{-1} = 0.1$, as expected.

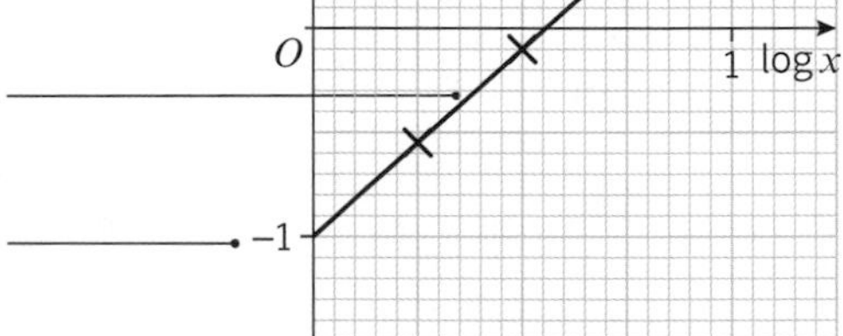

Example 1

The table shows some data collected on the temperature, in °C, of a colony of bacteria (t) and its growth rate (g).

Temperature, t (°C)	3	5	6	8	9	11
Growth rate, g	1.04	1.49	1.79	2.58	3.1	4.46

The data are coded using the changes of variable $x = t$ and $y = \log g$. The regression line of y on x is found to be $y = -0.2215 + 0.0792x$.

a Mika says that the constant −0.2215 in the regression line means that the colony is shrinking when the temperature is 0 °C. Explain why Mika is wrong.

b Given that the data can be modelled by an equation of the form $g = kb^t$ where k and b are constants, find the values of k and b.

a When $t = 0$, $x = 0$, so according to the model,

$y = -0.2215$

$\log g = -0.2215$

$g = 10^{-0.2215} = 0.600$ (3 s.f.).

This growth rate is positive: the colony is not shrinking.

Remember that the original data have been coded. Use the coding in reverse to find the corresponding value of g. You could also observe that a prediction based on $t = 0$ would be outside the range of the data so would be an example of **extrapolation**. ← Year 1, Section 4.2

b Substitute $x = t$ and $y = \log g$:

Use the change of variable to find an expression for $\log g$ in terms of t. You could also compare the equation of the regression line with $\log g = \log k + t \log b$. ← Pure Year 1, Section 14.6

$\log g = -0.2215 + 0.0792t$

$g = 10^{-0.2215 + 0.0792t}$

Remember log means log to the base 10. So $10^{\log g} = g$.

$g = 10^{-0.2215} \times (10^{0.0792})^t$

Use the laws of indices to write the expression in the form $g = kb^t$.

$g = 0.600 \times 1.20^t$ (both values given to 3 s.f.)

Exercise 1A

Online Explore the original and coded data graphically using technology.

1 Data are coded using $Y = \log y$ and $X = \log x$ to give a linear relationship. The equation of the regression line for the coded data is $Y = 1.2 + 0.4X$.

a State whether the relationship between y and x is of the form $y = ax^n$ or $y = kb^x$.

b Write down the relationship between y and x and find the values of the constants.

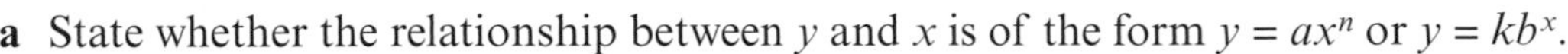

2 Data are coded using $Y = \log y$ and $X = x$ to give a linear relationship. The equation of the regression line for the coded data is $Y = 0.4 + 1.6X$.

a State whether the relationship between y and x is of the form $y = ax^n$ or $y = kb^x$.

b Write down the relationship between y and x and find the values of the constants.

(P) 3 The scatter diagram shows the relationship between two sets of coded data, X and Y, where $X = \log x$ and $Y = \log y$. The regression line of Y on X is shown, and passes through the points (0, 172) and (23, 109).

The relationship between the original data sets is modelled by an equation of the form $y = ax^n$. Find, correct to 3 decimal places, the values of a and n.

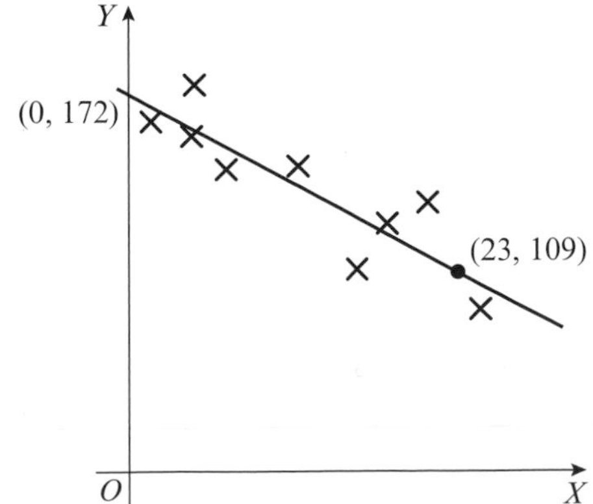

(P) 4 The size of a population of moles is recorded and the data are shown in the table. T is the time, in months, elapsed since the beginning of the study and P is the number of moles in the population.

T	2	3	5	7	8	9
P	72	86	125	179	214	257

a Plot a scatter diagram showing $\log P$ against T.

b Comment on the correlation between $\log P$ and T.

c State whether your answer to **b** supports the fact that the original data can be modelled by a relationship of the form $P = ab^T$.

d Approximate the values of a and b for this model.

e Give an interpretation of the value of b you calculated in part **d**.

Hint Think about what happens when the value of T increases by 1. When interpreting coefficients, refer in your answer to the context given in the question.

5 The time, t ms, needed for a computer algorithm to determine whether a number, n, is prime is recorded for different values of n. A scatter graph of t against n is drawn.

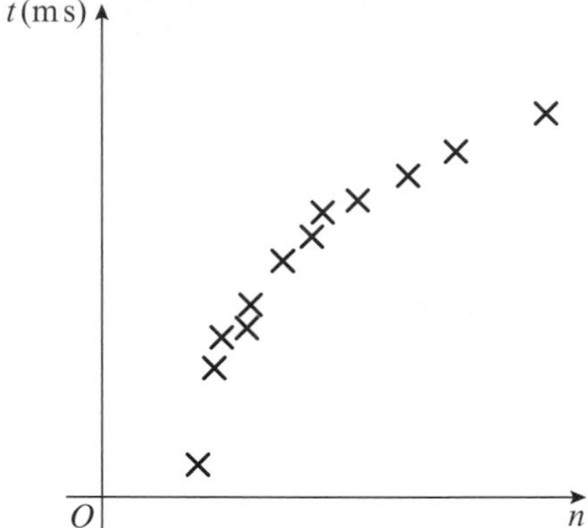

a Explain why a model of the form $t = a + bn$ is unlikely to fit these data.

The data are coded using the changes of variable $y = \log t$ and $x = \log n$. The regression line of y on x is found to be $y = -0.301 + 0.6x$.

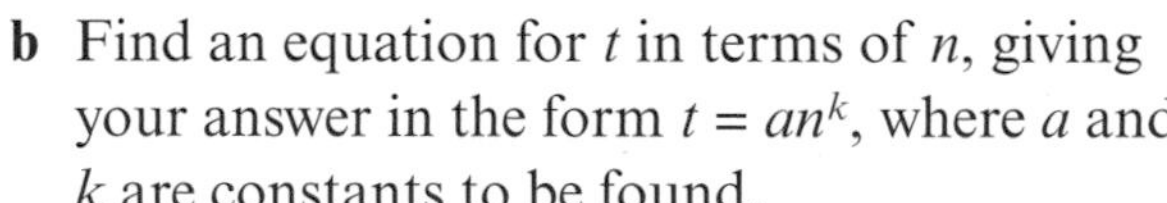

b Find an equation for t in terms of n, giving your answer in the form $t = an^k$, where a and k are constants to be found.

6 Data are collected on the number of units (c) of a catalyst added to a chemical process, and the rate of reaction (r).

The data are coded using $x = \log c$ and $y = \log r$. It is found that a linear relationship exists between x and y and that the equation of the regression line of y on x is $y = 1.31x - 0.41$.

Use this equation to determine an expression for r in terms of c.

7 The heights, h cm, and masses, m kg, of a sample of Galapagos penguins are recorded. The data are coded using $y = \log m$ and $x = \log h$ and it is found that a linear relationship exists between x and y. The equation of the regression line of y on x is $y = 0.0023 + 1.8x$.

Find an equation to describe the relationship between m and h, giving your answer in the form $m = ah^n$, where a and n are constants to be found.

E/P **8** The table shows some data collected on the temperature, t °C, of a colony of insect larvae and the growth rate, g, of the population.

Temp, t (°C)	13	17	21	25	26	28
Growth rate, g	5.37	8.44	13.29	20.91	23.42	29.38

The data are coded using the changes of variable $x = t$ and $y = \log g$. The regression line of y on x is found to be $y = 0.09 + 0.05x$.

a Given that the data can be modelled by an equation of the form $g = ab^t$ where a and b are constants, find the values of a and b. **(3 marks)**

b Give an interpretation of the constant b in this equation. **(1 mark)**

c Explain why this model is not reliable for estimating the growth rate of the population when the temperature is 35 °C. **(1 mark)**

Challenge

The table shows some data collected on the efficiency rating, E, of a new type of super-cooled engine when operating at a certain temperature, T.

Temp, T (°C)	1.2	1.5	2	3	4	6	8
Efficiency, E	9	5.5	3	1.4	0.8	0.4	0.2

It is thought that the relationship between E and t is of the form $E = aT^b$.

a By plotting an appropriate scatter diagram, verify that this relationship is valid for the data given.

b By drawing a suitable line on your scatter diagram and finding its equation, estimate the values of a and b.

c Give a reason why the model will not predict the efficiency of the engine when the temperature is 0 °C.

1.2 Measuring correlation

You can calculate quantitative measures for the strength and type of linear correlation between two variables. One of these measures is known as the **product moment correlation coefficient**.

- **The product moment correlation coefficient describes the linear correlation between two variables. It can take values between −1 and 1.**

If $r = 1$, there is perfect positive linear correlation.

If $r = -1$, there is perfect negative linear correlation.

The closer r is to −1 or 1, the stronger the negative or positive correlation, respectively.

If $r = 0$ (or is close to 0) there is no linear correlation. In this case there might still be a non-linear relationship between the variables.

Notation The product moment correlation coefficient, or PMCC, for a sample of data is denoted by the letter r.

Hint For $r = \pm1$, the points all lie on a straight line.

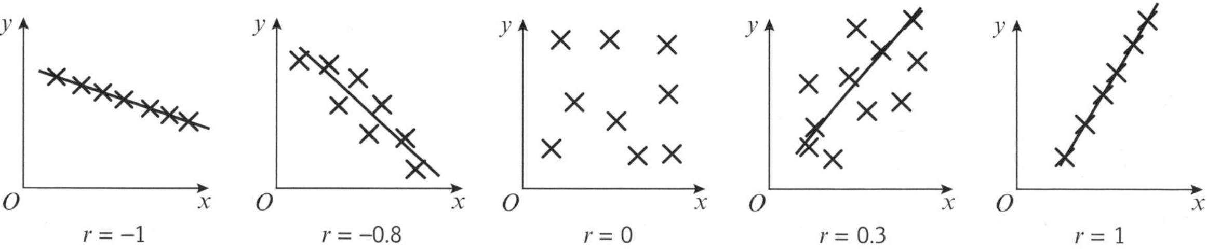

You need to know how to calculate the product moment correlation coefficient for bivariate data using your calculator.

Example 2

From the large data set, the daily mean windspeed, w knots, and the daily maximum gust, g knots, were recorded for the first 10 days in September in Hurn in 1987.

Day of month	1	2	3	4	5	6	7	8	9	10
w	4	4	8	7	12	12	3	4	7	10
g	13	12	19	23	33	37	10	n/a	n/a	23

a State the meaning of n/a in the table above.

b Calculate the product moment correlation coefficient for the remaining 8 days.

c With reference to your answer to part **b**, comment on the suitability of a linear regression model for these data.

a Data on daily maximum gust is not available for these days.

b $r = 0.9533$

c r is close to 1 so there is a strong positive correlation between daily mean windspeed and daily maximum gust. This means that the data points lie close to a straight line, so a linear regression model is suitable.

Online Use your calculator to calculate the PMCC.

r measures linear correlation. The closer r is to 1 or –1, the more closely a linear regression model will fit the data.

Exercise 1B

1 Suggest a value of r for each of these scatter diagrams:

a

b

c

2 The following table shows 10 observations from a bivariate data set.

v	50	70	60	82	45	35	110	70	35	30
m	140	200	180	210	120	100	200	180	120	60

a State what is measured by the product moment correlation coefficient.

b Use your calculator to find the value of the product moment correlation coefficient between v and m.

3 In a training scheme for young people, the average time taken for each age group to reach a certain level of proficiency was measured. The table below shows the data.

Age, x (years)	16	17	18	19	20	21	22	23	24	25
Average time, y (hours)	12	11	10	9	11	8	9	7	6	8

a Use your calculator to find the value of the product moment correlation coefficient for these data.

b Use your answer to part **a** to describe the correlation between the age and average time taken based on this sample.

E/P **4** The number of atoms of a radioactive substance, n, is measured at various times, t minutes after the start of an experiment. The table below shows the data.

Time, t	1	2	4	5	7
Atoms, n	231	41	17	7	2
log n					

The data is coded using $x = t$ and $y = \log n$.

a Copy and complete the table showing the values of $\log n$. **(2 marks)**

b Calculate the product moment correlation coefficient for the coded data. **(1 mark)**

c With reference to your answer to **b**, state whether an exponential model is a good fit for these data. **(2 marks)**

The equation of the regression line of y on x is found to be $y = 2.487 - 0.320x$.

d Find an expression for n in terms of t, giving your answer in the form $n = ab^t$, where a and b are constants to be found. **(3 marks)**

Hint For part **b** enter corresponding values of t and $\log n$ into your calculator.

E/P **5** The width, w cm, and the mass, m grams, of snowballs are measured. The table below shows the data.

Width, w	3	4	6	8	11
Mass, m	23	40	80	147	265
log w					
log m					

The data are coded using $x = \log w$ and $y = \log m$.

a Copy and complete the table showing the values of $\log w$ and $\log m$. **(3 marks)**

b Calculate the product moment correlation coefficient for the coded data: **(1 mark)**

c With reference to your answer to **b**, state whether a model in the form $y = kx^n$ where k and n are constants is a good fit for these data. **(2 marks)**

d Determine the values of k and n. **(3 marks)**

(E) 6 From the large data set, the daily mean air temperature, t °C, and the rainfall, f mm, were recorded for Perth on seven consecutive days in August 2015.

Temp, t	18.0	16.4	15.3	15.0	13.7	10.2	12.0
Rainfall, f	3.0	13.0	4.6	32.0	28.0	63.0	22.0

a Calculate the product moment correlation coefficient for these data. **(1 mark)**

b With reference to your answer to part **a**, comment on the suitability of a linear regression model for these data. **(2 marks)**

(E/P) 7 From the large data set, the daily total rainfall, x mm, and the daily total sunshine, y hours, were recorded for Camborne on seven consecutive days in May 2015.

Rainfall, x	2.2	tr	1.4	4.4	tr	0.2	0.6
Sunshine, y	5.2	7.7	5.6	0.3	5.1	0.1	8.9

a State the meaning of 'tr' in the table above. **(1 mark)**

b Calculate the product moment correlation coefficient for these 7 days, stating clearly how you deal with the entries marked 'tr'. **(2 marks)**

c With reference to your answer to part **b**, comment on the suitability of a linear regression model for these data. **(2 marks)**

Challenge

Data are recorded for two variables, x and y.

x	3.1	5.6	7.1	8.6	9.4	10.7
y	3.2	4.8	5.7	6.5	6.9	7.6

By calculating the product moment correlation coefficients for suitably coded values of x and y state, with reasons, whether these data are more closely modelled by a relationship of the form $y = ab^x$ or a relationship of the form $y = kx^n$, where a, b, k and n are constants.

1.3 Hypothesis testing for zero correlation

You can use a hypothesis test to determine whether the product moment correlation coefficient, r, for a particular sample indicates that there is likely to be a linear relationship within the whole population.

Notation r denotes the PMCC for a **sample**. ρ denotes the PMCC for a **whole population**. It is the Greek letter *rho*.

If you want to test for whether or not the population PMCC, ρ, is either greater than zero or less than zero you can use a **one-tailed test**:

For a one-tailed test use either:

- $\mathbf{H_0: \rho = 0, H_1: \rho > 0}$ **or**
- $\mathbf{H_0: \rho = 0, H_1: \rho < 0}$

If you want to test whether the population PMCC, ρ, is not equal to zero you need to use a **two-tailed test**:

For a two-tailed test use:

- $\mathbf{H_0 : \rho = 0, H_1 : \rho \neq 0}$

You can determine the critical region for r for your hypothesis test by using the table of critical values on page 192. This table will be given in the *Mathematical Formulae and Statistical Tables* booklet in your exam. The critical region depends on the **significance level** of the test and the **sample size**.

Product moment coefficient					
		Level			**Sample**
0.10	**0.05**	**0.025**	**0.01**	**0.005**	**size**
0.8000	0.9000	0.9500	0.9800	0.9900	4
0.6870	0.8054	0.8783	0.9343	0.9587	5
0.6084	0.7293	0.8114	0.8822	0.9172	6
0.5509	0.6694	0.7545	0.8329	0.8745	7
0.5067	0.6215	0.7067	0.7887	0.8343	8
0.4716	0.5822	0.6664	0.7498	0.7977	9

For a sample size of 8 you see from the table that the critical value of r to be significant at the 5% level on a one-tailed test is 0.6215. An observed value of r greater than 0.6215 from a sample of size 8 would provide sufficient evidence to reject the null hypothesis and conclude that $\rho > 0$. Similarly, an observed value of r less than −0.6215 would provide sufficient evidence to conclude that $\rho < 0$.

Example 3

A scientist takes 30 observations of the masses of two reactants in an experiment. She calculates a product moment correlation coefficient of $r = -0.45$.

The scientist believes there is no correlation between the masses of the two reactants. Test, at the 10% level of significance, the scientist's claim, stating your hypotheses clearly.

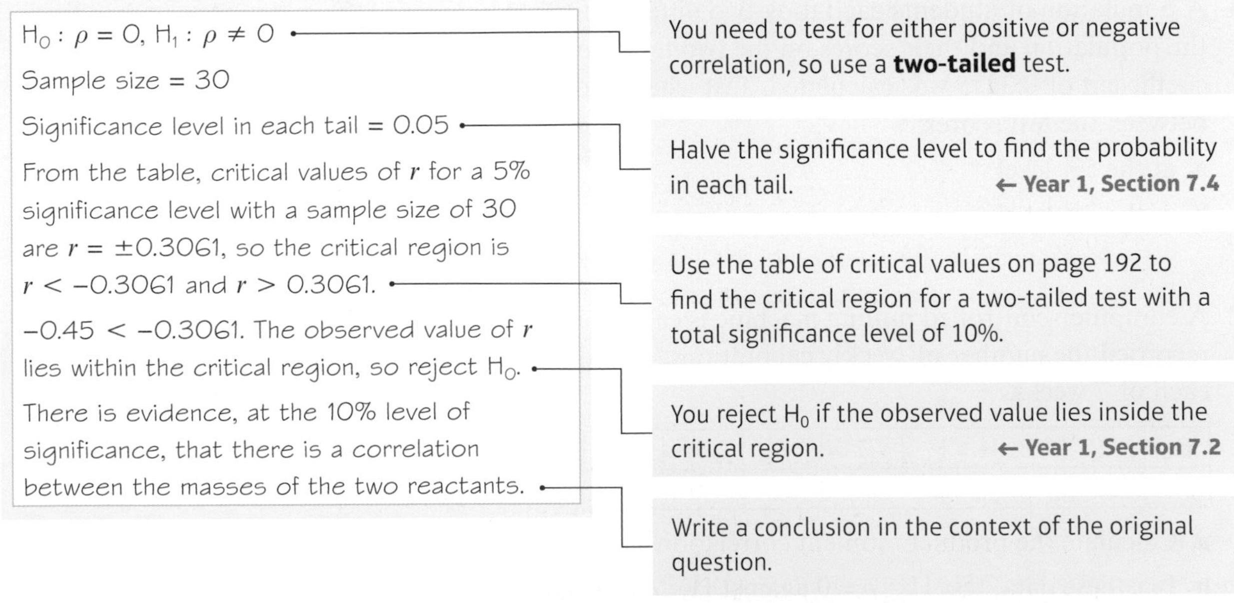

$H_0 : \rho = 0$, $H_1 : \rho \neq 0$

Sample size = 30

Significance level in each tail = 0.05

From the table, critical values of r for a 5% significance level with a sample size of 30 are $r = \pm 0.3061$, so the critical region is $r < -0.3061$ and $r > 0.3061$.

$-0.45 < -0.3061$. The observed value of r lies within the critical region, so reject H_0.

There is evidence, at the 10% level of significance, that there is a correlation between the masses of the two reactants.

You need to test for either positive or negative correlation, so use a **two-tailed** test.

Halve the significance level to find the probability in each tail. ← **Year 1, Section 7.4**

Use the table of critical values on page 192 to find the critical region for a two-tailed test with a total significance level of 10%.

You reject H_0 if the observed value lies inside the critical region. ← **Year 1, Section 7.2**

Write a conclusion in the context of the original question.

Example 4

The table from the large data set shows the daily maximum gust, x kn, and the daily maximum relative humidity, y%, in Leeming for a sample of eight days in May 2015.

x	31	28	38	37	18	17	21	29
y	99	94	87	80	80	89	84	86

a Find the product moment correlation coefficient for these data.

b Test, at the 10% level of significance, whether there is evidence of a positive correlation between daily maximum gust and daily maximum relative humidity. State your hypotheses clearly.

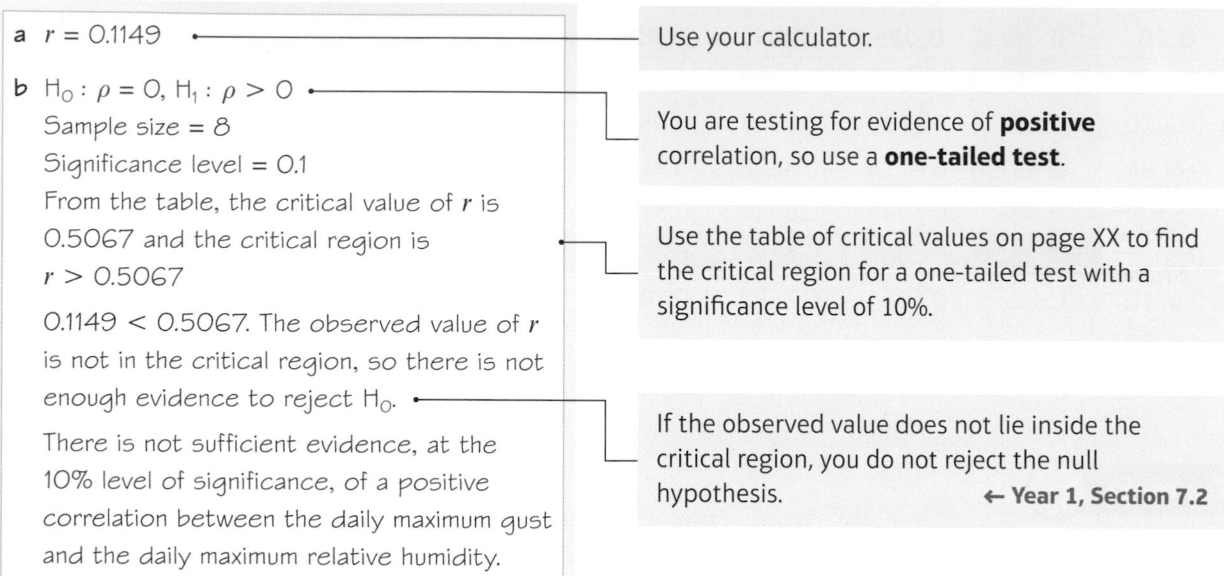

a $r = 0.1149$

Use your calculator.

b $H_0 : \rho = 0$, $H_1 : \rho > 0$

You are testing for evidence of **positive** correlation, so use a **one-tailed test**.

Sample size = 8

Significance level = 0.1

From the table, the critical value of r is 0.5067 and the critical region is $r > 0.5067$

Use the table of critical values on page XX to find the critical region for a one-tailed test with a significance level of 10%.

$0.1149 < 0.5067$. The observed value of r is not in the critical region, so there is not enough evidence to reject H_0.

If the observed value does not lie inside the critical region, you do not reject the null hypothesis. ← **Year 1, Section 7.2**

There is not sufficient evidence, at the 10% level of significance, of a positive correlation between the daily maximum gust and the daily maximum relative humidity.

Exercise 1C

1 A population of students each took two different tests. A sample of 40 students was taken from the population and their scores on the two tests were recorded. A product moment correlation coefficient of 0.3275 was calculated. Test whether or not this shows evidence of correlation between the test scores:

a at the 5% level

b at the 2% level.

Hint 'Evidence of correlation' could mean either positive or negative correlation, so you need to use a two-tailed test with $H_0 : \rho = 0$, $H_1 : \rho \neq 0$

2 A computer-controlled milling machine is calibrated between 1 and 7 times a week. A supervisor recorded the number of weekly calibrations, x, and the number of manufacturing errors, y, in each of 7 weeks.

x	2	3	1	7	6	5	4
y	53	55	62	19	35	40	41

a Calculate the product moment correlation coefficient for these data.

b For these data, test $H_0 : \rho = 0$ against $H_1 : \rho \neq 0$, using a 1% significance level.

(E/P) **3** **a** State what is measured by the product moment correlation coefficient. **(1 mark)**

Twelve students sat two Biology tests, one theoretical the other practical. Their marks are shown below.

Marks in theoretical test, *t*	5	9	7	11	20	4	6	17	12	10	15	16
Marks in practical test, *p*	6	8	9	13	20	9	8	17	14	8	17	18

b Find the product moment correlation coefficient for these data, correct to 3 significant figures. **(2 marks)**

A teacher claims that students who do well in their theoretical test also tend to do well in their practical test.

c Test this claim at a 0.05 significance level, stating your hypotheses clearly. **(3 marks)**

d Give an interpretation of the value 0.05 in your hypothesis test. **(1 mark)**

(E) **4** The following table shows the marks attained by 8 students in English and Mathematics tests.

Student	A	B	C	D	E	F	G	H
English	25	18	32	27	21	35	28	30
Mathematics	16	11	20	17	15	26	32	20

a Calculate the product moment correlation coefficient. **(1 mark)**

b Test, at the 5% significance level, whether these results show evidence of a linear relationship between English and Mathematics marks. State your hypotheses clearly. **(3 marks)**

5 A small company decided to import fine Chinese porcelain. They believed that in the long term this would prove to be an increasingly profitable arrangement with profits increasing proportionally to sales. Over the next 6 years their sales and profits were as shown in the table below.

Year	1994	1995	1996	1997	1998	1999
Sale in thousands	165	165	170	178	178	175
Profits in £1000	65	72	75	76	80	83

Using a 1% significance level, test to see if there is any evidence that the company's beliefs were correct, and that profit and sales were positively correlated.

(E) **6** A scientific researcher collects data on the amount of solvent in a solution and the rate of reaction. She calculates the product moment correlation coefficient between the two sets of data and finds it to be −0.43. Given that she collected data from 15 samples, test, at the 5% level of significance, the claim that there is a negative correlation between the amount of solvent and the rate of reaction. State your hypotheses clearly. **(3 marks)**

(P) **7** A safari ranger believes that there is a positive correlation between the amount of grass per square kilometre and the number of meerkats that graze there. He decides to carry out a hypothesis test to see if there is evidence for his claim. He takes a random sample of 10 equal-sized areas of grassland, records the amount of grass and the number of meerkats grazing in each, and finds that the correlation coefficient is 0.66.

Given that this result provided the ranger with sufficient evidence to reject his null hypothesis, suggest the least possible significance level for the ranger's test.

(P) **8** Data on the daily mean temperature and the daily total sunshine is taken from the large data set for Leuchars in May and June 1987. A meteorologist finds that the product moment correlation coefficient for these data is 0.715. Given that the researcher tests for positive correlation at the 2.5% level of significance, and concludes that the value is significant, find the smallest possible sample size.

(E) **9** An employee at a weather centre believes that there is a negative correlation between humidity and visibility. She takes a sample of data from Heathrow in August 1987.

Humidity (%)	92	93	91	82	91	100
Visibility (m)	2500	1500	2700	2900	2200	1000

© Crown Copyright Met Office

a Calculate the product moment correlation coefficient for these data. **(1 mark)**

b Test, at the 1% level of significance, the employee's claim. State your hypotheses clearly. **(3 marks)**

Mixed exercise 1

(E) **1** Conor uses a 3D printer to produce various pieces for a model. He records the time taken, t hours, to produce each piece, and its base area, $x\,\text{cm}^2$.

Base area, x (cm²)	1.1	1.3	1.9	2.2	2.5	3.7
Time, t (hours)	0.7	0.9	1.5	1.8	2.2	3.8

a Calculate the product moment correlation coefficient between $\log x$ and $\log t$. **(2 marks)**

b Use your answer to part **a** to explain why an equation of the form $t = ax^n$, where a and n are constants, is likely to be a good model for the relationship between x and t. **(1 mark)**

c The regression line of $\log t$ on $\log x$ is given as $\log t = -0.210 + 1.38 \log x$. Determine the values of the constants a and n in the equation given in part **b**. **(2 marks)**

(E) **2** The table shows some data collected on the temperature in °C of a chemical reaction (t) and the amount of dry residue produced (d grams).

Temperature, t (°C)	38	51	72	83	89	94
Dry residue, d (grams)	4.3	11.7	58.6	136.7	217.0	318.8

The data are coded using the changes of variable $x = t$ and $y = \log d$. The regression line of y on x is found to be $y = -0.635 + 0.0334x$.

a Given that the data can be modelled by an equation of the form $d = ab^t$ where a and b are constants, find the values of a and b. **(3 marks)**

b Explain why this model is not reliable for estimating the amount of dry residue produced when the temperature is 151 °C. **(1 mark)**

3 The product moment correlation coefficient for a person's age and his score on a memory test is −0.86. Interpret this value.

Ⓟ **4** Each of 10 cows was given an additive (x) every day for four weeks to see if it would improve the milk yield (y). At the beginning, the average milk yield per day was 4 gallons. The milk yield of each cow was measured on the last day of the four weeks. The data collected is shown in the table.

Cow	A	B	C	D	E	F	G	H	I	J
Additive, x (25 gm units)	1	2	3	4	5	6	7	8	9	10
Yield, y (gallons)	4.0	4.2	4.3	4.5	4.5	4.7	5.2	5.2	5.1	5.1

a By drawing a scatter diagram or otherwise, suggest the maximum amount of additive that should be given to the cows to maximise yield.

b Use your calculator to find the value of the product moment correlation coefficient for the first seven cows.

c Without further calculation, write down, with a reason, how the product moment correlation coefficient for all 10 cows would differ from your answer to **b**.

Ⓔ **5** The following table shows the engine size (c), in cubic centimetres, and the fuel consumption (f), in miles per gallon to the nearest mile, for 10 car models.

c (cm³)	1000	1200	1400	1500	1600	1800	2000	2200	2500	3000
f (mpg)	46	42	43	39	41	37	35	29	28	25

a Use your calculator to find the value of the product moment correlation coefficient between c and f. **(1 mark)**

b Interpret your answer to part **a**. **(2 marks)**

Ⓔ **6** As part of a survey in a particular profession, age, x years, and yearly salary, £y thousands, were recorded.

The values of x and y for a randomly selected sample of ten members of the profession are as follows:

x	30	52	38	48	56	44	41	25	32	27
y	22	38	40	34	35	32	28	27	29	41

a Calculate, to 3 decimal places, the product moment correlation coefficient between age and salary. **(1 mark)**

It is suggested that there is no correlation between age and salary.

b Test this suggestion at the 5% significance level, stating your null and alternative hypotheses clearly. **(3 marks)**

Ⓔ **7** A machine hire company kept records of the age, X months, and the maintenance costs, £Y, of one type of machine. The following table summarises the data for a random sample of 10 machines.

Machine	A	B	C	D	E	F	G	H	I	J
Age, X	63	12	34	81	51	14	45	74	24	89
Maintenance costs, Y	111	25	41	181	64	21	51	145	43	241

a Calculate, to 3 decimal places, the product moment correlation coefficient. **(1 mark)**

It is believed that there is a relationship between the age and maintenance cost of these machines.

b Using a 5% level of significance and quoting from the table of critical values, interpret your correlation coefficient. Use a two-tailed test and state clearly your null and alternative hypotheses. **(3 marks)**

Ⓔ **8** The data below show the height above sea level, x metres, and the temperature, y °C, at 7.00 a.m., on the same day in summer at nine places in Europe.

Height, x (m)	1400	400	280	790	390	590	540	1250	680
Temperature, y (°C)	6	15	18	10	16	14	13	7	13

The product moment correlation coefficient is −0.975. Use this value to test for negative correlation at the 5% significance level. Interpret your result in context. **(3 marks)**

Ⓔ **9** The ages, in months, and the weights, in kg, of a random sample of nine babies are shown in the table below.

Baby	A	B	C	D	E	F	G	H	I
Age, x	1	2	2	3	3	3	4	4	5
Weight, y	4.4	5.2	5.8	6.4	6.7	7.2	7.6	7.9	8.4

The product moment correlation coefficient between weight and age for these babies was found to be 0.972. By testing for positive correlation at the 5% significance level interpret this value. **(3 marks)**

Ⓔ **10** An ice-cream seller believes that there is a positive correlation between the amount of sunshine and sales of ice cream. He collects data on six days during June 2015 at his 'pitch' in Camborne:

Sunshine (hours)	4.2	7.9	13.8	8.7	6.2	0.7
Ice-cream sales (£100s)	7.0	8.3	12.4	8.1	7.9	6.2

a Calculate the product moment correlation coefficient for these data. **(1 mark)**

b Carry out a hypothesis test to determine, at the 5% level, if there is significant evidence in support of the ice-cream seller's belief. State your hypotheses clearly. **(3 marks)**

Ⓔ **11** A meteorologist believes that there is a positive correlation between daily mean windspeed and daily maximum gust. She collects data from the large data set for 5 days during August 2015 in the town of Hurn.

Mean windspeed (knots)	4	7	7	8	5
Daily maximum gust (knots)	14	22	18	20	17

By calculating the product moment correlation coefficient for these data, test at the 5% level of significance whether there is evidence to support the meteorologist's claim. State your hypotheses clearly. **(4 marks)**

(E) **12** The table shows data from the large data set on the daily mean air temperature and the daily mean pressure during May and June 2015 in Beijing.

Temperature (°C)	17.5	18.5	18.0	24.6	22.2	23.1	27.3
Pressure (hPa)	1010	1011	1012	997	1009	998	1002

Test at the 2.5% level of significance the claim that there is negative correlation between the daily mean air temperature and the daily mean pressure. State your hypotheses clearly.

(4 marks)

Large data set

You will need access to the large data set and spreadsheet software to answer these questions.

1 a Take a random sample of size 20 from the data for Heathrow in 2015, and record the daily mean air temperature and daily total rainfall.

b Calculate the product moment correlation coefficient between these variables for your sample.

c Test, at the 5% level of significance, the claim that there is a correlation between the daily mean air temperature and the daily total rainfall.

2 a State with a reason whether you would expect to find a relationship between daily mean total cloud cover and daily mean visibility.

b Use a random sample from the large data set to test for this relationship. You should state clearly:

- Your sample size and location
- Your sampling method
- The hypotheses and significance level for your test
- A conclusion in the context of the question

Hint You might be able to use the **Correl** or **CorrelationCoefficient** commands in your spreadsheet software to calculate the PMCC.

Summary of key points

1 If $y = ax^n$ for constants a and n then $\log y = \log a + n \log x$

2 If $y = kb^x$ for constants k and b then $\log y = \log k + x \log b$

3 The **product moment correlation coefficient** describes the linear correlation between two variables. It can take values between −1 and 1.

4 For a one-tailed test use either:

- $H_0: \rho = 0, H_1: \rho > 0$ or
- $H_0: \rho = 0, H_1: \rho < 0$

For a two-tailed test use:

- $H_0: \rho = 0, H_1: \rho \neq 0$

2 Conditional probability

Objectives

After completing this chapter you should be able to:

- Understand set notation in probability → pages 17–21
- Understand conditional probability → pages 21–24
- Solve conditional probability problems using two-way tables and Venn diagrams → pages 24–27
- Use probability formulae to solve problems → pages 27–30
- Solve conditional probability using tree diagrams → pages 30–31

Prior knowledge check

1 Events A and B are mutually exclusive. $P(A) = 0.3$ and $P(B) = 0.45$. Find:
  a P(A or B)
  b P(A and B)
  c P(neither A nor B). ← Year 1, Chapter 5

2 Events C and D are independent. $P(C) = 0.2$ and $P(D) = 0.6$.
  a Find P(C and D).
  b Draw a Venn diagram to show events C and D and the whole sample space.
  c Find P(neither C nor D). ← Year 1, Chapter 5

3 A bag contains seven counters numbered 1–7. Two counters are selected at random without replacement. Find the probability that:
  a Both counters are odd-numbered
  b At least one counter is odd-numbered.

← Year 1, Chapter 5

The outcome of one event can affect the probability for another event. If a football team scores a goal, the probability that they will win the match will increase.

→ Mixed exercise Q8

2.1 Set notation

You can use **set notation** to describe events within a sample space. This can help you abbreviate probability statements.

For example:

■ **The event A and B can be written as $A \cap B$. The '$\cap$' symbol is the symbol for intersection.**

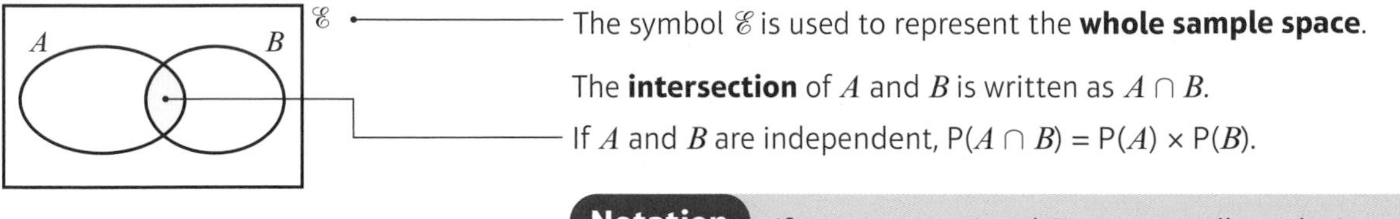

The symbol $\mathscr{E}$ is used to represent the **whole sample space**.

The **intersection** of A and B is written as $A \cap B$.

If A and B are independent, $P(A \cap B) = P(A) \times P(B)$.

Notation If two events, A and B, are mutually exclusive, then their intersection is the **empty set**, $\varnothing$. You can write $A \cap B = \varnothing$.

■ **The event A or B can be written as $A \cup B$. The '$\cup$' symbol is the symbol for union.**

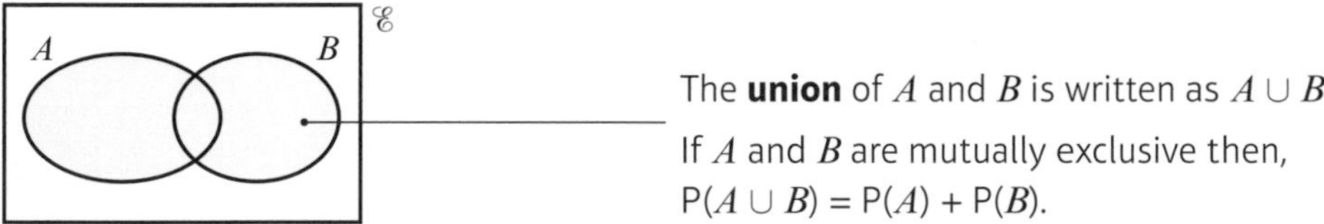

The **union** of A and B is written as $A \cup B$.

If A and B are mutually exclusive then,
$P(A \cup B) = P(A) + P(B)$.

■ **The event not A can be written as A'. This is also called the complement of A.**

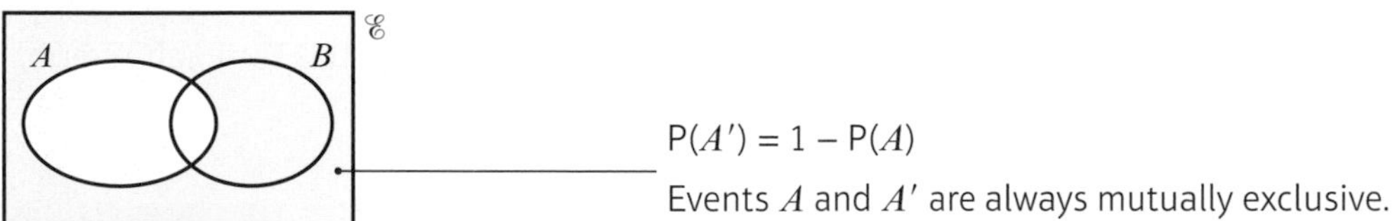

$P(A') = 1 - P(A)$

Events A and A' are always mutually exclusive.

Online Explore set notation on a Venn diagram using GeoGebra.

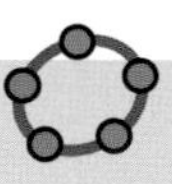

Example 1

A card is selected at random from a pack of 52 playing cards. Let A be the event that the card is an ace and D the event that the card is a diamond. Find:

a $P(A \cap D)$ **b** $P(A \cup D)$ **c** $P(A')$ **d** $P(A' \cap D)$

Draw a Venn diagram:

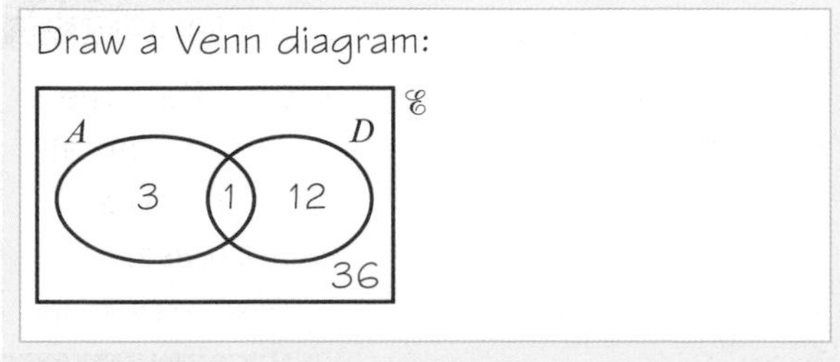

Notation Venn diagrams can show either probabilities or the number of outcomes in each event.
$n(A)$ is the notation used to indicate the number of outcomes. For example there are four aces so $n(A) = 4$ and there is one ace of diamonds so $n(A \cap D) = 1$.

a $A \cap D$ is the event 'the card chosen is the ace of diamonds'.

$P(A \cap D) = \frac{1}{52}$

There is one outcome in $A \cap D$ and 52 outcomes in $\mathscr{E}$ so probability is $\frac{1}{52}$.

b $A \cup D$ is the event 'the card chosen is an ace or a diamond or both'.

$P(A \cup D) = \frac{16}{52} = \frac{4}{13}$

$n(A \cup D) = 3 + 12 + 1 = 16$

c A' is the event 'the card chosen is not an ace'.

$P(A') = \frac{48}{52} = \frac{12}{13}$

d $A' \cap D$ is the event 'the card chosen is not an ace and is a diamond'.

$P(A' \cap D) = \frac{12}{52} = \frac{3}{13}$

This is the set of all outcomes that are not in A but **are** in D.

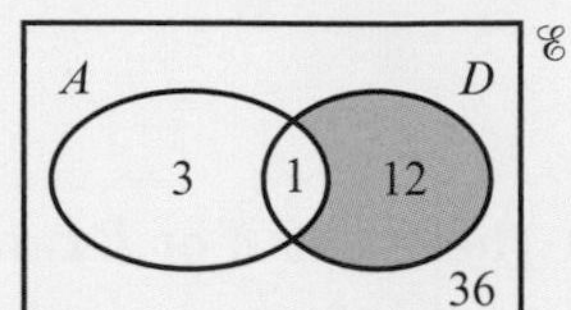

Example 2

Given that $P(A) = 0.3$, $P(B) = 0.4$ and $P(A \cap B) = 0.25$,

a explain why events A and B are not independent.

Given also that $P(C) = 0.2$, that events A and C are mutually exclusive and that events B and C are independent,

b draw a Venn diagram to illustrate the events A, B and C, showing the probabilities for each region.

c Find $P((A \cap B') \cup C)$.

a $P(A) \times P(B) = 0.3 \times 0.4 = 0.12$

$P(A) \times P(B) \neq P(A \cap B)$ so A and B are not independent.

b

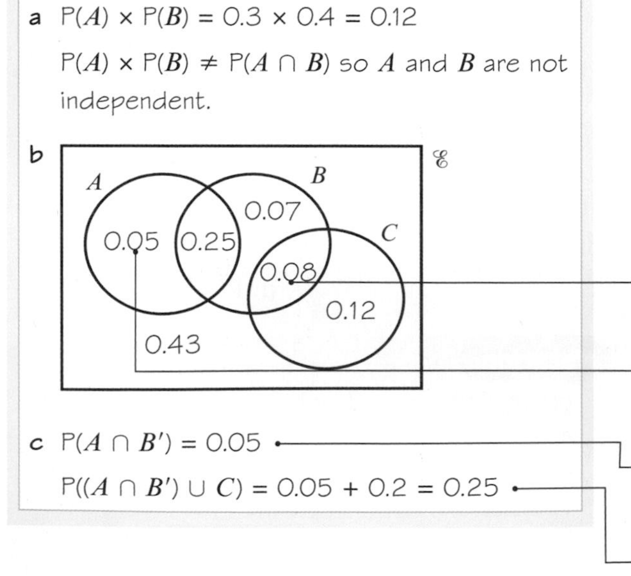

c $P(A \cap B') = 0.05$

$P((A \cap B') \cup C) = 0.05 + 0.2 = 0.25$

Problem-solving

When transferring information onto a Venn diagram, work from the intersections outwards if possible.

Since B and C are independent, $P(B \cap C) = 0.4 \times 0.2 = 0.08$.

Since A and C are mutually exclusive, A overlaps only with B. This region representing just A is $0.3 - 0.25$.

This is the region inside set A but outside set B.

Add the two probabilities, since it is a union relationship and there is no overlap.

Exercise 2A

1 Use set notation to describe the shaded area in each of these Venn diagrams:

a
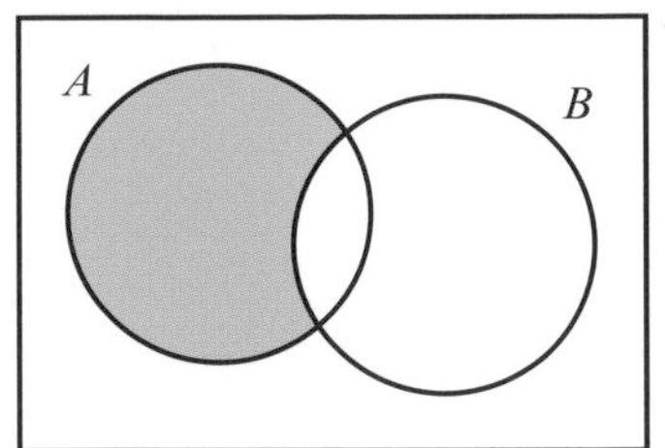

b
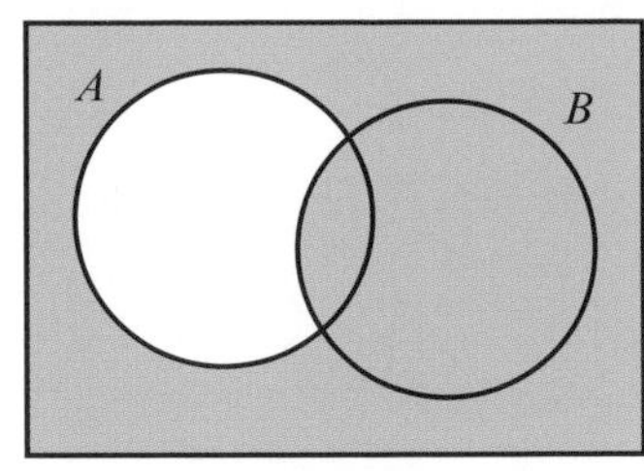

c
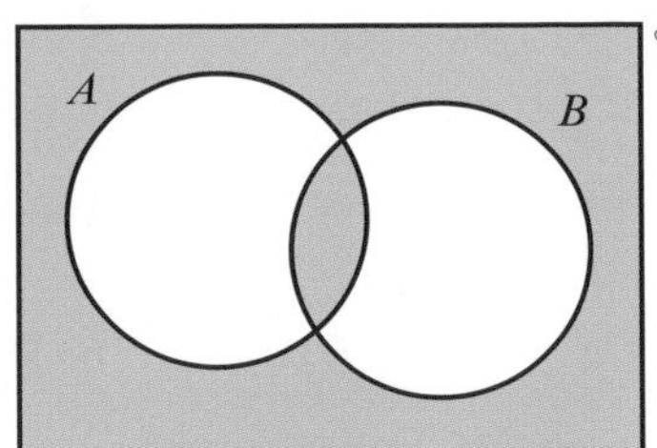

d
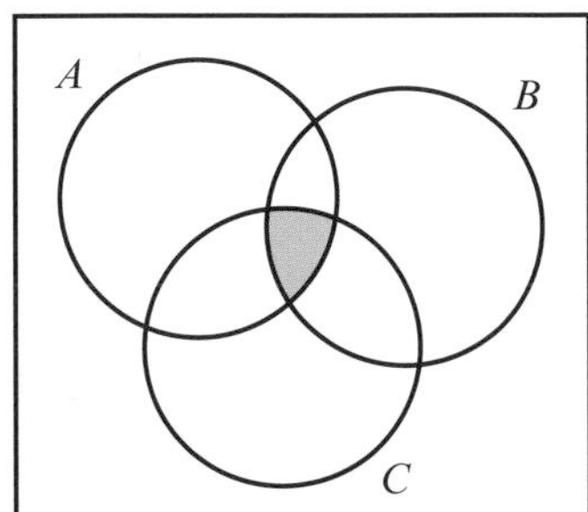

e
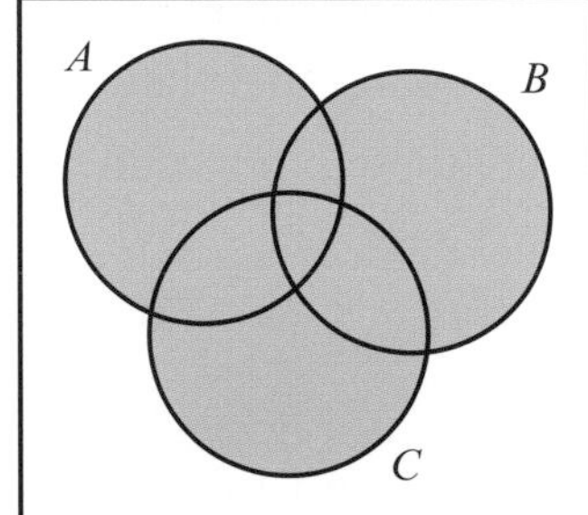

f
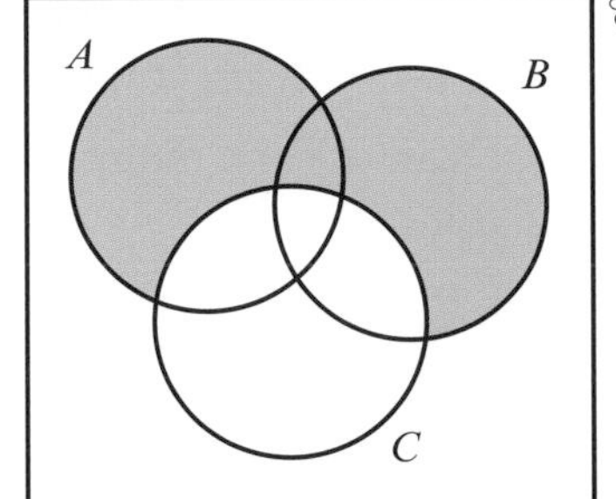

2 On copies of this Venn diagram, shade:

a $A \cup B'$

b $A' \cap B'$

c $(A \cap B)'$

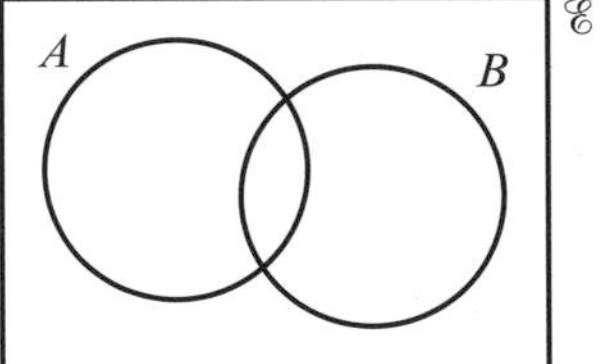

3 On copies of this Venn diagram, shade:

a $(A \cap B) \cup C$

b $(A' \cup B') \cap C$

c $(A \cap B \cap C')'$

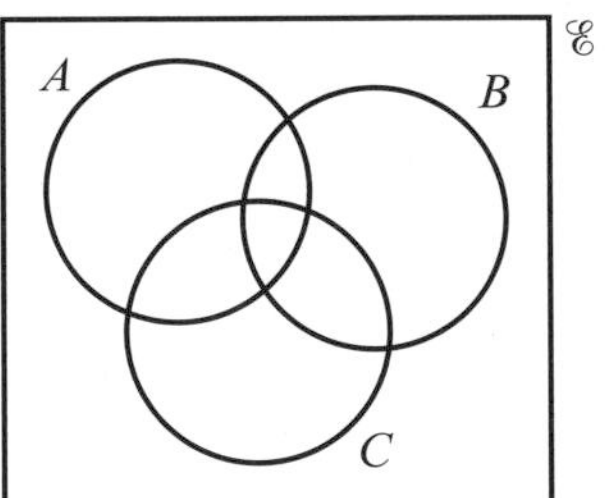

4 A card is chosen at random from a pack of 52 playing cards. C is the event 'the card chosen is a club' and K is the event 'the card chosen is a King'.
The Venn diagram shows the number of outcomes for each event.

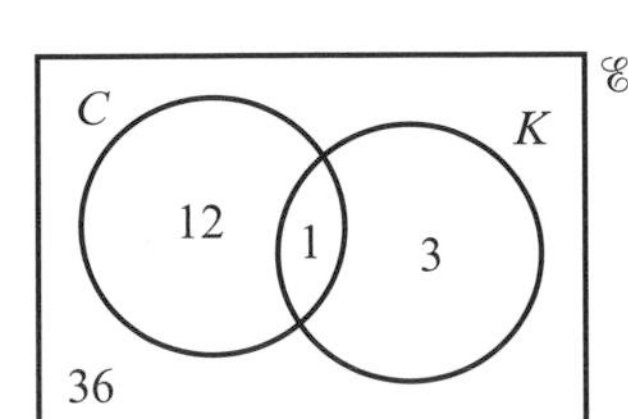

Find:

a $P(K)$ **b** $P(C)$ **c** $P(C \cap K)$

d $P(C \cup K)$ **e** $P(C')$ **f** $(K' \cap C)$

5 A and B are two events. $P(A) = 0.5$, $P(B) = 0.2$ and $P(A \cap B) = 0.1$.
Find:

a $P(A \cup B)$ **b** $P(B')$
c $P(A \cap B')$ **d** $P(A \cup B')$

Hint Draw a Venn diagram.

6 C and D are two events. $P(D) = 0.4$, $P(C \cap D) = 0.15$ and $P(C' \cap D') = 0.1$.
Find:

a $P(C' \cap D)$ **b** $P(C \cap D')$ **c** $P(C)$ **d** $P(C' \cap D')$

7 The probability that a member of a sports club plays hockey (H) is 0.5 and the probability that they play cricket (C) is 0.4. The probability that they play both is 0.25.

a Draw a Venn diagram to illustrate these probabilities.

b Find:
i $P(H \cup C)$ **ii** $P(H' \cap C)$ **iii** $P(H \cup C')$

(P) 8 A bag contains 50 counters numbered from 1 to 50. The counters are either red or blue. A counter is picked at random. The two events R and E are the events 'counter is red' and 'counter is even-numbered' respectively. Given that $n(R) = 17$, $n(E) = 30$ and $n(R \cup E) = 40$,

a draw a Venn diagram to illustrate the outcomes.

b Find:
i $n(R \cap E)$
ii $P(R' \cap E')$
iii $P((R \cap E)')$

Watch out $n(R)$ represents the **number** of outcomes in the event R, whereas $P(R)$ represents the **probability** that the event R occurs.

(E/P) 9 A, B and C are three events with $P(A) = 0.55$, $P(B) = 0.35$ and $P(C) = 0.4$. $P(A \cap C) = 0.2$.
Given that A and B are mutually exclusive and B and C are independent,

a draw a Venn diagram to illustrate the probabilities. **(4 marks)**

b Find:
i $P(A' \cap B')$ **(1 mark)**
ii $P(A \cup (B \cap C'))$ **(1 mark)**
iii $P((A \cap C)' \cup B')$ **(1 mark)**

(E/P) 10 A, B and C are three events with $P(A) = 0.25$, $P(B) = 0.4$, $P(C) = 0.45$ and $P(A \cap B \cap C) = 0.1$.
Given that A and B are independent, B and C are independent, and $A \cap B' \cap C = \emptyset$,

Problem-solving $\emptyset$ is the empty set. $P(\emptyset) = 0$.

a draw a Venn diagram to illustrate the probabilities. **(4 marks)**

b Find:
i $P(A' \cap (B' \cup C))$ **(1 mark)**
ii $P((A \cup B) \cap C)$ **(1 mark)**

c State, with reasons, whether events A' and C independent. **(2 marks)**

E/P **11** Members of a school book club read either murder mysteries (M), ghost stories (G) or epic fiction (E). $P(M) = 0.5$, $P(G) = 0.4$ and $P(E) = 0.6$. Given that no one reads both ghost stories and epic fiction and that $P(M \cap G) = 0.3$,

a draw a Venn diagram to illustrate these probabilities. **(4 marks)**

b Find:

i $P(M \cup G)$ **ii** $P((M \cap G) \cup (M \cap E))$ **(2 marks)**

c Are the events G' and M independent? You must justify your answer. **(2 marks)**

E/P **12** Given that events A and B are independent and that $P(A) = x$ and $P(B) = y$, find, in terms of x and y:

a $P(A \cap B)$ **(2 marks)**

b $P(A \cup B)$ **(2 marks)**

c $P(A \cup B')$ **(2 marks)**

Challenge

Given that events A, B and C are all independent and that $P(A) = x$, $P(B) = y$ and $P(C) = z$, find, in terms of x, y and z:

a $P(A \cap B \cap C)$ **b** $P(A \cup B \cup C)$ **c** $P((A \cup B') \cap C)$

2.2 Conditional probability

The probability of an event can change depending on the outcome of a previous event. For example, the probability of your being late for work may change depending on whether you oversleep or not.

Situations like this can be modelled using **conditional probability**. You use a vertical line to indicate conditional probabilities.

- **The probability that B occurs given that A has already occurred is written as $P(B|A)$.**

Similarly, $P(B|A')$ describes the probability of B occurring given that A has not occurred.

- **For independent events, $P(A|B) = P(A|B') = P(A)$, and $P(B|A) = P(B|A') = P(B)$.**

You can use this condition to determine independence.

You can solve some problems involving conditional probability by considering a **restricted sample space** of the outcomes where one event has already occurred.

Example 3

A school has 75 students in year 12. Of these students, 25 study only humanities subjects (H) and 37 study only science subjects (S). 11 students study both science and humanities subjects.

a Draw a two-way table to show this information.

b Find:

i $P(S' \cap H')$ **ii** $P(S|H)$ **iii** $P(H|S')$

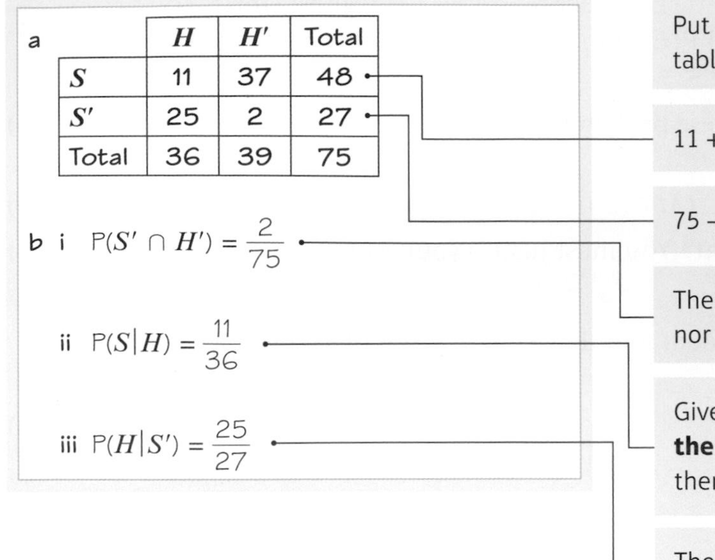

a

	H	H'	Total
S	11	37	48
S'	25	2	27
Total	36	39	75

b i $P(S' \cap H') = \frac{2}{75}$

ii $P(S|H) = \frac{11}{36}$

iii $P(H|S') = \frac{25}{27}$

Put the information from the question in the table. These values are shown in bold.

$11 + 37 = 48$

$75 - 48 = 27$

There are two students who study neither science nor humanities out of a total of 75.

Given that H is already true, you need to **restrict the sample space** to those 36 students. 11 of them also study science.

There are 25 humanities students out of the 27 students who do not study science.

Example 4

Two four-sided dice are thrown together, and the sum of the numbers shown is recorded.

a Draw a sample-space diagram showing the possible outcomes.

b Given that at least one dice lands on a 3, find the probability that the sum on the two dice is exactly 5.

c State one modelling assumption used in your calculations.

a

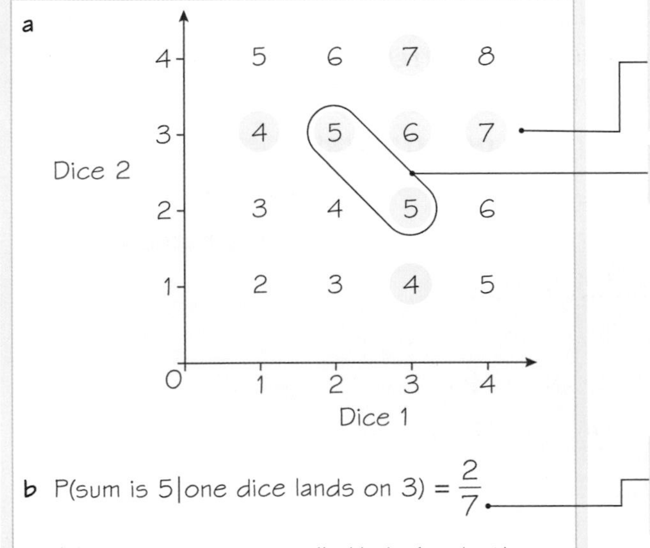

b $P(\text{sum is } 5|\text{one dice lands on } 3) = \frac{2}{7}$

c All outcomes are equally likely (i.e. both dice are fair).

There are seven outcomes where one of the dice lands on a 3. The yellow circles show the restricted sample space.

Two of these outcomes have a total of 5.

There are 7 equally likely outcomes in the restricted sample space.

Exercise 2B

1 The two-way table shows the fast-food preferences of 60 students in a sixth-form.

	Pizza	**Curry**	**Total**
Male	11	18	29
Female	14	17	31
Total	25	35	60

Find:

a P(Male) **b** P(Curry|Male) **c** P(Male|Curry) **d** P(Pizza|Female)

2 In a sports club, there are 75 members of whom 32 are female. Of the female members, 15 play badminton and 17 play squash. There are 22 men who play squash and the rest play badminton.

a Draw a two-way table to illustrate this situation.

b Find:

i P(Male|Squash) **ii** P(Female|Badminton) **iii** P(Squash|Female)

3 A group of 80 children are asked about their favourite ice-cream flavour. Of the 45 girls, 13 like vanilla, 12 like chocolate and the rest like strawberry. Of the boys, 2 like vanilla and 23 like strawberry. The rest like chocolate.

a Draw a two-way table to show this situation.

b Find:

i P(Boy|Strawberry) **ii** P(Girl|Vanilla) **iii** P(Chocolate|Boy)

4 A red and a blue spinner each have four equally likely outcomes, numbered 1 to 4. The two spinners are spun at the same time, and the sum of the numbers shown, X, is recorded.

a Draw a sample space diagram for X.

b Find:

i $P(X = 5)$ **ii** $P(X = 3|\text{Red spinner is 2})$ **iii** $P(\text{Blue spinner is 3}|X = 5)$

5 Two fair six-sided dice are thrown and the product is recorded.

a Draw a sample-space diagram to illustrate the possible outcomes.

b Given that the first dice shows a 5, find the probability that the product is 20.

c Given that the product is 12, find the probability that the second dice shows a 6.

d Explain the importance of the word ‘fair’ in this context.

6 A card is drawn at random from a pack of 52 playing cards. Given that the card is a diamond, find the probability that the card is an ace.

7 Two coins are flipped and the results are recorded. Given that one coin lands on a head, find the probability of:

a two heads b a head and a tail.

c State one modelling assumption used in your calculations.

(E) 8 120 students are asked about their viewing habits. 56 say they watch sports (S) and 77 say they watch dramas (D). Of those who watch dramas, 18 also watch sports.

a Draw a two-way table to show this information. **(2 marks)**

b One student is chosen at random. Find:

i $P(D')$ **(1 mark)**

ii $P(S' \cap D')$ **(1 mark)**

iii $P(S|D)$ **(1 mark)**

iv $P(D'|S)$ **(1 mark)**

(E) 9 A rambling group is made up of 63 women and 47 men. 26 of the women and 18 of the men use a walking stick.

a Draw a two-way table to show this information. **(2 marks)**

b One rambler is chosen at random. Find:

i P(Uses a stick) **(1 mark)**

ii P(Uses a stick|Female) **(1 mark)**

iii P(Male|Uses a stick) **(1 mark)**

(P) 10 A veterinary surgery has 750 registered pet owners. Of these 450 are female. 320 of the pet owners own a cat and 250 own a dog. Of the remaining pet owners, 25 are males who own another type of pet. No one owns more than one type of pet. 175 female owners have a cat. One owner is chosen at random. Given that:

F is the event that an owner is female

D is the event that an owner has a dog

C is the event that an owner has a cat.

Find:

a $P(D' \cap C')$ b $P(D|F')$ c $P(F'|C)$ d $P((D' \cap C')|F)$

2.3 Conditional probabilities in Venn diagrams

You can find conditional probabilities from a Venn diagram by considering the section of the Venn diagram that corresponds to the restricted sample space.

Example 5

A and B are two events such that $P(A) = 0.55$, $P(B) = 0.4$ and $P(A \cap B) = 0.15$.

a Draw a Venn diagram showing the probabilities for events A and B.

b Find:

i $P(A|B)$ ii $P(B|(A \cup B))$ iii $P(A'|B')$

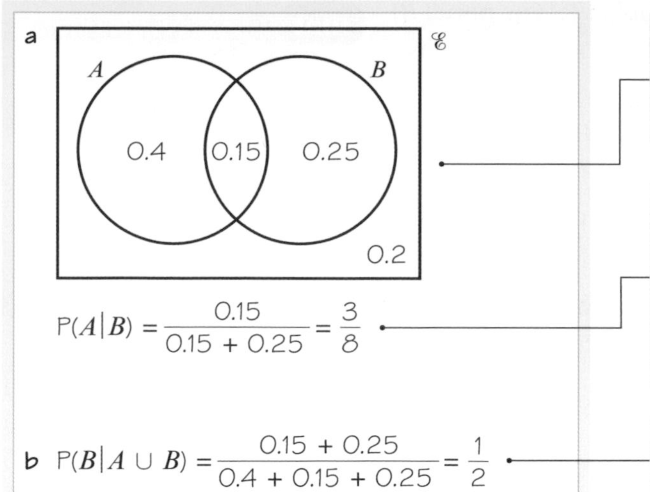

a $\text{P}(A|B) = \dfrac{0.15}{0.15 + 0.25} = \dfrac{3}{8}$

Use the information given to fill in the probabilities on each of the four regions in the Venn diagram.

The sample space is restricted to just circle B. The part of circle A inside B has probability 0.15.

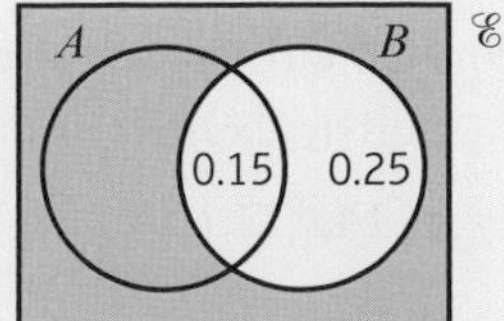

b $\text{P}(B|A \cup B) = \dfrac{0.15 + 0.25}{0.4 + 0.15 + 0.25} = \dfrac{1}{2}$

The sample space is restricted to just the union of A and B.

c $\text{P}(A'|B') = \dfrac{0.2}{0.4 + 0.2} = \dfrac{1}{3}$

Consider the restricted sample space first. This is everything **not** inside circle B.

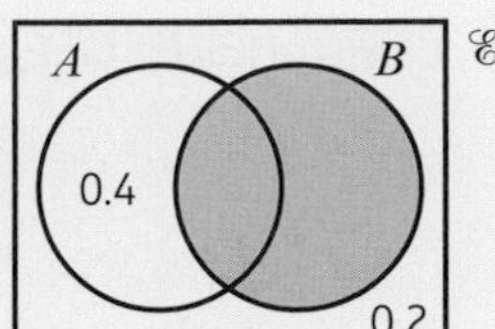

Exercise 2C

1 The Venn diagram shows the probabilities for two events, A and B.

Find:

a $\text{P}(A \cup B)$ **b** $\text{P}(A|B)$

c $\text{P}(B|A')$ **d** $\text{P}(B|A \cup B)$

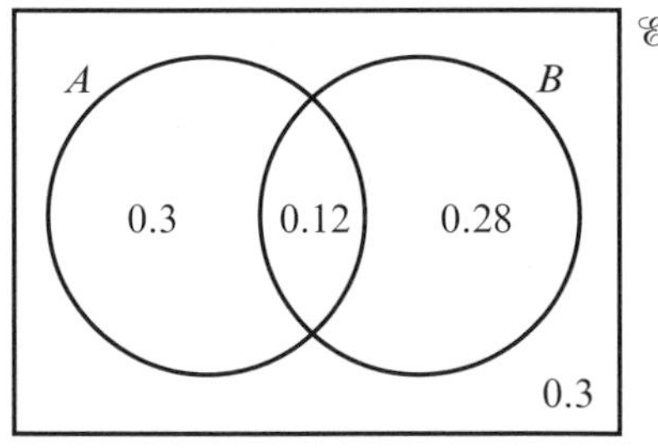

2 C and D are two events such that $\text{P}(C) = 0.8$, $\text{P}(D) = 0.4$ and $\text{P}(C \cap D) = 0.25$.

a Draw a Venn diagram showing the probabilities for events C and D.

b Find:

i $\text{P}(C \cup D)$ **ii** $\text{P}(C|D)$ **iii** $\text{P}(D|C)$ **iv** $\text{P}(D'|C')$

3 S and T are two events such that $\text{P}(S) = 0.5$ and $\text{P}(T) = 0.7$. Given that S and T are independent,

a draw a Venn diagram showing the probabilities for events S and T.

b Find:

i $\text{P}(S \cap T)$ **ii** $\text{P}(S|T)$ **iii** $\text{P}(T|S')$ **iv** $\text{P}(S|S' \cup T')$

4 120 members of a youth club play either snooker (A) or pool (B) or neither. Given that 65 play snooker, 50 play pool and 20 play both, find:

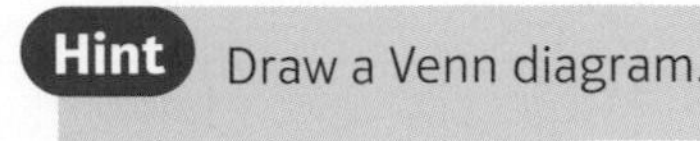

a $P(A \cap B')$ **b** $P(A|B)$ **c** $P(B|A')$ **d** $P(A|A \cup B)$

5 The eating tastes of 80 cats are recorded. 45 like Feskers (F) and 32 like Whilix (W). 12 like neither. One cat is chosen at random. Find:

a $P(F \cap W)$ **b** $P(F|W)$ **c** $P(W|F)$ **d** $P(W'|F')$

6 The Venn diagram shows the probabilities of three events, A, B and C.

Find:

a $P(A|B)$ **b** $P(C|A')$

c $P((A \cap B)|C')$ **d** $P(C|(A' \cup B'))$

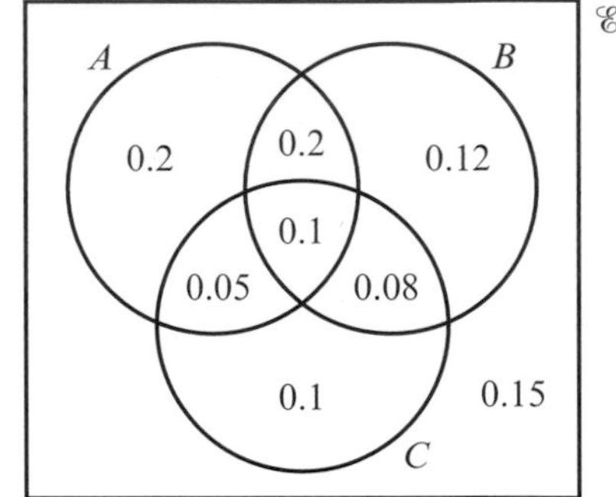

E/P 7 The Venn diagram shows the number of students in a class who watch any of 3 popular TV programmes A, B and C.

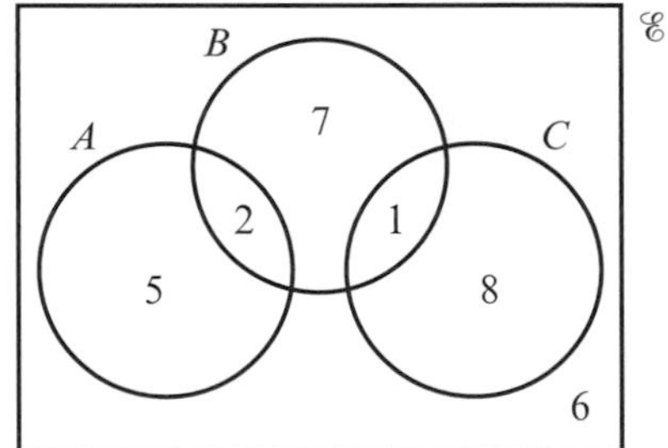

One of these students is selected at random. Given that the student watches at least one of the TV programmes, find the probability that the student watches:

a programme C **(2 marks)**

b exactly two of the programmes. **(2 marks)**

c Determine whether or not watching programme B and watching programme C are statistically independent. **(3 marks)**

Problem-solving

If $P(A|B) = P(A)$ then events A and B are independent.

E/P 8 Three events, A, B and C are such that A and B are mutually exclusive and B and C are independent. $P(A) = 0.2$, $P(B) = 0.6$ and $P(C) = 0.5$. Given that $P(A' \cap B' \cap C') = 0.1$,

a draw a Venn diagram to show the probabilities for events A, B and C. **(4 marks)**

b Find:

i $P(A|C)$ **(1 mark)**

ii $P(B|C')$ **(1 mark)**

iii $P(C|(A \cup B))$ **(1 mark)**

(E/P) **9** A doctor completes a medical study of 100 people, 5 of whom are known to have an illness and 95 of whom are known not to. A diagnostic test is applied. All 5 of the people with the illness test positive, and 10 people without the illness also test positive. Given that event A = person has the disease and event B = person tests positive,

a draw a Venn diagram to represent this situation. **(3 marks)**

b Calculate $P(A|B)$. **(2 marks)**

c With reference to your answer to part **b**, comment on the usefulness of the diagnostic test. **(2 marks)**

(P) **10** Events A and B are such that $P(A) = 0.6$ and $P(B) = 0.7$. Given that $P(A' \cap B') = 0.12$, find:

a $P(B|A')$ **b** $P(B|A)$

c Explain what your answers to parts **a** and **b** tell you about events A and B.

(E/P) **11** The Venn diagram shows the probabilities for two events, A and B. Given that $P(A|B) = P(B')$, find the values of x and y.

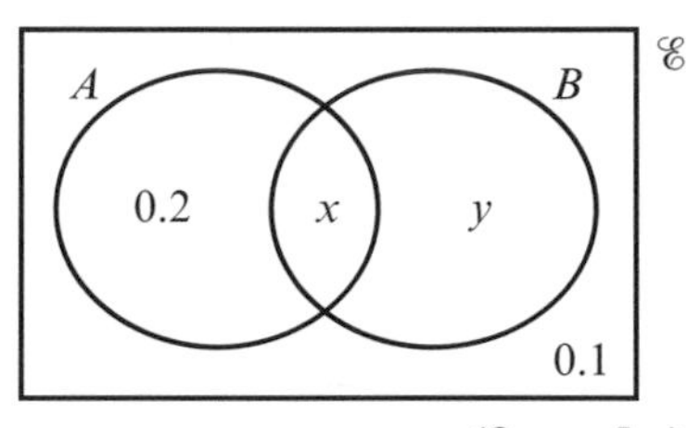

(3 marks)

(E/P) **12** The Venn diagram shows the probabilities for two events, A and B. Given that $P(A|B) = P(A')$, find the values of c and d.

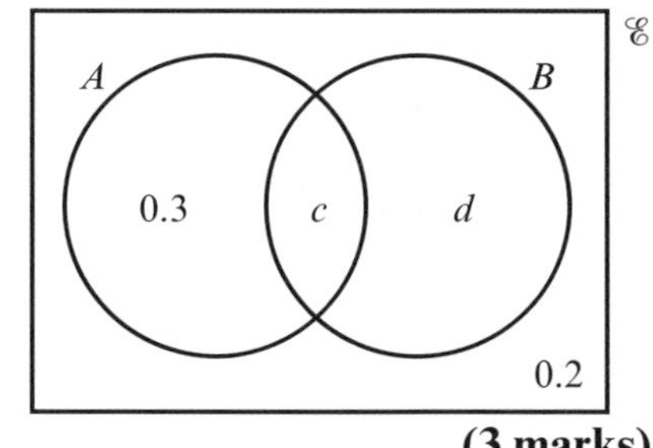

(3 marks)

2.4 Probability formulae

There is a formula you can use for two events that links the probability of the union and the probability of the intersection.

If $P(A) = a$ and $P(B) = b$

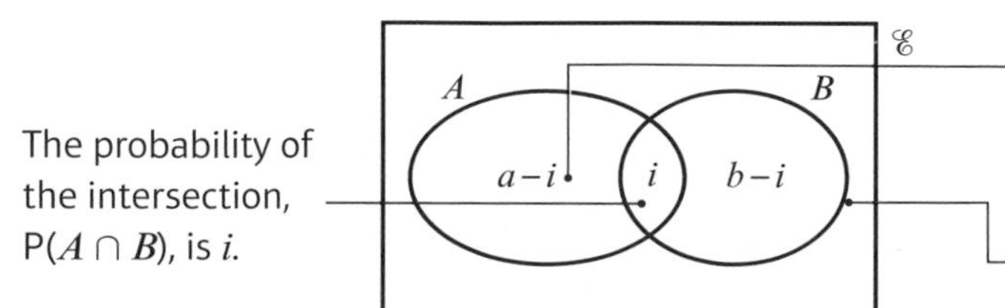

Since $i = P(A \cap B)$ you can write the following **addition formula** for two events A and B:

- $\mathbf{P(A \cup B) = P(A) + P(B) - P(A \cap B)}$

Example 6

A and B are two events, with $P(A) = 0.6$, $P(B) = 0.7$ and $P(A \cup B) = 0.9$.
Find $P(A \cap B)$.

Watch out You do not know whether A and B are independent so you can't use $(A \cap B) = P(A) \times P(B)$. Use the addition formula.

$P(A \cup B) = P(A) + P(B) - P(A \cap B)$
So $P(A \cap B) = P(A) + P(B) - P(A \cup B)$
$= 0.6 + 0.7 - 0.9$
$= 0.4$

Rearrange the addition formula to make $P(A \cap B)$ the subject.

You can also use the Venn diagram in the explanation above to find a formula for $P(B|A)$:

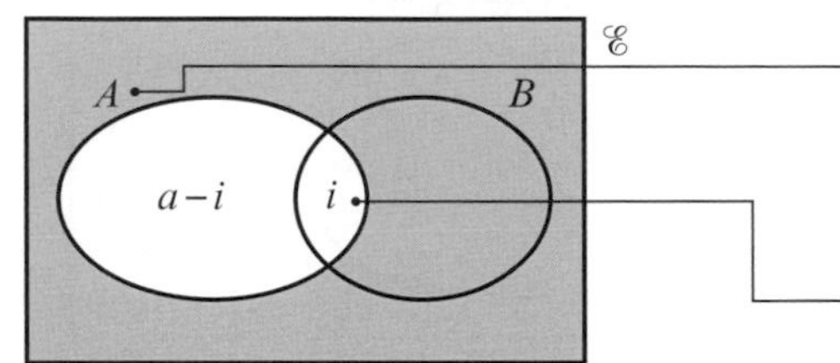

To find $P(B|A)$ restrict the sample space to the set of outcomes in which A has already occurred.

This is the subset of outcomes in the restricted sample space in which B occurs.

So $P(B|A) = \frac{i}{(a - i) + i} = \frac{i}{a}$

Since $P(B \cap A) = i$ and $P(A) = a$, you can write the following **multiplication formula** for conditional probability.

- $\mathbf{P(B|A) = \frac{P(B \cap A)}{P(A)}}$ **so** $\mathbf{P(B \cap A) = P(B|A) \times P(A)}$

Example 7

C and D are two events such that $P(C) = 0.2$, $P(D) = 0.6$ and $P(C|D) = 0.3$.
Find:

a $P(C \cap D)$ **b** $P(D|C)$ **c** $P(C \cup D)$

a $P(C \cap D) = P(C|D) \times P(D)$
$= 0.3 \times 0.6 = 0.18$

Use the multiplication formula.

b $P(D|C) = \frac{P(D \cap C)}{P(C)}$

$= \frac{0.18}{0.2} = 0.9$

c $P(C \cup D) = P(C) + P(D) - P(C \cap D)$
$= 0.2 + 0.6 - 0.18 = 0.62$

Problem-solving

If you wanted to draw a Venn diagram to show these events it would help to find $P(C \cap D)$ first using the multiplication formula.

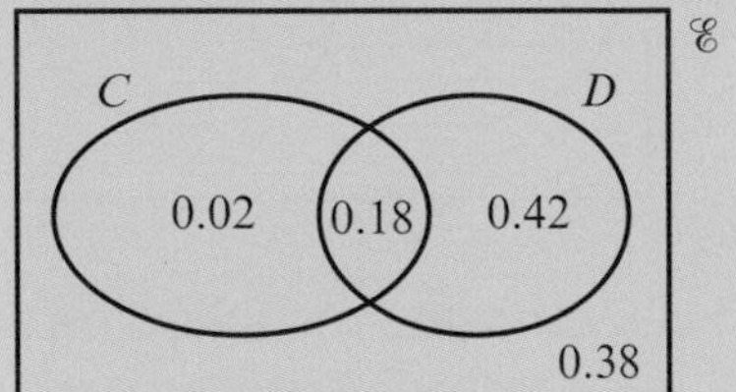

Exercise 2D

1 A and B are two events where $P(A) = 0.4$, $P(B) = 0.5$ and $P(A \cup B) = 0.6$.
Find:

a $P(A \cap B)$ **b** $P(A')$ **c** $P(A \cup B')$ **d** $P(A' \cup B)$

(P) 2 C and D are two events where $P(C) = 0.55$, $P(D) = 0.65$ and $P(C \cap D) = 0.4$.

a Find $P(C \cup D)$.

b Draw a Venn diagram and use it to find:
i $P(C' \cap D')$ **ii** $P(C|D)$ **iii** $P(C|D')$

c Explain why events C and D are not statistically independent.

3 E and F are two events where $P(E) = 0.7$, $P(F) = 0.8$ and $P(E \cap F) = 0.6$.

a Find $P(E \cup F)$.

b Draw a Venn diagram and use it to find:
i $P(E \cup F')$ **ii** $P(E' \cap F)$ **iii** $P(E|F')$

(P) 4 There are two events T and Q where $P(T) = P(Q) = 3P(T \cap Q)$ and $P(T \cup Q) = 0.75$.
Find:

a $P(T \cap Q)$ **b** $P(T)$ **c** $P(Q')$ **d** $P(T' \cap Q')$ **e** $P(T \cap Q')$

5 A survey of all the households in the town of Bury was carried out. The survey showed that 70% have a freezer and 20% have a dishwasher and 80% have either a dishwasher or a freezer or both appliances. Find the probability that a randomly chosen household in Bury has both appliances.

6 A and B are two events such that $P(A) = 0.4$, $P(B) = 0.5$ and $P(A|B) = 0.4$. Find:

a $P(B|A)$ **b** $P(A' \cap B')$ **c** $P(A' \cap B)$.

7 Let A and B be events such that $P(A) = \frac{1}{4}$, $P(B) = \frac{1}{2}$ and $P(A \cup B) = \frac{3}{5}$.
Find:

a $P(A|B)$ **b** $P(A' \cap B)$ **c** $P(A' \cap B')$

8 C and D are two events where $P(C|D) = \frac{1}{3}$, $P(C|D') = \frac{1}{5}$ and $P(D) = \frac{1}{4}$. Find:

a $P(C \cap D)$ **b** $P(C \cap D')$ **c** $P(C)$
d $P(D|C)$ **e** $P(D'|C)$ **f** $P(D'|C')$

(E) 9 Given that $P(A) = 0.42$, $P(B) = 0.37$ and $P(A \cap B) = 0.12$. Find:

a $P(A \cup B)$ **(2 marks)**

b $P(A \mid B')$ **(2 marks)**

The event C has $\text{P}(C) = 0.3$.

The events B and C are mutually exclusive and the events A and C are independent.

c Find $\text{P}(A \cap C)$. **(2 marks)**

d Draw a Venn diagram to illustrate the events A, B and C, giving the probabilities for each region. **(4 marks)**

e Find $\text{P}((A' \cup C)')$. **(2 marks)**

(E/P) **10** Three events A, B and C are such that $\text{P}(A) = 0.4$, $\text{P}(B) = 0.7$, $\text{P}(C) = 0.4$ and $\text{P}(A \cap B) = 0.3$. Given that A and C are mutually exclusive and that B and C are independent, find:

a $\text{P}(B \cap C)$ **(1 mark)**

b $\text{P}(B|C)$ **(1 mark)**

c $\text{P}(A|B')$ **(1 mark)**

d $\text{P}((B \cap C)|A')$ **(1 mark)**

(E/P) **11** Anna and Bella are sometimes late for school. The events A and B are defined as follows:
A is the event that Anna is late for school
B is the event that Bella is late for school
$\text{P}(A) = 0.3$, $\text{P}(B) = 0.7$ and $\text{P}(A' \cap B') = 0.1$. On a randomly selected day, find the probability that:

a both Anna and Bella are late to school **(1 mark)**

b Anna is late to school given that Bella is late to school. **(2 marks)**

Their teacher suspects that Anna and Bella being late for school is linked in some way.

c Comment on his suspicion, showing your working. **(2 marks)**

(E/P) **12** John and Kayleigh play darts in the same team. The events J and K are defined as follows:
J is the event that John wins his match
K is the event that Kayleigh wins her match
$\text{P}(J) = 0.6$, $\text{P}(K) = 0.7$ and $\text{P}(J \cup K) = 0.8$.
Find the probability that:

a both John and Kayleigh win their matches **(1 mark)**

b John wins his match given that Kayleigh loses hers **(2 marks)**

c Kayleigh wins her match given that John wins his. **(2 marks)**

d Determine whether the events J and K are statistically independent.
You must show all your working. **(2 marks)**

Challenge

The discrete random variable X has probability function:

$\text{P}(X = x) = kx$, $x = 1, 2, 3, 4, 5$

Find:

a the value of k

b $\text{P}(X = 5|X > 2)$

c $\text{P}(X \text{ is odd}|X \text{ is prime})$

2.5 Tree diagrams

Conditional probabilities can be represented on a tree diagram.

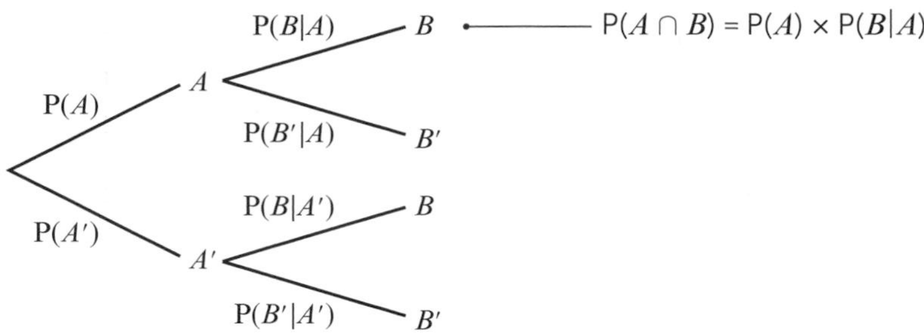

The probabilities on the second set of branches represent the conditional probabilities of B given that A has, or has not, happened.

Example 8

A bag contains 6 green beads and 4 yellow beads. A bead is taken from the bag at random, the colour is recorded and it is not replaced. A second bead is then taken from the bag and its colour recorded. Given that both balls are the same colour, find the probability that they are both yellow.

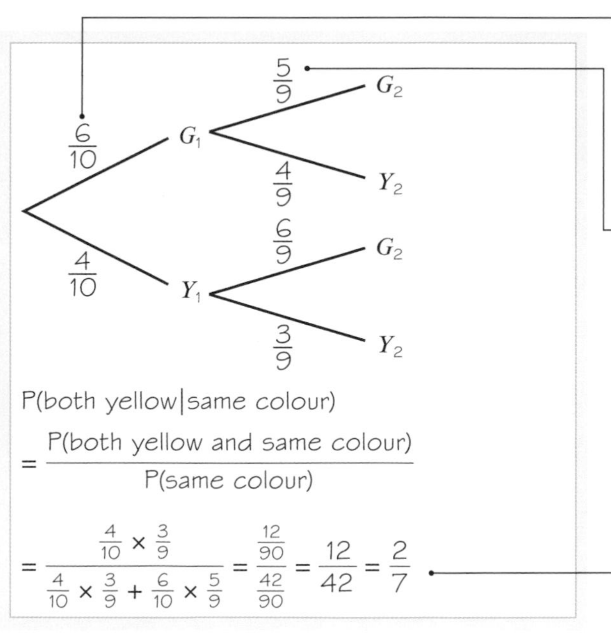

P(both yellow|same colour)

$$= \frac{\text{P(both yellow and same colour)}}{\text{P(same colour)}}$$

$$= \frac{\frac{4}{10} \times \frac{3}{9}}{\frac{4}{10} \times \frac{3}{9} + \frac{6}{10} \times \frac{5}{9}} = \frac{\frac{12}{90}}{\frac{42}{90}} = \frac{12}{42} = \frac{2}{7}$$

Initially there are 10 beads in the bag and 6 are green. $P(G_1) = \frac{6}{10}$.

Since a green bead is removed and not replaced, the total number of beads is reduced to 9 and there are just 5 green beads remaining.

Use $P(B|A) = \frac{P(B \cap A)}{P(A)}$

Exercise 2E

1 A bag contains five red and four blue tokens. A token is chosen at random, the colour recorded and the token is not replaced. A second token is chosen and the colour recorded.

a Draw a tree diagram to illustrate this situation.

Find the probability that:

b the second token is red given that the first token is blue

c the first token is red given that the second token is blue

d the first token is blue given that the tokens are different colours

e the tokens are the same colour given that the second token is red.

2 A and B are two events such that $\mathrm{P}(B|A) = 0.45$, $\mathrm{P}(B|A') = 0.35$ and $\mathrm{P}(A) = 0.7$.

a Copy and complete the tree diagram representing this information.

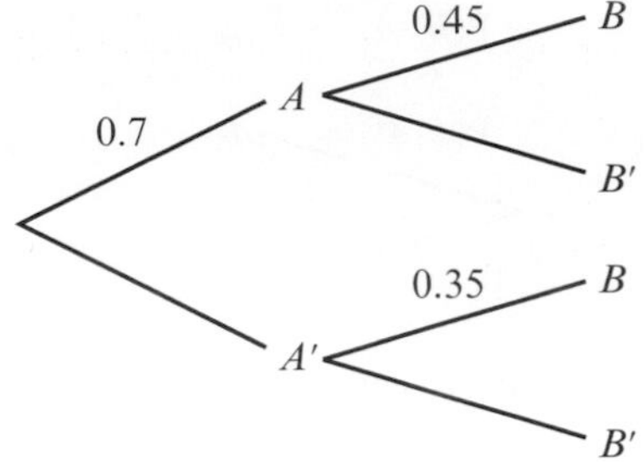

b Find:

i $\mathrm{P}(A \cap B)$

ii $\mathrm{P}(A' \cap B')$

iii $\mathrm{P}(A|B)$

3 A box of 24 chocolates contains 10 dark and 14 milk chocolates. Linda chooses a chocolate at random and eats it, followed by another one.

a Draw a tree diagram to represent this information.

Find the probability that Linda eats:

b two dark chocolates

c one dark and one milk chocolate

d two dark chocolates given that she eats at least one dark chocolate.

(P) **4** Jean always goes to work by bus or takes a taxi. If one day she goes to work by bus, the probability she goes to work by taxi the next day is 0.4. If one day she goes to work by taxi, the probability she goes to work by bus the next day is 0.7.

Given that Jean takes the bus to work on Monday, find the probability that she takes a taxi to work on Wednesday.

(P) **5** Sue has two coins. One is fair, with a head on one side and a tail on the other.
The second is a trick coin and has a tail on both sides. Sue picks up one of the coins at random and flips it.

a Find the probability that it lands heads up.

b Given that it lands tails up, find the probability that she picked up the fair coin.

(E) **6** A bag contains 4 blue balls and 7 green balls. A ball is selected at random from the bag and its colour is recorded. The ball is not replaced. A second ball is selected at random and its colour is recorded.

a Draw a tree diagram to represent the information. **(3 marks)**

Find the probability that:

b the second ball selected is green **(2 marks)**

c both balls selected are green, given that the second ball selected is green. **(2 marks)**

(E) **7** In an engineering company, factories A, B and C are all producing tin sheets of the same type. Factory A produces 25% of the sheets, factory B produces 45% and the rest are produced by factory C. Factories A, B and C produce flawed sheets with probabilities 0.02, 0.07 and 0.04 respectively.

a Draw a tree diagram to represent this information. **(3 marks)**

b Find the probability that a randomly selected sheet is:

i produced by factory B and flawed **(2 marks)**

ii flawed. **(3 marks)**

c Given that a randomly selected sheet is flawed, find the probability that it was produced by factory A. **(3 marks)**

(E/P) **8** A genetic condition is known to be present in 4% of a population. A test is developed to help determine whether or not someone has the genetic condition.
If a person has the condition, the test is positive with probability 0.9.
If a person does not have the condition, the test is positive with probability 0.02.

a Draw a tree diagram to represent this information. **(3 marks)**

A person is selected at random from the population and tested for this condition.

b Find the probability that the test is negative. **(3 marks)**

A doctor randomly selects a person from the population and tests them for the condition.
Given that the test is negative,

c find the probability that they do have the condition. **(2 marks)**

d Comment on the effectiveness of this test. **(1 mark)**

(E) **9** On a randomly chosen day the probabilities that Bill travels to work by car, by bus or by train are 0.1, 0.6 and 0.3 respectively. The probabilities of being late when using these methods of travel are 0.55, 0.3 and 0.05 respectively.

a Draw a tree diagram to represent this information. **(3 marks)**

b Find the probability that on a randomly chosen day,

i Bill travels by train and is late **(2 marks)**

ii Bill is late. **(2 marks)**

c Given that Bill is late, find the probability that he did not travel by car. **(4 marks)**

(E/P) **10** A box A contains 7 counters of which 4 are green and 3 are blue.
A box B contains 5 counters of which 2 are green and 3 are blue.
A counter is drawn at random from box A and placed in box B. A second counter is drawn at random from box A and placed in box B.
A third counter is then drawn at random from the counters in box B.

a Draw a tree diagram to show this situation. **(4 marks)**

The event C occurs when the 2 counters drawn from box A are of the same colour.
The event D occurs when the counter drawn from box B is blue.

b Find P(C). **(3 marks)**

c Show that $P(D) = \frac{27}{49}$ **(3 marks)**

d Show that $P(C \cap D) = \frac{11}{49}$ **(2 marks)**

e Hence find $P(C \cup D)$. **(2 marks)**

f Given that all three counters drawn are the same colour, find the probability that they are all green. **(3 marks)**

(E/P) **11** A box of jelly beans contains 7 sweet flavours and 3 sour flavours. Two of the jelly beans are taken one after the other and eaten. Emilia wants to find the probability that both jelly beans eaten are sweet given that at least one of them is. Her solution is shown below:

> P(both jelly beans are sweet) $= \frac{7}{10} \times \frac{7}{10} = \frac{49}{100}$
>
> P(at least one jelly bean is sweet)
>
> $= 1 - \text{P(neither are sweet)} = 1 - \frac{3}{10} \times \frac{3}{10} = \frac{91}{100}$
>
> P(both are sweet given at least one is sweet)
>
> $= \dfrac{\frac{49}{100}}{\frac{91}{100}} = \frac{49}{91}$

Identify Emilia's mistake and find the correct probability. **(4 marks)**

Mixed exercise 2

(E) **1** A and B are two events such that $P(A) = 0.4$ and $P(B) = 0.35$. If $P(A \cap B) = 0.2$, find:

a $P(A \cup B)$ **(1 mark)**

b $P(A' \cap B')$ **(1 mark)**

c $P(B|A)$ **(2 marks)**

d $P(A'|B)$ **(2 marks)**

(E/P) **2** J, K and L are three events such that $P(J) = 0.25$, $P(K) = 0.45$ and $P(L) = 0.15$. Given that K and L are independent, J and L are mutually exclusive and $P(J \cap K) = 0.1$

a draw a Venn diagram to illustrate this situation. **(2 marks)**

b Find:

i $P(J \cup K)$ **(1 mark)**

ii $P(J' \cap L')$ **(1 mark)**

iii $P(J|K)$ **(2 marks)**

iv $P(K|J' \cap L')$ **(2 marks)**

(E/P) **3** Of 60 students in a high-school sixth form, 35 study French and 45 study Spanish. If 27 students study both, find the probability that a student chosen at random:

a studies only one subject **(1 mark)**

b studies French given that they study Spanish **(2 marks)**

c studies Spanish given that they do not study French. **(2 marks)**

It is found that 75% of the students who study just French wear glasses and half of the students who study just Spanish wear glasses. Find the probability that a student chosen at random:

d studies one language and wears glasses **(2 marks)**

e wears glasses given that they study one language. **(2 marks)**

(E/P) **4** A bag contains 6 red balls and 9 green balls. A ball is chosen at random from that bag, its colour noted and the ball placed to one side. A second ball is chosen at random and its colour noted.

a Draw a tree diagram to illustrate this situation. **(2 marks)**

b Find the probability that:

i both balls are green **(1 mark)**

ii the balls are different colours. **(2 marks)**

Further balls are drawn from the bag and not replaced. Find the probability that:

c the third ball is red **(2 marks)**

d it takes just four selections to get four green balls. **(2 marks)**

(E) **5** In a tennis match, the probability that Anne wins the first set against Colin is 0.7. If Anne wins the first set, the probability that she wins the second set is 0.8. If Anne loses the first set, the probability that she wins the second set is 0.4. A match is won when one player wins two sets.

a Find the probability that the game is over after two sets. **(2 marks)**

b Find the probability that Anne wins given that the game is over after two sets. **(2 marks)**

If the game is tied at one set all, a tiebreaker is played and the probability of Anne winning it is 0.55.

c Find the probability of Anne winning the entire match. **(3 marks)**

(E/P) **6** The colours of the paws of 75 kittens are recorded. 26 kittens have all black paws and 14 kittens have all white paws. 15 have a combination of black and white paws. One kitten is chosen at random. Find the probability that the kitten has:

a neither white nor black paws **(1 mark)**

b a combination of black and white paws given that they have some black paws. **(2 marks)**

Two kittens are now chosen. Find the probability that:

c both kittens have all black paws **(2 marks)**

d both kittens have some white paws. **(2 marks)**

(E/P) **7** Two events A and B are such that $P(A) = 0.4$ and $P(A \cap B) = 0.12$. If A and B are independent, find:

a $P(B)$ **(1 mark)**

b $P(A' \cap B')$ **(1 mark)**

A third event C has $P(C) = 0.4$. Given that A and C are mutually exclusive and $P(B \cap C) = 0.1$,

c draw a Venn diagram to illustrate this situation. **(2 marks)**

d Find:

i $P(B|C)$ **(2 marks)**

ii $P(A \cap (B' \cup C))$ **(2 marks)**

8 In a football match, the probability that team A scores first is 0.6, and the probability that team B scores first is 0.35.

a Suggest a reason why these probabilities do not add up to 1. **(1 mark)**

The probability that team A scores first and wins the match is 0.48.

b Find the probability that team A scores first and does not win the match. **(3 marks)**

If team B scores first, the probability that team A will win the match is 0.3.

c Given that team A won the match, find the probability that they did not score first. **(3 marks)**

Challenge

$P(A) = 0.6$ and $P(B) = 0.2$

a Given that $P(A \cap B') = p$, find the range of possible values of p.

$P(C) = 0.7$ and $P(A \cap B \cap C) = 0.1$

b Given $P(A \cap B' \cap C) = q$, find the range of possible values of q.

Summary of key points

1 The event A and B can be written as $A \cap B$. The '$\cap$' symbol is the symbol for **intersection**.
The event A or B can be written as $A \cup B$. The '$\cup$' symbol is the symbol for **union**.
The event not A can be written as A'. This is also called the **complement** of A.

2 The probability that B occurs given that A has already occurred is written as $P(B|A)$.
For independent events, $P(A|B) = P(A|B') = P(A)$, and $P(B|A) = P(B|A') = P(B)$.

3 $P(A \cup B) = P(A) + P(B) - P(A \cap B)$

4 $P(B|A) = \dfrac{P(B \cap A)}{P(A)}$ so $(B \cap A) = P(B|A) \times P(A)$

The normal distribution 3

Objectives

After completing this chapter you should be able to:

- Understand the normal distribution and the characteristics of a normal distribution curve → pages 38–41
- Find percentage points on a standard normal curve → pages 41–47
- Calculate values on a standard normal curve → pages 47–49
- Find unknown means and/or standard deviations for a normal distribution → pages 49–53
- Approximate a binomial distribution using a normal distribution → pages 53–55
- Select appropriate distributions and solve real-life problems in context → pages 53–60
- Carry out a hypothesis test for the mean of a normal distribution → pages 53–60

Biologists use the normal distribution to model the distributions of physical characteristics, such as height and mass, in large populations.

→ Exercise 3E, Q13

Prior knowledge check

1 The probability that a one-month old Labrador puppy weighs under 2 kg is 0.735. Two puppies are chosen at random from different litters. Find:

a P(both weigh under 2 kg)

b P(exactly one weighs under 2 kg)

← Year 1, Chapter 1, Chapter 5

2 $X \sim \mathrm{B}(20, 0.4)$. Find:

a $\mathrm{P}(X = 6)$

b $\mathrm{P}(X \geqslant 8)$

c $\mathrm{P}(3 \leqslant X \leqslant 10)$

← Year 1, Chapter 6

3 The probability that a plate made using a particular production process is faulty is given as 0.16. A sample of 20 plates is taken. Find:

a the probability that exactly two plates are faulty

b the probability that no more than three plates are faulty.

← Year 1, Chapter 6

3.1 The normal distribution

A **continuous random variable** can take any one of infinitely many values. The probability that a continuous random variable takes any one specific value is 0, but you can write the probability that it takes values within a given range. If ten coins are flipped:

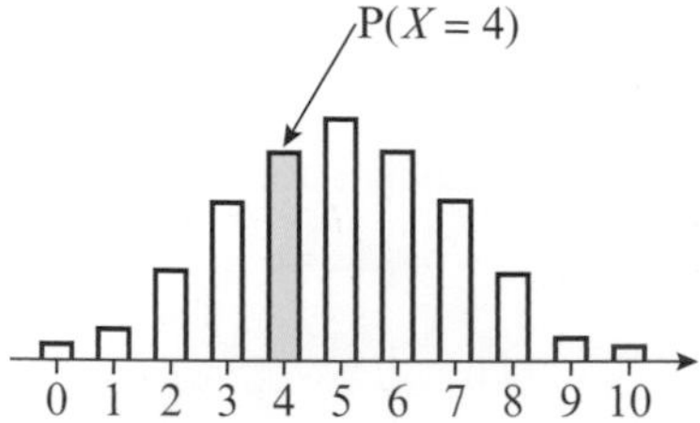

X = number of heads

Probability of getting 4 heads is written as $P(X = 4)$

X is a **discrete** random variable

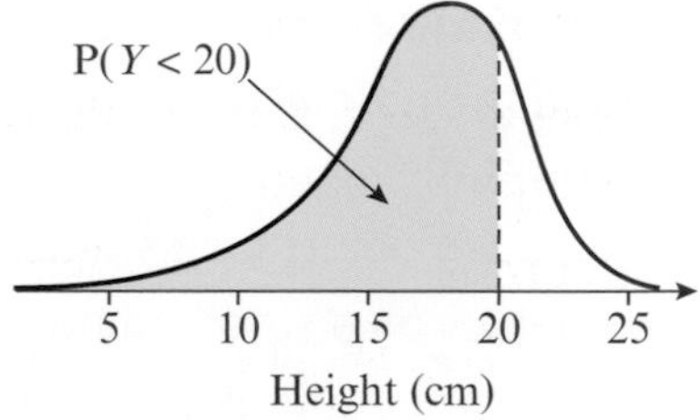

Y = average height of flipped coin

Probability that the average height is less than 20 cm is written as $P(Y < 20)$

Y is a **continuous** random variable

A continuous random variable has a **continuous probability distribution**. This can be shown as a curve on a graph.

- **The area under a continuous probability distribution is equal to 1.**

Links A discrete random variable can only take certain distinct values. The sum of the probabilities in a discrete probability distribution is equal to 1.

← Year 1, Chapter 6

The continuous variables generally encountered in real life are more likely to take values grouped around a central value than to take extreme values. The **normal distribution** is a continuous probability distribution that can be used to model many naturally occurring characteristics that behave in this way. Examples of continuous variables that can be modelled using the normal distribution are:

- heights of people within a given population
- weights of tigers in a jungle
- errors in scientific measurements
- size variations in manufactured objects

These histograms show the distribution of heights of adult males in a particular city. As the class width reduces, the distribution gets smoother.

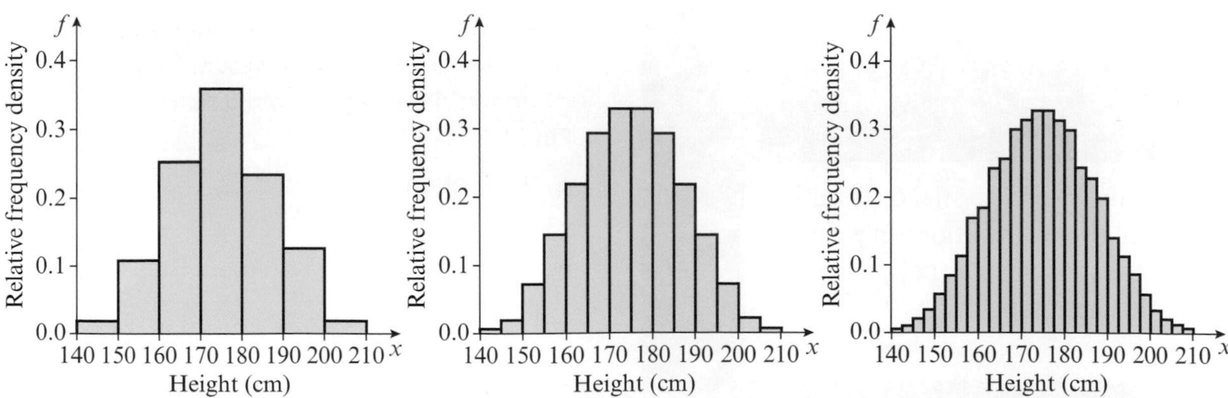

The distribution becomes bell-shaped and is symmetrical about the mean. You can model the heights of adult males in this city using a normal distribution, with mean 175 cm and standard deviation 12 cm.

Notation If X is a normally distributed random variable, you write $X \sim \text{N}(\mu, \sigma^2)$ where μ is the population mean and σ^2 is the population variance.

- **The normal distribution**
  - **has parameters μ, the population mean and σ^2, the population variance**
  - **is symmetrical (mean = median = mode)**
  - **has a bell-shaped curve with asymptotes at each end**
  - **has total area under the curve equal to 1**
  - **has points of inflection at $\mu + \sigma$ and $\mu - \sigma$**

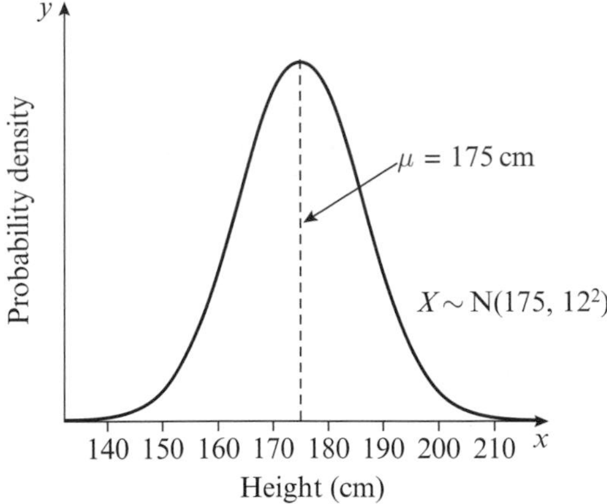

For a normally distributed variable:

- approximately 68% of the data lies within one standard deviation of the mean
- 95% of the data lies within two standard deviations of the mean
- nearly all of the data (99.7%) lies within three standard deviations of the mean

Watch out Although a normal random variable could take any value, in practice observations a long way (more than 5 standard deviations) from the mean have probabilities close to 0.

Example 1

The diameters of a rivet produced by a particular machine, X mm, is modelled as $X \sim \text{N}(8, 0.2^2)$. Find:

a $\text{P}(X > 8)$

b $\text{P}(7.8 < X < 8.2)$

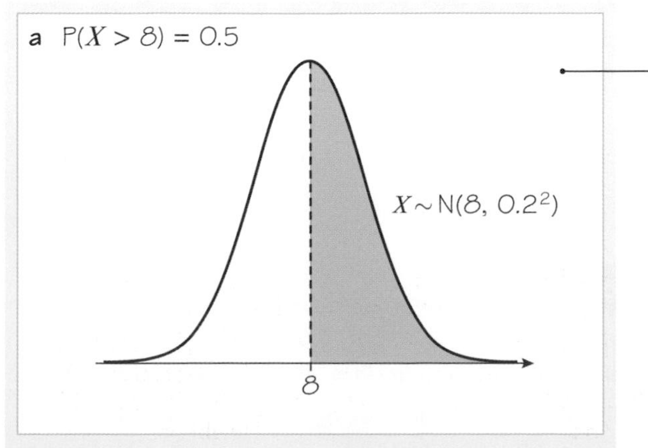

8 is the mean of the distribution. The normal distribution is **symmetrical**, so for any normally distributed random variable $\text{P}(X > \mu) = 0.5$.

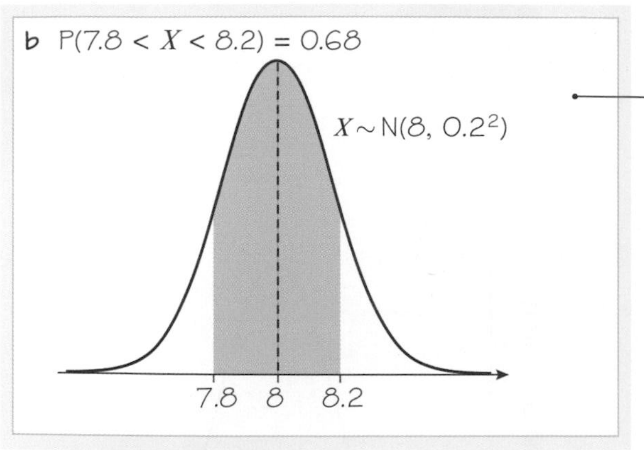

7.8 and 8.2 are each one standard deviation from the mean. For a normally distributed random variable, 68% of the data lies within one standard deviation of the mean. You can also write $P(\mu - \sigma < X < \mu + \sigma) = 0.68$.

Online Explore the normal distribution curve using technology.

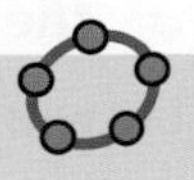

Exercise 3A

1 State, with a reason, whether these random variables are discrete or continuous:

a X, the lengths of a random sample of 100 sidewinder snakes in the Sahara desert

b Y, the scores achieved by 250 students in a university entrance exam

c C, the masses of honey badgers in a random sample of 1000

d Q, the shoe sizes of 200 randomly selected women in a particular town.

2 The lengths, X mm, of a bolt produced by a particular machine are normally distributed with mean 35 mm and standard deviation 0.4 mm. Sketch the distribution of X.

3 The distribution of incomes, in £000s per year, of employees of a bank is shown on the right.

State, with reasons, why the normal distribution is not a suitable model for this data.

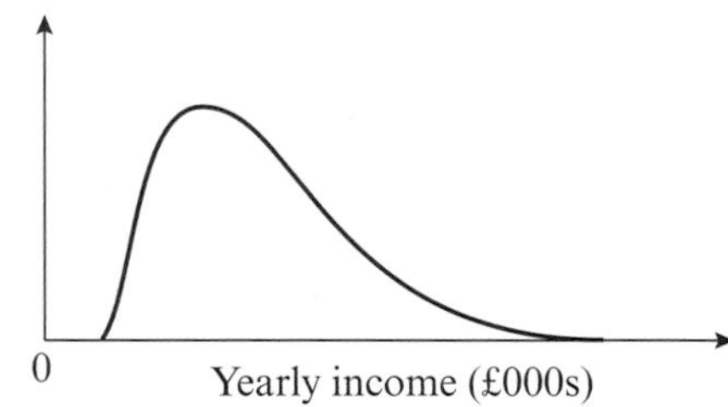

4 The armspans of a group of Year 5 pupils, X cm, are modelled as $X \sim N(120, 16)$.

a State the proportion of pupils that have an armspan between 116 cm and 124 cm.

b State the proportion of pupils that have an armspan between 112 cm and 128 cm.

5 The lengths of a colony of adders, Y cm, are modelled as $Y \sim N(100, \sigma^2)$. If 68% of the adders have a length between 93 cm and 107 cm, find σ^2.

(P) **6** The weights of a group of dormice, D grams, are modelled as $D \sim N(\mu, 25)$. If 97.5% of dormice weigh less than 70 grams, find μ.

Problem-solving

Draw a sketch of the distribution. Use the symmetry of the distribution and the fact that 95% of the data lies within 2 standard deviations of the mean.

(P) **7** The masses of the pigs, M kg, on a farm are modelled as $M \sim N(\mu, \sigma^2)$. If 84% of the pigs weigh more than 52 kg and 97.5% of the pigs weigh more than 47.5 kg, find μ and σ^2.

(P) **8** The percentage scores of a group of students in a test, S, are modelled as a normal distribution with mean 45 and standard deviation 15. Find:

a $P(S > 45)$ **b** $P(30 < S < 60)$ **c** $P(15 < S < 75)$

Alexia states that since it is impossible to score above 100%, this is not a suitable model.

d State, with a reason, whether Alexia is correct.

(E) **9** The diagram shows the distribution of heights, in cm, of barn owls in the UK.

An ornithologist notices that the distribution is approximately normal.

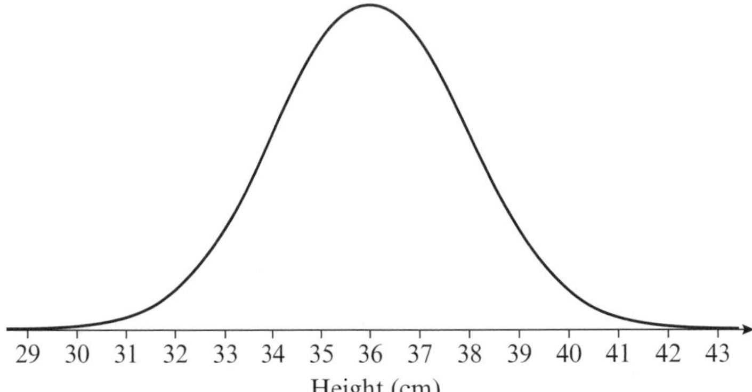

Hint The points of inflection on a normal distribution curve occur at $\mu \pm \sigma$.

a State the value of the mean height. **(1 mark)**

b Estimate the standard deviation of the heights. **(2 marks)**

3.2 Finding probabilities for normal distributions

You can find probabilities for a normal distribution using the **normal cumulative distribution** function on your calculator.

Example 2

$X \sim N(30, 4^2)$. Find:

a $P(X < 33)$ **b** $P(X \geq 24)$ **c** $P(33.5 < X < 38.2)$ **d** $P(X < 27 \text{ or } X > 32)$

a $P(X < 33) = 0.7734$ (4 d.p.)

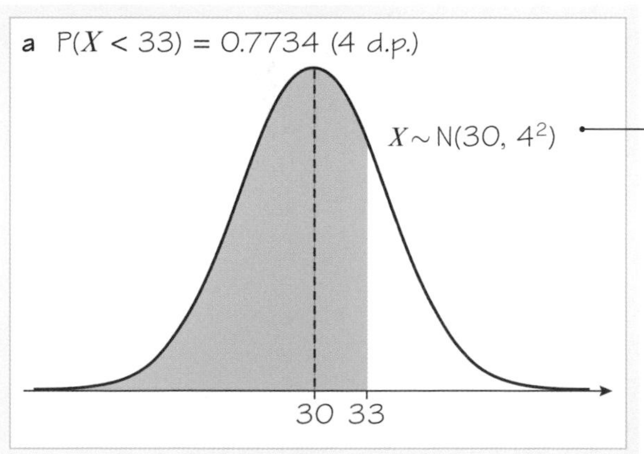

You should always draw a sketch to check your answer makes sense. 33 is larger than the mean so the probability should be greater than 0.5.

Watch out You need to enter a lower limit into your calculator. Choose a value at least 5 standard deviations away from the mean. For example 10, or –100. Because $P(X < 10)$ is very close to 0, $P(10 < X < 33) \approx P(X < 33)$.

Online Use the Normal CD function on your calculator to find probabilities from a normal distribution.

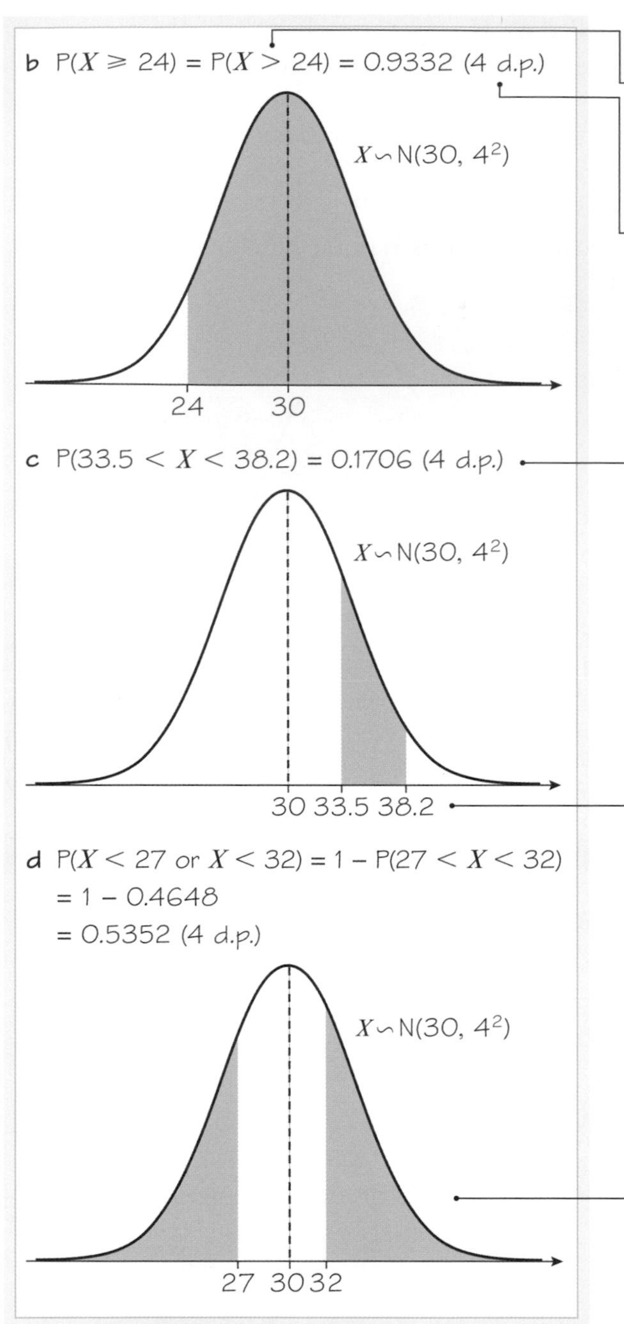

You can use either $>$ or $\geqslant$ interchangeably with a continuous distribution. This is because $P(X = 24) = 0$.

Set the upper limit on your calculator to any large value greater than 5 standard deviations above the mean. You could use 50, 100 or 1000.

Enter both the upper and the lower limits in your calculator.

Both 33.5 and 38.2 are above the mean, so the probability should be less than 0.5.

Use the fact that the total probability is equal to 1. Sketch the two 'tails' of the required area.

Example 3

An IQ test is applied to a population of adults. The scores, X, on the test are found to be normally distributed with $X \sim N(100, 15^2)$. Adults scoring more than 140 on the test are classified as 'genius'.

a Find the probability that an adult chosen at random achieves a 'genius' classification. Give your answer to three significant figures.

b Twenty adults take the test. Find the probability that two or more are classified as 'genius'.

a $P(X > 140) = 0.00383$ (3 s.f.)

Use your calculator and choose a large upper limit, such as 200.

b Let Y be the number of adults who classify as 'genius'.

$Y \sim B(20, 0.00383)$

This is 20 trials, each with probability of success 0.00383. You can model the number of successful trials as a binomial random variable.

$P(Y \geqslant 2) = 1 - P(Y \leqslant 1)$
$= 1 - 0.9973379...$
$= 0.00266$ (3 s.f.)

Use the binomial cumulative distribution function on your calculator to find $P(Y \leqslant 1)$. You could also find $P(Y = 1) + P(Y = 0)$ which will have the same value. Then subtract this result from 1 to find $P(Y \geqslant 2)$.

← Year 1, Chapter 6

Exercise 3B

1 The random variable $X \sim N(30, 2^2)$.
Find: **a** $P(X < 33)$ **b** $P(X > 26)$ **c** $P(X \geqslant 31.6)$

2 The random variable $X \sim N(40, 9)$.
Find: **a** $P(X > 45)$ **b** $P(X \leqslant 38)$ **c** $P(41 \leqslant X \leqslant 44)$

Watch out In the normal distribution N(40, 9) the second parameter is the **variance**. The standard deviation in this normal distribution is $\sqrt{9} = 3$.

3 The random variable $X \sim N(25, 25)$.
Find: **a** $P(Y < 20)$ **b** $P(18 < Y < 26)$ **c** $P(Y > 23.8)$

4 The random variable $X \sim N(18, 10)$.
Find: **a** $P(X \geqslant 20)$ **b** $P(X < 15)$ **c** $P(18.4 < X < 18.7)$

5 The random variable $M \sim N(15, 1.5^2)$.
a Find: **i** $P(M > 14)$ **ii** $P(M < 14)$
b Calculate the sum of your answers to **a i** and **ii** and comment on your answer.

6 The random variable $T \sim N(4.5, 0.4)$.
a Find $P(T < 4.2)$.
b Without further calculation, write down $P(T > 4.2)$.

(P) **7** The random variable $Y \sim N(45, 2^2)$. Find:
a $P(Y < 41 \text{ or } Y > 47)$ **b** $P(Y < 44 \text{ or } 46.5 < Y < 47.5)$

(E) **8** The volume of soap dispensed by a soap-dispenser on each press, X ml, is modelled as $X \sim N(6, 0.8^2)$.
a Find: **i** $P(X > 7)$ **ii** $P(X < 5)$ **(2 marks)**
The soap dispenser is pressed three times.
b Find the probability that on all three presses, less than 5 ml of soap is dispensed. **(2 marks)**

(E) **9** The amount of mineral water, W ml, in a bottle produced by a certain manufacturer is modelled as $W \sim N(500, 14^2)$.

a Find: **i** $P(W > 505)$ **ii** $P(W < 490)$ **(2 marks)**

A sample of 4 bottles is taken.

b Find the probability that all of the bottles contain more than 490 ml. **(2 marks)**

(P) **10** The heights of a large group of women are normally distributed with a mean of 165 cm and a standard deviation of 3.5 cm. A woman is selected at random from this group.

a Find the probability that she is shorter than 160 cm.

Problem-solving

For part **c**, formulate a binomial random variable to represent the number of women in the sample who meet Steven's criteria.

Steven is looking for a woman whose height is between 168 cm and 174 cm for a part in his next film.

b Find the proportion of women from this group who meet Steven's criteria.

A sample of 20 women is taken from the group.

c Find the probability that at least 5 of the women meet Steven's criteria.

(E/P) **11** The diameters of bolts, D mm, made by a particular machine are modelled as $D \sim N(13, 0.1^2)$.

a Find the probability that a bolt, chosen at random, has a diameter less than 12.8 mm. **(1 mark)**

Bolts are considered to be 'perfect' if the diameter lies between 12.9 mm and 13.1 mm. A random sample of 40 bolts is taken.

b Find the probability that more than 25 of the bolts are 'perfect'. **(4 marks)**

(E/P) **12** The masses, X grams, of a large population of squirrels are modelled as a normal distribution with $X \sim N(480, 40^2)$.

a Find the probability that a squirrel chosen at random has a mass greater than 490 g. **(1 mark)**

A naturalist takes a random sample of 30 squirrels from the population.

b Find the probability that at least 15 of the squirrels have a mass between 470 g and 490 g. **(4 marks)**

3.3 The inverse normal distribution function

For a given probability, p, you can use your calculator to find a value of a such that $P(X < a) = p$. This function is usually called the **inverse normal distribution** function on your calculator.

Example 4

$X \sim N(20, 3^2)$. Find, correct to two decimal places, the values of a such that:

a $P(X < a) = 0.75$

b $P(X > a) = 0.4$

c $P(16 < X < a) = 0.3$

a $a = 22.02$ (2 d.p.)

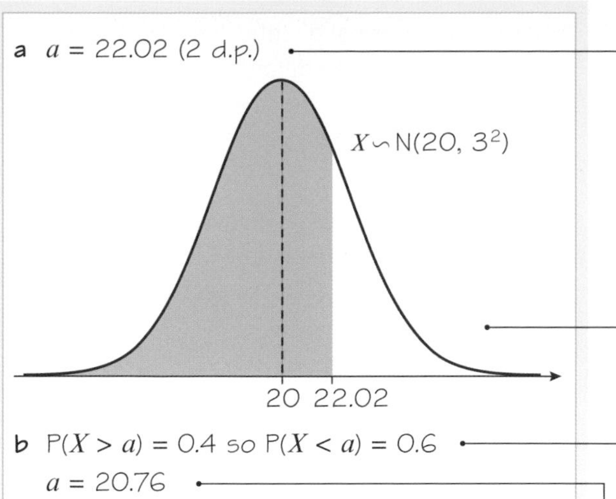

Enter $\mu = 20$, $\sigma = 3$ and $p = 0.75$ into your calculator. The value for p might be labelled 'Area' on your calculator because it represents the area under the curve to the left of a.

This means that for $X \sim N(20, 3^2)$, $P(X < 22.02) = 0.75$. You can check this result using your calculator.

Draw a sketch to check that your answer makes sense. 0.75 is more than 0.5 so the value should be greater than the mean.

b $P(X > a) = 0.4$ so $P(X < a) = 0.6$

$a = 20.76$

Use the fact that $P(X > a) + P(X < a) = 1$ to find the area to the **left** of a before using your calculator.

$P(X < 20.76) = 0.6$, so $P(X > 20.76) = 0.4$ as needed.

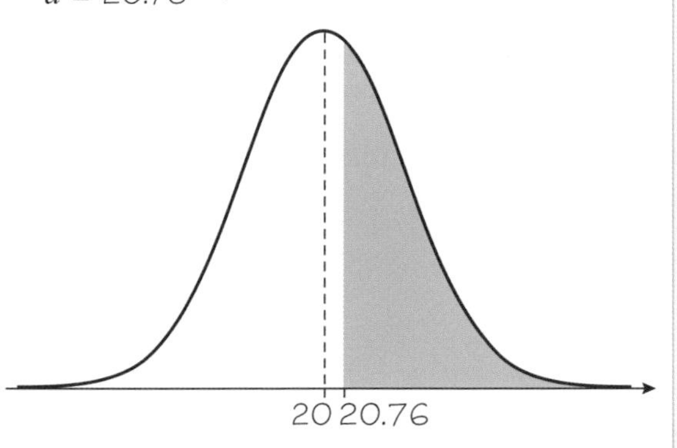

c $P(16 < X < a) = 0.3$

So $P(X < a) = 0.3 + P(X < 16)$

$= 0.3 + 0.09121 = 0.39121$

So $a = 19.17$

Watch out You can't use your calculator to find a directly. Use the fact that $P(16 < X < a) = P(X < a) - P(X < 16)$

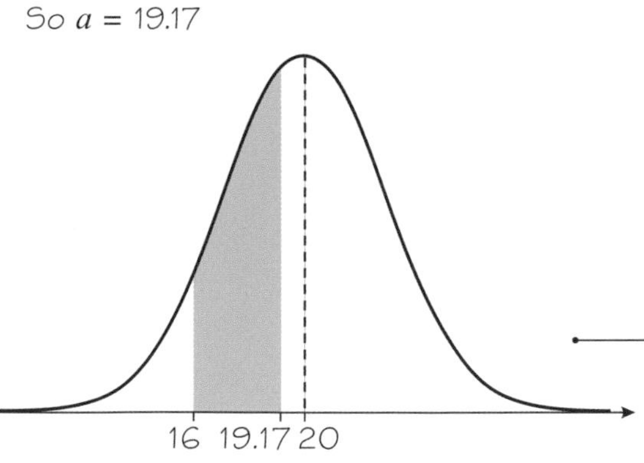

You can check your answer using your calculator by working out $P(16 < X < 19.17)$.

Online Use the Inverse Normal function on your calculator to calculate values which satisfy given probability statements for the normal distribution.

Example 5

Plates made using a particular manufacturing process have a diameter, D cm, which can be modelled using a normal distribution, $D \sim N(20, 1.5^2)$.

a Given that 60% of plates are less than x cm, find x.

b Find the interquartile range of the plate diameters.

a $P(D < x) = 0.6$
$\Rightarrow x = 20.38$ cm

b $P(D < Q_1) = 0.25$
$\Rightarrow Q_1 = 18.99$ cm
$P(D < Q_3) = 0.75$
$\Rightarrow Q_3 = 21.01$ cm
The interquartile range is
$21.01 - 18.99 = 2.02$ cm (2 d.p.)

You can use a normal distribution to determine the **proportion** of the data that lie within a certain interval. Use the inverse normal distribution function on your calculator.

25% of the data values lie below the lower quartile, Q_1, and 75% lie below the upper quartile, Q_3. ← Year 1, Chapter 2

The distribution is symmetrical so Q_1 and Q_3 should be the same distance away from the mean.

Exercise 3C

1 The random variable $X \sim N(30, 5^2)$. Find the value of a, to 2 decimal places, such that:

a $P(X < a) = 0.3$ b $P(X < a) = 0.75$ c $P(X > a) = 0.4$ d $P(32 < X < a) = 0.2$

2 The random variable $X \sim N(12, 3^2)$. Find the value of a, to 2 decimal places, such that:

a $P(X < a) = 0.1$ b $P(X > a) = 0.65$

c $P(10 \leqslant X \leqslant a) = 0.25$ d $P(a < X < 14) = 0.32$

3 The random variable $X \sim N(20, 12)$.

a Find the value of a and the value of b such that:

i $P(X < a) = 0.40$ ii $P(X > b) = 0.6915$

b Find $P(b < X < a)$.

4 The random variable $Y \sim N(100, 15^2)$.

a Find the value of a and the value of b such that:

i $P(Y > a) = 0.975$ ii $P(Y < b) = 0.10$

b Find $P(a < Y < b)$.

5 The random variable $X \sim N(80, 16)$.

a Find the value of a and the value of b such that:

i $P(X > a) = 0.40$ ii $P(X < b) = 0.5636$

b Find $P(b < X < a)$.

(P) 6 The masses, M kg, of a population of badgers are modelled as $M \sim N(4.5, 0.6^2)$.

For this population, find:

a the lower quartile

b the 80th percentile

c Explain without calculation why $Q_2 = 4.5$ kg.

(E) 7 The percentage scores, X, of a group of learner drivers in a theory test is modelled as a normal distribution with $X \sim N(72, 6^2)$.

a Find the value of a such that $P(X < a) = 0.6$. **(1 mark)**

b Find the interquartile range of the scores. **(2 marks)**

E/P **8** The masses, Y grams, of a brand of chocolate bar are modelled as $Y \sim N(60, 2^2)$.

a Find the value of y such that $P(Y > y) = 0.2$. **(1 mark)**

b Find the 10% to 90% interpercentile range of masses. **(2 marks)**

c Tom says that the median is equal to the mean. State, with a reason, whether Tom is correct. **(1 mark)**

E/P **9** The distribution of heights, H cm, of a large group of men is modelled using $H \sim N(170, 10^2)$. A frock coat is a coat that goes from the neck of a person to near the floor. A clothing manufacturer uses the information to make three different lengths of frock coats. The table below shows the proportion of each size they will make.

Short	Regular	Long
30%	50%	20%

a The company wants to advertise a range of heights for which the regular frock coat is suitable. Use the model to suggest suitable heights for the advertisement. **(4 marks)**

b State one assumption you have made in deciding these values. **(1 mark)**

3.4 The standard normal distribution

It is often useful to **standardise** normally distributed random variables. You do this by coding the data so that it can be modelled by the **standard normal distribution**.

- **The standard normal distribution has mean 0 and standard deviation 1.**

Notation The standard normal variable is written as $Z \sim N(0, 1^2)$.

If $X \sim N(\mu, \sigma^2)$ is a normal random variable with mean μ and standard deviation σ, then you can code X using the formula:

$$Z = \frac{X - \mu}{\sigma}$$

Links If $X = x$ then the corresponding value of Z will be $z = \frac{x - \mu}{\sigma}$. The mean of the coded data will be $\frac{\mu - \mu}{\sigma} = 0$ and the standard deviation will be $\frac{\sigma}{\sigma} = 1$. ← Year 1, Section 2.5

The resulting **z-values** will be normally distributed with mean 0 and standard deviation 1.

For the standard normal curve $Z \sim N(0, 1^2)$, the probability $P(Z < a)$ is sometimes written as $\Phi(a)$. You can find it by entering $\mu = 0$ and $\sigma = 1$ into the normal cumulative distribution function on your calculator.

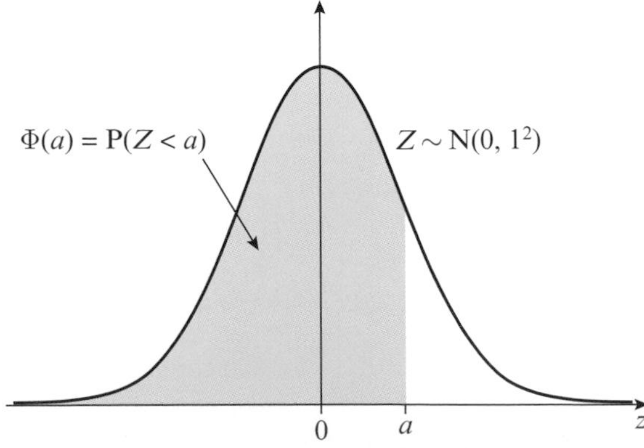

Example 6

The random variable $X \sim \text{N}(50, 4^2)$. Write in terms of $\Phi(z)$ for some value z:

a $\text{P}(X < 53)$ **b** $\text{P}(X \geqslant 55)$

a $z = \dfrac{53 - 50}{4} = 0.75$

$\text{P}(X < 53) = \text{P}(Z < 0.75)$

$= \Phi(0.75)$

b $\text{P}(X \geqslant 55) = 1 - \text{P}(X < 55)$

$z = \dfrac{55 - 50}{4} = 1.25$

$\text{P}(X \geqslant 55) = 1 - \text{P}(Z < 1.25)$

$= 1 - \Phi(1.25)$

Code the data so that it is modelled by the standard normal distribution $\text{N}(0, 1^2)$. Use $Z = \dfrac{X - \mu}{\sigma}$

The distribution is continuous, so you can use $<$ and $\leqslant$ interchangeably.

You sometimes need to find z-values that correspond to given probabilities. You can find these probabilities for some standard values of p by using the percentage points of the normal distribution table on page 191. This table will be given in the *Mathematics Formulae and Statistical Tables* booklet in your exam. It gives values of z and p such that $\text{P}(Z > z) = p$.

p	z	p	z
0.5000	0.0000	0.0500	1.6449
0.4000	0.2533	0.0250	1.9600
0.3000	0.5244	0.0100	2.3263
0.2000	0.8416	0.0050	2.5758
0.1500	1.0364	0.0010	3.0902
0.1000	1.2816	0.0005	3.2905

So $\text{P}(Z > 1.96) = 0.025$. You can use the symmetry of the distribution to find corresponding negative z-values. $\text{P}(Z < -1.96) = 0.025$ so $\text{P}(Z > -1.96) = 0.975$.

Example 7

The systolic blood pressure of an adult population, S mmHg, is modelled as a normal distribution with mean 127 and standard deviation 16.

A medical researcher wants to study adults with blood pressures higher than the 95th percentile. Find the minimum blood pressure for an adult included in her study.

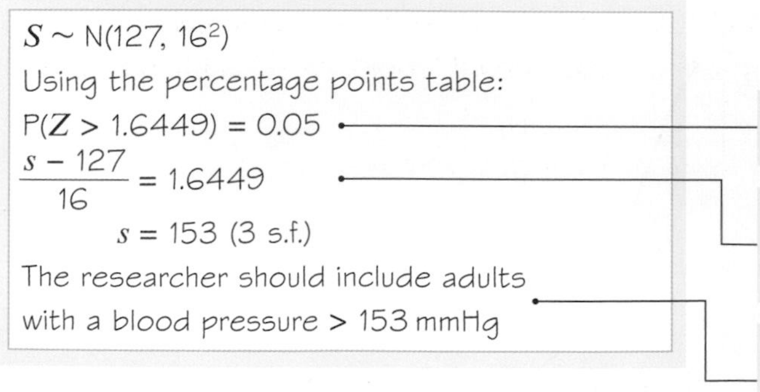

$S \sim \text{N}(127, 16^2)$

Using the percentage points table:

$\text{P}(Z > 1.6449) = 0.05$

$\dfrac{s - 127}{16} = 1.6449$

$s = 153$ (3 s.f.)

The researcher should include adults with a blood pressure > 153 mmHg

Use the percentage points table with $p = 0.05$.

Convert the value for Z back into a value for S. Remember that the denominator is σ, not σ^2.

You could also find the inverse normal function on your calculator with $\mu = 127$, $\sigma = 16$ and $p = 0.95$.

Exercise 3D

1 For the standard normal distribution $Z \sim N(0, 1^2)$, find:

a $P(Z < 2.12)$ b $P(Z < 1.36)$ c $P(Z > 0.84)$ d $P(Z < -0.38)$
e $P(-2.30 < Z < 0)$ f $P(Z < -1.63)$ g $P(-2.16 < Z < -0.85)$ h $P(-1.57 < Z < 1.57)$

2 For the standard normal distribution $Z \sim N(0, 1^2)$, find values of a such that:

a $P(Z < a) = 0.9082$ b $P(Z > a) = 0.0314$
c $P(Z > a) = 0.1500$ d $P(Z > a) = 0.9500$
e $P(0 < Z < a) = 0.3554$ f $P(0 < Z < a) = 0.4946$
g $P(-a < Z < a) = 0.80$ h $P(-a < Z < a) = 0.40$

Hint For parts **g** and **h** you will need to use the symmetry properties of the distribution.

3 The random variable $X \sim N(0.8, 0.05^2)$. For each of the following values of X, write down the corresponding value of the standardised normal distribution, $Z \sim N(0, 1^2)$.

a $x = 0.8$ b $x = 0.792$ c $x = 0.81$ d $x = 0.837$

4 The normal distribution $X \sim N(154, 12^2)$. Write in terms of $\Phi(z)$:

a $P(X < 154)$ b $P(X < 160)$
c $P(X > 151)$ d $P(140 < X < 155)$

Hint Write your answer to part **d** in the form $\Phi(z_1) - \Phi(z_2)$.

(E) 5 a Use the percentage points table to find a value of z such that $P(Z > z) = 0.025$. **(1 mark)**

b A fighter jet training programme takes only the top 2.5% of candidates on a test. Given that the scores can be modelled using a normal distribution with mean 80 and standard deviation 4, use your answer to part **a** to find the score necessary to get on the programme. **(2 marks)**

(E) 6 a Use the percentage points table to find a value of z such that $P(Z < z) = 0.15$. **(1 mark)**

b A hat manufacturer makes a special ‘petite’ hat which should fit 15% of its customers. Given that hat sizes can be modelled using a normal distribution with mean 57 cm and standard deviation 2 cm, use your answer to part **a** to find the size of a ‘petite’ hat. **(2 marks)**

(E) 7 a Use the percentage points table to find the values of z that correspond to the 10% to 90% interpercentile range. **(2 marks)**

A particular brand of light bulb has a life modelled as a normal distribution with mean 1175 hours and standard deviation 56 hours. The bulb life is considered ‘standard’ if its life falls into the 10% to 90% interpercentile range.

b Use your answer to part **a** to find the range of life to the nearest hour for a ‘standard’ bulb. **(2 marks)**

3.5 Finding μ and σ

You might need to find an unknown mean or standard deviation for a normally distributed variable.

Example 8

The random variable $X \sim N(\mu, 3^2)$.

Given that $P(X > 20) = 0.20$, find the value of μ.

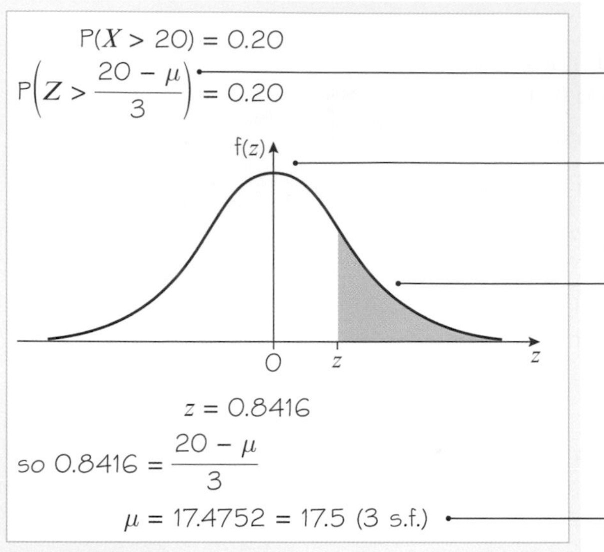

Use $Z = \frac{X - \mu}{\sigma}$

Draw a diagram for Z.

$p = 0.20$

Problem-solving

You don't know μ, so you need to use the **standard normal distribution**. Use your calculator with $\mu = 0$, $\sigma = 1$ and $p = 0.8$ to find the value of z such that $P(Z > z) = 0.2$. You could also use the percentage points table.

You know one value of X and the corresponding value of Z so use the coding formula to find μ.

Example 9

A machine makes metal sheets with width, X cm, modelled as a normal distribution such that $X \sim N(50, \sigma^2)$.

a Given that $P(X < 46) = 0.2119$, find the value of σ.

b Find the 90th percentile of the widths.

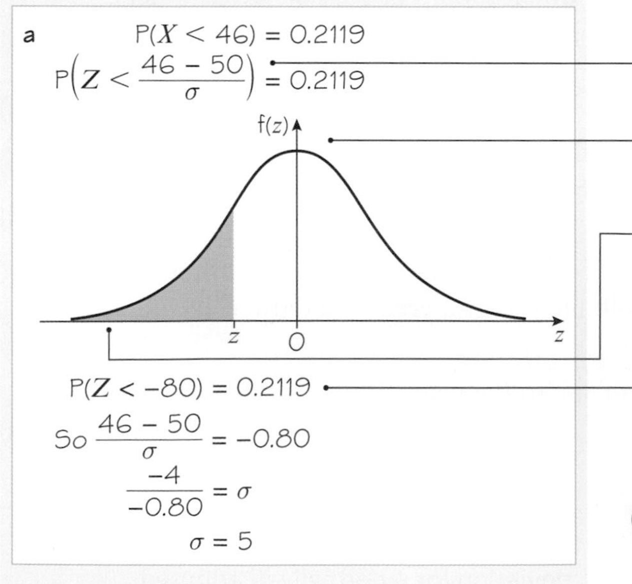

Use $Z = \frac{X - \mu}{\sigma}$

Draw a diagram for Z.

$p = 0.2119$

Use the inverse normal distribution function on your calculator with $\mu = 0$, $\sigma = 1$ and $p = 0.2119$. You can also find this value from the percentage points table by noting that $P(Z > 0.80) = 0.2119$.

Online Use the Inverse Normal function on your calculator with the standard normal distribution.

b $X \sim N(50, 5^2)$.
Let a be the 90th percentile.
$P(X < a) = 0.9$
$a = 56.4$ cm (1 d.p.)

Now that you have calculated σ you can write out the distribution.

Use the inverse normal distribution function of your calculator with $\mu = 50$, $\sigma = 5$ and $p = 0.9$.

Example 10

The random variable $X \sim N(\mu, \sigma^2)$.
Given that $P(X > 35) = 0.025$ and $P(X < 15) = 0.1469$, find the value of μ and the value of σ.

$P(Z > z_1) = 0.025 \Rightarrow z_1 = 1.96$
$P(Z < z_2) = 0.1469 \Rightarrow z_2 = -1.05$

So $-1.05 = \dfrac{15 - \mu}{\sigma}$

$-1.05\sigma + \mu = 15$ (1)

and $1.96 = \dfrac{35 - \mu}{\sigma}$

$1.96\sigma + \mu = 35$ (2)

(2) − (1): $3.01\sigma = 20$

$\sigma = 6.6445...$

Substituting into (2):
$\mu = 35 - 1.96 \times 6.6445... = 21.976...$
So $\sigma = 6.64$ and $\mu = 22.0$ (3 s.f.)

Find z-values corresponding to a 'right-tail' of 0.025 and a 'left-tail' of 0.1469:

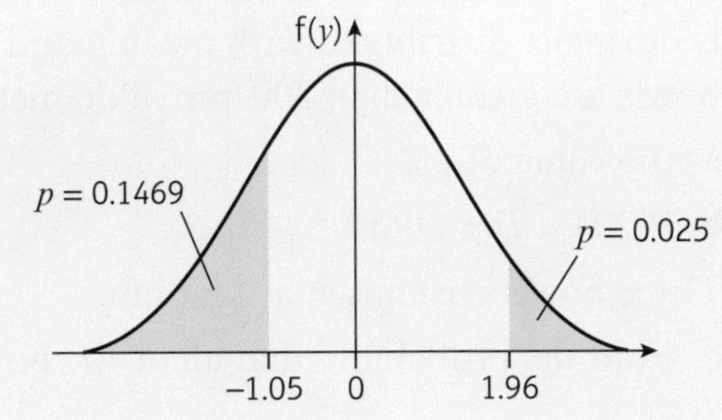

Use $\dfrac{X - \mu}{\sigma}$ to link X and Z values and form two simultaneous equations in μ and σ.

Exercise 3E

1 The random variable $X \sim N(\mu, 5^2)$ and $P(X < 18) = 0.9032$.
Find the value of μ.

2 The random variable $X \sim N(11, \sigma^2)$ and $P(X > 20) = 0.01$.
Find the value of σ.

3 The random variable $Y \sim N(\mu, 40)$ and $P(Y < 25) = 0.15$.
Find the value of μ.

4 The random variable $Y \sim N(50, \sigma^2)$ and $P(Y > 40) = 0.6554$.
Find the value of σ.

(P) 5 The random variable $X \sim N(\mu, \sigma^2)$.
Given that $P(X < 17) = 0.8159$ and $P(X < 25) = 0.9970$, find the value of μ and the value of σ.

(P) 6 The random variable $Y \sim N(\mu, \sigma^2)$.
Given that $P(Y < 25) = 0.10$ and $P(Y > 35) = 0.005$, find the value of μ and the value of σ.

(P) **7** The random variable $X \sim N(\mu, \sigma^2)$.
Given that $P(X > 15) = 0.20$ and $P(X < 9) = 0.20$,
find the value of μ and the value of σ.

Hint Draw a diagram and use symmetry to find μ.

(P) **8** The random variable $X \sim N(\mu, \sigma^2)$.
The lower quartile of X is 25 and the upper quartile of X is 45.
Find the value of μ and the value of σ.

(P) **9** The random variable $X \sim N(0, \sigma^2)$.
Given that $P(-4 < X < 4) = 0.6$, find the value of σ.

(P) **10** The random variable $X \sim N(2.68, \sigma^2)$.
Given that $P(X > 2a) = 0.2$ and $P(X < a) = 0.4$, find the value of σ and the value of a.

(E) **11** An automated pottery wheel is used to make bowls. The diameter of the bowls, D mm, is normally distributed with mean μ and standard deviation 5 mm. Given that 75% of bowls are greater than 200 mm in diameter, find:

a the value of μ **(2 marks)**

b $P(204 < D < 206)$ **(1 mark)**

Three bowls are chosen at random.

c Find the probability that all of the bowls are greater than 205 mm in diameter. **(3 marks)**

(E/P) **12** A loom makes table cloths with an average thickness of 2.5 mm. The thickness, T mm, can be modelled using a normal distribution. Given that 65% of table cloths are less than 2.55 mm thick, find:

a the standard deviation of the thickness **(2 marks)**

b the proportion of table cloths with thickness between 2.4 mm and 2.6 mm. **(1 mark)**

A table cloth can be sold if the thickness is between 2.4 mm and 2.6 mm. A sample of 20 table cloths is taken.

c Find the probability that at least 15 table cloths can be sold. **(3 marks)**

(E/P) **13** The masses of the penguins on an island are found to be normally distributed with mean μ, and standard deviation σ. Given that 10% of the penguins have a mass less than 18 kg and 5% of the penguins have a mass greater than 30 kg,

a sketch a diagram to represent this information **(2 marks)**

b find the value of μ and the value of σ. **(6 marks)**

10 penguins are chosen at random.

c Find the probability that at least 4 of them have a mass greater than 25 kg. **(4 marks)**

(E/P) **14** The length of an adult Dachshund is found to be normally distributed with mean μ and standard deviation σ. Given that 20% of Dachshunds have a length less than 16 inches and 10% have a length greater than 18 inches, find:

a the value of μ and the value of σ **(6 marks)**

b the interquartile range. **(2 marks)**

Challenge

A normally distributed random variable $X \sim N(\mu, \sigma^2)$ has interquartile range q.

a Show that $\sigma = 0.742q$, where the coefficient of q is correct to 3 s.f.

b Explain why it is not possible to write μ in terms of q only.

3.6 Approximating a binomial distribution

Consider the binomial random variable $X \sim B(n, p)$. It can be difficult to calculate probabilities for X when n is large. In certain circumstances you can use a normal distribution to **approximate** a binomial distribution.

Links The cumulative binomial tables in the formulae booklet only go up to $n = 50$.
← **Year 1, Chapter 6**

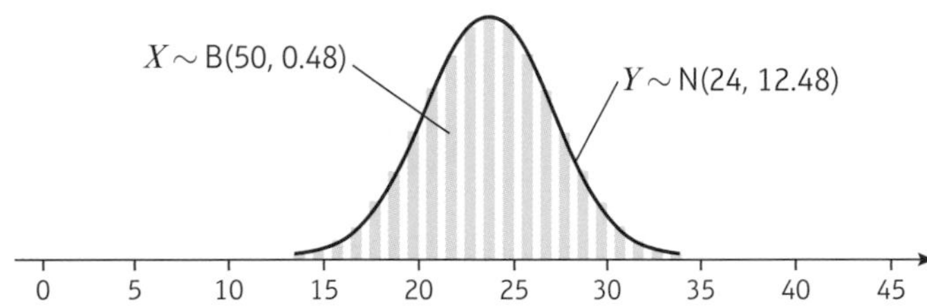

You need to understand the conditions under which this approximation is valid, and learn the relationship between the values of n and p in $B(n, p)$ and the values of μ and σ in the normal approximation $N(\mu, \sigma^2)$.

- **If n is large and p is close to 0.5, then the binomial distribution $X \sim B(n, p)$ can be approximated by the normal distribution $N(\mu, \sigma^2)$ where**
  - $\mu = np$
  - $\sigma = \sqrt{np(1-p)}$

Hint The approximation is only valid when p is close is to 0.5 because the normal distribution is **symmetrical**.

Example 11

A biased coin has P(Head) = 0.53. The coin is tossed 100 times and the number of heads, X, is recorded.

a Write down a binomial model for X.

b Explain why X can be approximated with a normal distribution, $Y \sim N(\mu, \sigma^2)$.

c Find the values of μ and σ in this approximation.

a $X \sim B(100, 0.53)$

b The distribution can be approximated with a normal distribution since n is large and p is close to 0.5.

c $\mu = 100 \times 0.53 = 53$ — Use $\mu = np$

$\sigma = \sqrt{100 \times 0.53 \times (1 - 0.53)} = 4.99$ (3 s.f.) — Use $\sigma = \sqrt{np(1-p)}$

The binomial distribution is a discrete distribution but the normal distribution is continuous.

- **If you are using a normal approximation to a binomial distribution, you need to apply a continuity correction when calculating probabilities.**

The diagrams show $X \sim B(14, 0.5)$ being approximated by $Y \sim N(7, 1.87^2)$:

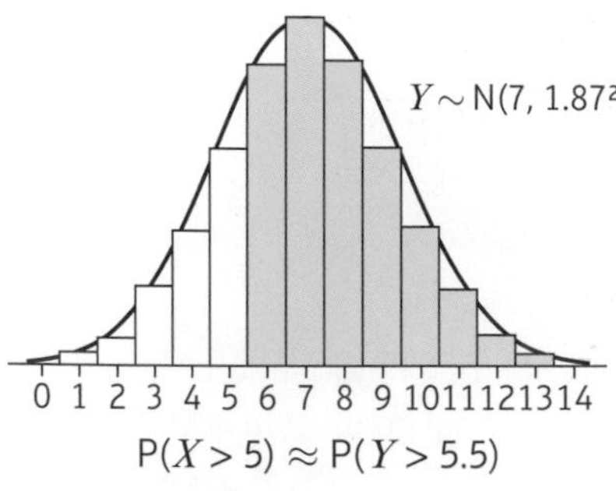

$P(X > 5) \approx P(Y > 5.5)$

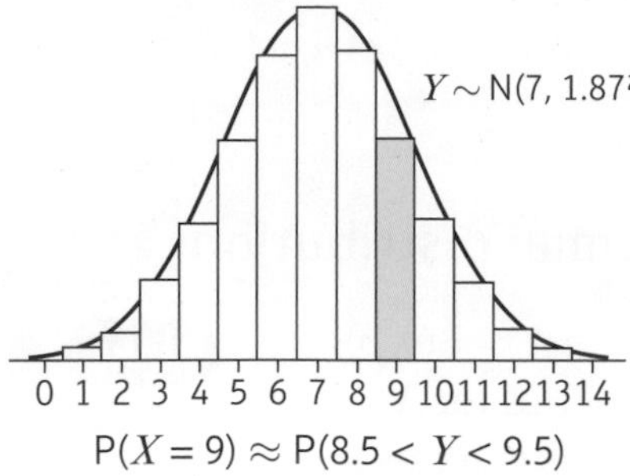

$P(X = 9) \approx P(8.5 < Y < 9.5)$

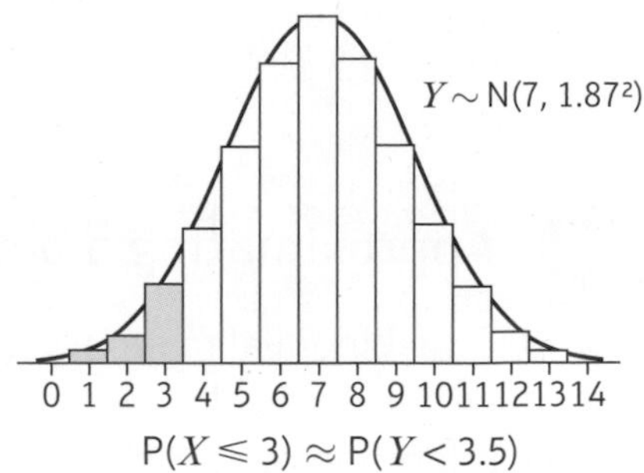

$P(X \leqslant 3) \approx P(Y < 3.5)$

Example 12

The binomial random variable $X \sim B(150, 0.48)$ is approximated by the normal random variable $Y \sim N(72, 6.12^2)$.

Use this approximation to find:

a $P(X \leqslant 70)$ **b** $P(80 \leqslant X < 90)$

Watch out Remember to apply the continuity correction. You are interested in values of the **discrete** random variable X that are less than **or equal to** 70, so you need to consider values less than 70.5 for the continuous random variable Y.

a $P(X \leqslant 70) \approx P(Y < 70.5) = 0.4032$ (4 d.p.)

b $P(80 \leqslant X < 90) \approx P(79.5 < Y < 89.5)$
$= 0.9979 - 0.8898$
$= 0.1081$ (4 d.p.)

For values of X less than 90 consider values of Y less than 89.5.

Example 13

For a particular type of flower bulb, 55% will produce yellow flowers. A random sample of 80 bulbs is planted.

Calculate the percentage error incurred when using a normal approximation to estimate the probability that there are exactly 50 yellow flowers.

Let X = the number of bulbs producing yellow flowers in a sample of 80.

Define a suitable binomial random variable.

Then $X \sim B(80, 0.55)$

$P(X = 50) = \binom{80}{50} 0.55^{50} 0.45^{30} = 0.0365$

Use your calculator to find the exact probability using a binomial distribution. ← **Year 1, Chapter 6**

X can be approximated by the normal distribution $Y \sim N(\mu, \sigma^2)$, where $\mu = 80 \times 0.55 = 44$

Use $\mu = np$

$\sigma = \sqrt{80 \times 0.55 \times (1 - 0.55)} = \sqrt{19.8}$ (3 s.f.)

$Y \sim N(44, 19.8)$

Write down the normal approximation.

$P(X = 50) \approx P(49.5 < Y < 50.5)$
$= 0.9280 - 0.8918 = 0.0362$ (4 d.p.)

To estimate the probability that X takes a single value, apply a continuity correction by considering values half a unit below and half a unit above.

Percentage error $= \dfrac{0.0365 - 0.0362}{0.0365} \times 100 = 0.82\%$

Exercise 3F

1 For each of the following binomial random variables, X:
  i state, with reasons, whether X can be approximated by a normal distribution.
  ii if appropriate, write down the normal approximation to X in the form $N(\mu, \sigma^2)$, giving the values of μ and σ.

a $X \sim B(120, 0.6)$ b $X \sim B(20, 0.5)$ c $X \sim B(250, 0.52)$
d $X \sim B(300, 0.85)$ e $X \sim B(400, 0.48)$ f $X \sim B(1000, 0.58)$

2 The random variable $X \sim B(150, 0.45)$. Use a suitable approximation to estimate:
a $P(X \leqslant 60)$ b $P(X > 75)$ c $P(65 \leqslant X \leqslant 80)$

3 The random variable $X \sim B(200, 0.53)$. Use a suitable approximation to estimate:
a $P(X < 90)$ b $P(100 \leqslant X < 110)$ c $P(X = 105)$

4 The random variable $X \sim B(100, 0.6)$. Use a suitable approximation to estimate:
a $P(X > 58)$ b $P(60 < X \leqslant 72)$ c $P(X = 70)$

5 A fair coin is tossed 70 times. Use a suitable approximation to estimate the probability of obtaining more than 45 heads.

6 The probability of a roulette ball landing on red when the wheel is spun is $\frac{50}{101}$.
On one day in a casino, the wheel is spun 1200 times.
Estimate the probability that the ball lands on red in at least half of these spins.

(E) 7 a Write down two conditions under which the normal distribution may be used as an approximation to the binomial distribution. **(2 marks)**

A company sells orchids of which 45% produce pink flowers.
A random sample of 20 orchids is taken and X produce pink flowers.

b Find $P(X = 10)$. **(1 mark)**

A second random sample of 240 orchids is taken.

c Using a suitable approximation, find the probability that fewer than 110 orchids produce pink flowers. **(3 marks)**

d The probability that at least q orchids produce pink flowers is 0.2. Find q. **(3 marks)**

(E) 8 A drill bit manufacturer claims that 52% of its bits last longer than 40 hours.
A random sample of 30 bits is taken and X last longer than 40 hours.

a Find $P(X < 17)$. **(1 mark)**

A second random sample of 600 drill bits is taken.

b Using a suitable approximation, find the probability that between 300 and 350 bits last longer than 40 hours. **(3 marks)**

(E/P) 9 A particular breakfast cereal has prizes in 56% of the boxes. A random sample of 100 boxes is taken.

a Find the exact value of the probability that exactly 55 boxes contain a prize. **(1 mark)**

b Find the percentage error when using a normal approximation to calculate the probability that exactly 55 boxes contain prizes. **(4 marks)**

3.7 Hypothesis testing with the normal distribution

You can test hypotheses about the mean of a normally distributed random variable by looking at the mean of a **sample** taken from the whole population.

- **For a random sample of size n taken from a random variable $X \sim \mathrm{N}(\mu, \sigma^2)$, the sample mean, $\overline{X}$, is normally distributed with $\overline{X} \sim \mathrm{N}\left(\mu, \frac{\sigma^2}{n}\right)$.**

Hint If you took lots of different random samples of size n from the population, their means would be normally distributed.

You can use the distribution of the sample mean to determine whether the mean from one particular sample, $\overline{x}$, is statistically significant.

Example 14

A certain company sells fruit juice in cartons. The amount of juice in a carton has a normal distribution with a standard deviation of 3 ml.

The company claims that the mean amount of juice per carton, μ, is 60 ml. A trading inspector has received complaints that the company is overstating the mean amount of juice per carton and he wishes to investigate this complaint. The trading inspector takes a random sample of 16 cartons and finds that the mean amount of juice per carton is 59.1 ml.

Using a 5% level of significance, and stating your hypotheses clearly, test whether or not there is evidence to uphold this complaint.

$\mathrm{H_0}: \mu = 60$

$\mathrm{H_1}: \mu < 60$

Let X represent the amount of juice in a carton and assume $\mathrm{H_0}$, so that $X \sim \mathrm{N}(60, 3^2)$.

$\overline{X} \sim \mathrm{N}\left(60, \frac{3^2}{16}\right)$ or $\overline{X} \sim \mathrm{N}(60, 0.75^2)$

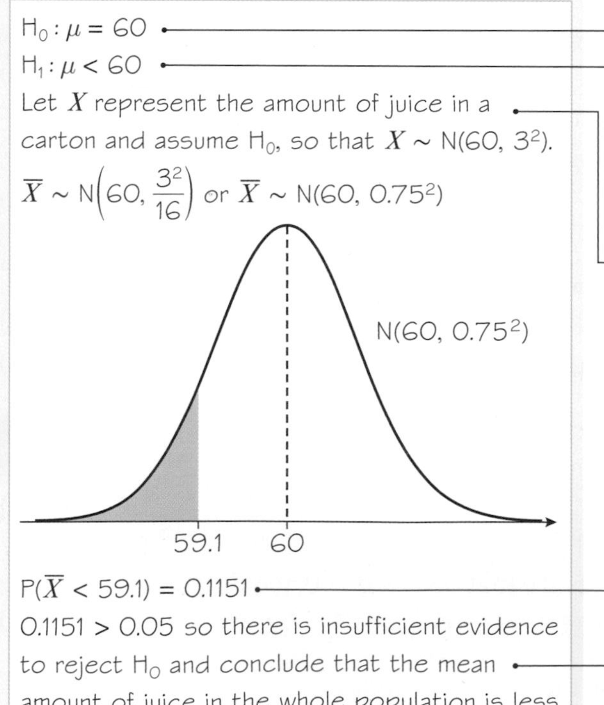

$\mathrm{P}(\overline{X} < 59.1) = 0.1151$

$0.1151 > 0.05$ so there is insufficient evidence to reject $\mathrm{H_0}$ and conclude that the mean amount of juice in the whole population is less than 60 ml.

The null hypothesis, $\mathrm{H_0}$, is that the population mean is equal to the claimed value.

The inspector is investigating whether the population mean is **less** than 60, so this is a **one-tailed test**.

Write out the population distribution assuming that $\mathrm{H_0}$ is true.

Watch out Your test statistic will be the **sample mean**, $\overline{X}$. This will have the same mean as X, but you need to divide the variance by the sample size. The new variance is $\frac{3^2}{16}$ so the new standard deviation is $\sqrt{\frac{3^2}{16}} = 0.75$.

Use your calculator to find $\mathrm{P}(\overline{X} < \overline{x})$.

Compare $\mathrm{P}(\overline{X} < 59.1)$ with the significance level of the test. The probability of obtaining this value of $\overline{x}$ is greater than 5%, so you do not reject the null hypothesis. Make sure your conclusion refers to the context given in the problem.

If you need to find a **critical region** or **critical value** for a hypothesis test for the mean of a normal distribution you can standardise your test statistic:

- **For the sample mean of a normally distributed random variable, $\overline{X} \sim \mathrm{N}\left(\mu, \frac{\sigma^2}{n}\right)$, $Z = \frac{\overline{X} - \mu}{\frac{\sigma}{\sqrt{n}}}$ is a normally distributed random variable with $Z \sim \mathrm{N}(0, 1)$.**

Coding the test statistic in this way allows you to use the percentage points table to determine critical values and critical regions.

You could also use the **inverse normal distribution** function on your calculator to determine critical values and critical regions for $\overline{X}$ directly.

Example 15

A machine produces bolts of diameter D where D has a normal distribution with mean 0.580 cm and standard deviation 0.015 cm.

The machine is serviced and after the service a random sample of 50 bolts from the next production run is taken to see if the mean diameter of the bolts has changed from 0.580 cm. The distribution of the diameters of bolts after the service is still normal with a standard deviation of 0.015 cm.

a Find, at the 1% level, the critical region for this test, stating your hypotheses clearly.

The mean diameter of the sample of 50 bolts is calculated to be 0.587 cm.

b Comment on this observation in light of the critical region.

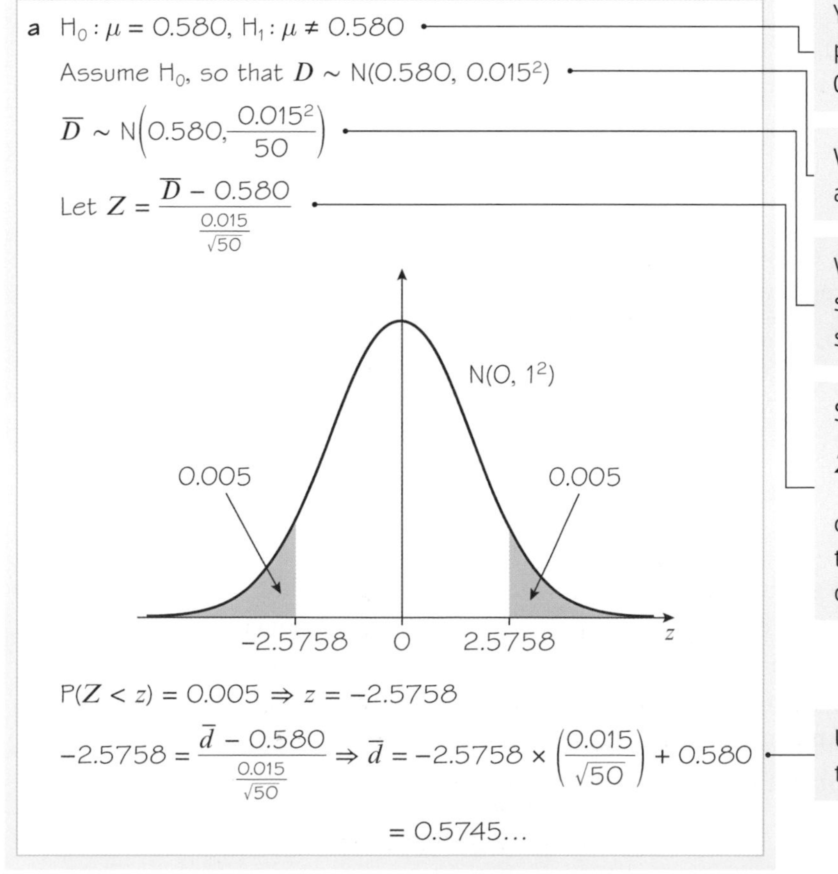

a $H_0 : \mu = 0.580$, $H_1 : \mu \neq 0.580$

Assume H_0, so that $D \sim \mathrm{N}(0.580, 0.015^2)$

$\overline{D} \sim \mathrm{N}\left(0.580, \frac{0.015^2}{50}\right)$

Let $Z = \frac{\overline{D} - 0.580}{\frac{0.015}{\sqrt{50}}}$

$\mathrm{P}(Z < z) = 0.005 \Rightarrow z = -2.5758$

$-2.5758 = \frac{\overline{d} - 0.580}{\frac{0.015}{\sqrt{50}}} \Rightarrow \overline{d} = -2.5758 \times \left(\frac{0.015}{\sqrt{50}}\right) + 0.580$

$= 0.5745\ldots$

You want to test whether the population mean differs from 0.580 so this is a two-tailed test.

Write out the population distribution assuming that H_0 is true.

Write out the distribution of the sample means, $\overline{D}$, for a sample of size 50.

Standardise $\overline{D}$ using the coding $Z = \frac{\overline{D} - \mu}{\frac{\sigma}{\sqrt{n}}}$. $Z \sim \mathrm{N}(0, 1)$, so you can use the percentage points table for the standard normal distribution.

Use the coding in reverse to find the corresponding value of $\overline{d}$.

$P(Z > z) = 0.005 \Rightarrow z = 2.5758$

$$2.5758 = \frac{\bar{d} - 0.580}{\frac{0.015}{\sqrt{50}}} \Rightarrow \bar{d} = 2.5758 \times \left(\frac{0.015}{\sqrt{50}}\right) + 0.580$$

$$= 0.5854...$$

So the critical region is $\bar{D} \leqslant 0.575$ or $\bar{D} \geqslant 0.585$ (3 s.f.)

b The observed value of $\bar{D}$ (0.587 cm) falls inside the critical region so there is sufficient evidence, at the 1% level, that the mean diameter has changed from 0.580 cm.

Problem-solving

This is a **two-tailed test** with total significance 1%, so you need the probability in each tail to be 0.5% = 0.005. Use the percentage points table to find z-values corresponding to the critical value for each tail.

Remember to refer to the context of the question. Don't just state whether H_0 is rejected or not.

Online Use the inverse normal distribution function on your calculator with $\mu = 0.580$ and $\sigma = \sqrt{\frac{0.015^2}{50}} = 0.002121$ to find the critical region for $\bar{D}$ directly. Set the percentages (or areas) equal to 0.005 and 0.995.

Exercise 3G

1 In each part, a random sample of size n is taken from a population having a normal distribution with mean μ and variance σ^2. Test the hypotheses at the stated levels of significance.

a $H_0: \mu = 21$, $H_1: \mu \neq 21$, $n = 20$, $\bar{x} = 21.2$, $\sigma = 1.5$, at the 5% level

b $H_0: \mu = 100$, $H_1: \mu < 100$, $n = 36$, $\bar{x} = 98.5$, $\sigma = 5.0$, at the 5% level

c $H_0: \mu = 5$, $H_1: \mu \neq 5$, $n = 25$, $\bar{x} = 6.1$, $\sigma = 3.0$, at the 5% level

d $H_0: \mu = 15$, $H_1: \mu > 15$, $n = 40$, $\bar{x} = 16.5$, $\sigma = 3.5$, at the 1% level

e $H_0: \mu = 50$, $H_1: \mu \neq 50$, $n = 60$, $\bar{x} = 48.9$, $\sigma = 4.0$, at the 1% level

2 In each part, a random sample of size n is taken from a population having a $N(\mu, \sigma^2)$ distribution. Find the critical regions for the test statistic $\bar{X}$ in the following tests.

a $H_0: \mu = 120$, $H_1: \mu < 120$, $n = 30$, $\sigma = 2.0$, at the 5% level

b $H_0: \mu = 12.5$, $H_1: \mu > 12.5$, $n = 25$, $\sigma = 1.5$, at the 1% level

c $H_0: \mu = 85$, $H_1: \mu < 85$, $n = 50$, $\sigma = 4.0$, at the 10% level

d $H_0: \mu = 0$, $H_1: \mu \neq 0$, $n = 45$, $\sigma = 3.0$, at the 5% level

e $H_0: \mu = -8$, $H_1: \mu \neq -8$, $n = 20$, $\sigma = 1.2$, at the 1% level

3 The times taken for a capful of stain remover to remove a standard chocolate stain from a baby's bib are normally distributed with a mean of 185 seconds and a standard deviation of 15 seconds. The manufacturers of the stain remover claim to have developed a new formula which will shorten the time taken for a stain to be removed. A random sample of 25 capfuls of the new formula are tested and the mean time for the sample is 179 seconds.

Test, at the 5% level, whether or not there is evidence that the new formula is an improvement.

Hint You are testing for an improvement, so use a **one-tailed** test.

4 The IQ scores of a population are normally distributed with a mean of 100 and a standard deviation of 15. A psychologist wishes to test the theory that eating chocolate before sitting an IQ test improves your score. A random sample of 80 people are selected and they are each given an identical bar of chocolate to eat before taking an IQ test.

a Find, at the 2.5% level, the critical region for this test, stating your hypotheses clearly.

The mean score on the test for the sample of 80 people was 102.5.

b Comment on this observation in light of the critical region.

5 The diameters of circular cardboard drinks mats produced by a certain machine are normally distributed with a mean of 9 cm and a standard deviation of 0.15 cm. After the machine is serviced a random sample of 30 mats is selected and their diameters are measured to see if the mean diameter has altered.

The mean of the sample was 8.95 cm.
Test, at the 5% level, whether there is significant evidence of a change in the mean diameter of mats produced by the machine.

Hint You are testing for an alteration in either direction, so use a **two-tailed** test.

6 A machine produces metal bolts of diameter D mm, where D is normally distributed with standard deviation 0.1 mm. Bolts with diameter either less than 5.1 mm or greater than 5.6 mm cannot be sold.

Given that 5% of bolts have a diameter in excess of 5.62 mm,

a find the probability that a randomly chosen bolt can be sold. **(5 marks)**

Twelve bolts are chosen.

b Find the probability that fewer than three cannot be sold. **(2 marks)**

A second machine produces bolts of diameter Y mm, where Y is normally distributed with standard deviation 0.08 mm.

A random sample of 20 bolts produced by this machine is taken and the sample mean of the diameters is found to be 5.52 mm.

c Stating your hypotheses clearly, and using a 2.5% level of significance, test whether the mean diameter of all the bolts produced by the machine is less than 5.7 mm. **(4 marks)**

7 The mass of European water voles, M grams, is normally distributed with standard deviation 12 grams.

Given that 2.5% of water voles have a mass greater than 160 grams,

a find the mean mass of a European water vole. **(3 marks)**

Eight water voles are chosen at random.

b Find the probability that at least 4 have a mass greater than 150 grams. **(3 marks)**

European water rats have mass, N grams, which is normally distributed with standard deviation 85 grams.

A random sample of 15 water rats is taken and the sample mean mass is found to be 875 grams.

c Stating your hypotheses clearly, and using a 10% level of significance, test whether the mean mass of all water rats is different from 860 grams. **(4 marks)**

(E) **8** Daily mean windspeed is modelled as being normally distributed with a standard deviation of 3.1 knots.

A random sample of 25 recorded daily mean windspeeds is taken at Heathrow in 2015.

Given that the mean of the sample is 12.2 knots, test at the 2.5% level of significance whether the mean of the daily mean windspeeds is greater than 9.5 knots.
State your hypotheses clearly. **(4 marks)**

Mixed exercise 3

(E) **1** The heights of a large group of men are normally distributed with a mean of 178 cm and a standard deviation of 4 cm. A man is selected at random from this group.

a Find the probability that he is taller than 185 cm. **(2 marks)**

b Find the probability that three men, selected at random, are all less than 180 cm tall. **(3 marks)**

A manufacturer of door frames wants to ensure that fewer than 0.005 men have to stoop to pass through the frame.

c On the basis of this group, find the minimum height of a door frame to the nearest centimetre. **(2 marks)**

(E) **2** The weights of steel sheets produced by a factory are known to be normally distributed with mean 32.5 kg and standard deviation 2.2 kg.

a Find the percentage of sheets that weigh less than 30 kg. **(1 mark)**

Bob requires sheets that weigh between 31.6 kg and 34.8 kg.

b Find the percentage of sheets produced that satisfy Bob's requirements. **(3 marks)**

(E/P) **3** The time a mobile phone battery lasts before needing to be recharged is assumed to be normally distributed with a mean of 48 hours and a standard deviation of 8 hours.

a Find the probability that a battery will last for more than 60 hours. **(2 marks)**

b Find the probability that the battery lasts less than 35 hours. **(1 mark)**

A random sample of 30 phone batteries is taken.

c Find the probability that 3 or fewer last less than 35 hours. **(2 marks)**

(E) **4** The random variable $X \sim N(24, \sigma^2)$. Given that $P(X > 30) = 0.05$, find:

a the value of σ **(2 marks)**

b $P(X < 20)$ **(1 mark)**

c the value of d so that $P(X > d) = 0.01$. **(2 marks)**

(E) **5** A machine dispenses liquid into plastic cups in such a way that the volume of liquid dispensed is normally distributed with a mean of 120 ml. The cups have a capacity of 140 ml and the probability that the machine dispenses too much liquid so that the cup overflows is 0.01.

a Find the standard deviation of the volume of liquid dispensed. **(2 marks)**

b Find the probability that the machine dispenses less than 110 ml. **(1 mark)**

Ten percent of customers complain that the machine has not dispensed enough liquid.

c Find the largest volume of liquid, to the nearest millilitre, that will lead to a complaint. **(2 marks)**

E/P **6** The random variable $X \sim N(\mu, \sigma^2)$. The lower quartile of X is 20 and the upper quartile is 40.

a Find μ and σ. **(3 marks)**

b Find the 10% to 90% interpercentile range. **(3 marks)**

E/P **7** The heights of seedlings are normally distributed. Given that 10% of the seedlings are taller than 15 cm and 5% are shorter than 4 cm, find the mean and standard deviation of the heights. **(4 marks)**

E **8** A psychologist gives a student two different tests. The first test has a mean of 80 and a standard deviation of 10 and the student scores 85.

a Find the probability of scoring 85 or more on the first test. **(2 marks)**

The second test has a mean of 100 and a standard deviation of 15. The student scores 105 on the second test.

b Find the probability of a score of 105 or more on the second test. **(2 marks)**

c State, giving a reason, which of the student's two test scores was better. **(2 marks)**

E/P **9** Jam is sold in jars and the mean weight of the contents is 108 grams. Only 3% of jars have contents weighing less than 100 grams. Assuming that the weight of jam in a jar is normally distributed, find:

a the standard deviation of the weight of jam in a jar **(2 marks)**

b the proportion of jars where the contents weigh more than 115 grams. **(2 marks)**

A random sample of 25 jars is taken.

c Find the probability that 2 or fewer jars have contents weighing more than 115 grams. **(3 marks)**

E **10** The waiting time at a doctor's surgery is assumed to be normally distributed with standard deviation of 3.8 minutes. Given that the probability of waiting more than 15 minutes is 0.0446, find:

a the mean waiting time **(2 marks)**

b the probability of waiting less than 5 minutes. **(2 marks)**

E/P **11** The thickness of some plastic shelving produced by a factory is normally distributed. As part of the production process the shelving is tested with two gauges. The first gauge is 7 mm thick and 98.61% of the shelving passes through this gauge. The second gauge is 5.2 mm thick and only 1.02% of the shelves pass through this gauge.

Find the mean and standard deviation of the thickness of the shelving. **(4 marks)**

12 A fair coin is spun 60 times. Use a suitable approximation to estimate the probability of obtaining fewer than 25 heads.

E **13** The owner of a local corner shop calculates that the probability of a customer buying a newspaper is 0.40.

A random sample of 100 customers is recorded.

a Give two reasons why a normal approximation may be used in this situation. **(2 marks)**

b Write down the parameters of the normal distribution used. **(2 marks)**

c Use this approximation to estimate the probability that at least half the customers bought a newspaper. **(2 marks)**

(E) **14** The random variable $X \sim B(120, 0.46)$.

a Find $P(X = 65)$. **(1 mark)**

b State why a normal distribution can be used to approximate X, and write down the parameters of such a normal distribution. **(4 marks)**

c Find the percentage error in using the normal approximation to calculate $P(X = 65)$. **(3 marks)**

(E/P) **15** The random variable $Y \sim B(300, 0.6)$.

a Give two reasons why a normal distribution can be used to approximate Y. **(2 marks)**

b Find, using the normal approximation, $P(150 < Y \leqslant 180)$. **(4 marks)**

c Find the largest value of y such that $P(Y < y) < 0.05$. **(3 marks)**

16 Past records from a supermarket show that 40% of people who buy chocolate bars buy the family-size bar. A random sample of 80 people is taken from those who bought chocolate bars. Use a suitable approximation to estimate the probability that more than 30 of these 80 people bought family-size bars.

(E/P) **17** A horticulture company sells apple-tree seedlings. It is claimed that 55% of these seedlings will produce apples within three years.

A random sample of 20 seedlings is taken and X produce apples within three years.

a Find $P(X > 10)$. **(2 marks)**

A second random sample of 200 seedlings is taken. 95 produce apples within three years.

b Assuming the company's claim is correct, use a suitable approximation to find the probability that 95 or fewer seedlings produce apples within three years. **(4 marks)**

c Using your answer to part **b**, comment on the company's claim. **(1 mark)**

(E/P) **18** A herbalist claims that a particular remedy is successful in curing a particular disease in 52% of cases.

A random sample of 25 people who took the remedy is taken.

a Find the probability that more than 12 people in the sample were cured. **(2 marks)**

A second random sample of 300 people was taken and 170 were cured.

b Assuming the herbalist's claim is true, use a suitable approximation to find the probability that at least 170 people were cured. **(4 marks)**

c Using your answer to part **b**, comment on the herbalist's claim. **(1 mark)**

(E) **19** The random variable X has a normal distribution with mean μ and standard deviation 2.

A random sample of 25 observations is taken and the sample mean $\overline{X}$ is calculated in order to test the null hypothesis $\mu = 7$ against the alternative hypothesis $\mu > 7$ using a 5% level of significance.

Find the critical region for $\overline{X}$. **(4 marks)**

(E) **20** A certain brand of mineral water comes in bottles. The amount of water in a bottle, in millilitres, follows a normal distribution of mean μ and standard deviation 2. The manufacturer claims that μ is 125. In order to maintain standards the manufacturer takes a sample of 15 bottles and calculates the mean amount of water per bottle to be 124.2 millilitres.

Test, at the 5% level, whether or not there is evidence that the value of μ is lower than the manufacturer's claim. State your hypotheses clearly. **(4 marks)**

(E) **21** Climbing rope produced by a manufacturer is known to be such that one-metre lengths have breaking strengths that are normally distributed with mean 170.2 kg and standard deviation 10.5 kg. Find, to 3 decimal places, the probability that:

a a one-metre length of rope chosen at random from those produced by the manufacturer will have a breaking strength of 175 kg to the nearest kg **(2 marks)**

b a random sample of 50 one-metre lengths will have a mean breaking strength of more than 172.4 kg. **(3 marks)**

A new component material is added to the ropes being produced. The manufacturer believes that this will increase the mean breaking strength without changing the standard deviation. A random sample of 50 one-metre lengths of the new rope is found to have a mean breaking strength of 172.4 kg.

c Perform a significance test at the 5% level to decide whether this result provides sufficient evidence to confirm the manufacturer's belief that the mean breaking strength is increased. State clearly the null and alternative hypotheses that you are using. **(3 marks)**

(E/P) **22** A machine fills 1 kg packets of sugar. The actual weight of sugar delivered to each packet can be assumed to be normally distributed. The manufacturer requires that,

i the mean weight of the contents of a packet is 1010 g, and

ii 95% of all packets filled by the machine contain between 1000 g and 1020 g of sugar.

a Show that this is equivalent to demanding that the variance of the sampling distribution, to 2 decimal places, is equal to 26.03 g^2. **(3 marks)**

A sample of 8 packets was selected at random from those filled by the machine. The weights, in grams, of the contents of these packets were

1012.6 1017.7 1015.2 1015.7 1020.9 1005.7 1009.9 1011.4

Assuming that the variance of the actual weights is 26.03 g^2,

b test at the 2% significance level (stating clearly the null and alternative hypotheses that you are using) to decide whether this sample provides sufficient evidence to conclude that the machine is not fulfilling condition **i**. **(4 marks)**

(E/P) **23** The diameters of eggs of the little-gull are approximately normally distributed with mean 4.11 cm and standard deviation 0.19 cm.

a Calculate the probability that an egg chosen at random has a diameter between 3.9 cm and 4.5 cm. **(3 marks)**

A sample of 8 little-gull eggs was collected from a particular island and their diameters, in cm, were

4.4, 4.5, 4.1, 3.9, 4.4, 4.6, 4.5, 4.1

b Assuming that the standard deviation of the diameters of eggs from the island is also 0.19 cm, test, at the 1% level, whether the results indicate that the mean diameter of little-gull eggs on this island is different from elsewhere. **(4 marks)**

(E/P) **24** The random variable X is normally distributed with mean μ and variance σ^2.

a Write down the distribution of the sample mean $\overline{X}$ of a random sample of size n. **(1 mark)**

A construction company wishes to determine the mean time taken to drill a fixed number of holes in a metal sheet.

b Determine how large a random sample is needed so that the expert can be 95% certain that the sample mean time will differ from the true mean time by less than 15 seconds. Assume that it is known from previous studies that $\sigma = 40$ seconds. **(4 marks)**

Challenge

A football manager claims to have the support of 48% of all the club's fans.
A random sample of 15 fans is taken.

a Find the probability that more than 8 of these fans supported the manager.

A second random sample of 250 fans was taken, and is used to test the football manager's claim at the 5% significance level.

b Use a suitable approximation to find the critical regions for this test.

It was found that 102 fans said they supported the manager.

c Using your answer to part **b**, comment on the manager's claim.

Summary of key points

1 The area under a continuous probability distribution is equal to 1.

2 If X is a normally distributed random variable, you write $X \sim \text{N}(\mu, \sigma^2)$ where μ is the population mean and σ^2 is the population variance.

3 The normal distribution

- has parameters μ, the population mean, and σ^2, the population variance
- is symmetrical (mean = median = mode)
- has a bell-shaped curve with asymptotes at each end
- has total area under the curve equal to 1
- has points of inflection at $\mu + \sigma$ and $\mu - \sigma$

4 The standard normal distribution has mean 0 and standard deviation 1.
The standard normal variable is written as $Z \sim \text{N}(0, 1^2)$.

5 If n is large and p is close to 0.5, then the binomial distribution $X \sim \text{B}(n, p)$ can be approximated by the normal distribution $\text{N}(\mu, \sigma^2)$ where

- $\mu = np$
- $\sigma = \sqrt{np(1 - p)}$

6 If you are using a normal approximation to a binomial distribution, you need to apply a **continuity correction** when calculating probabilities.

7 For a random sample of size n taken from a random variable $X \sim \text{N}(\mu, \sigma^2)$, the sample mean, $\overline{X}$, is normally distributed with $\overline{X} \sim \text{N}\left(\mu, \frac{\sigma^2}{n}\right)$.

8 For the sample mean of a normally distributed random variable, $\overline{X} \sim \text{N}\left(\mu, \frac{\sigma^2}{n}\right)$,
$Z = \frac{\overline{X} - \mu}{\frac{\sigma}{Rn}}$ is a normally distributed random variable with $Z \sim \text{N}(0, 1)$.

Review exercise

(E) **1** An engineer is developing a new production process to assemble microswitches. The time taken, t minutes, to produce each switch, and its size, s mm^3, are recorded in the table.

Size, s mm^3	2.0	4.5	6.2	7.3	9.1
Time, t minutes	0.33	1.15	1.76	2.15	2.95

a Calculate the product moment correlation coefficient between $\log s$ and $\log t$. **(2)**

b Use your answer to part **a** to explain why an equation of the form $t = as^n$, where a and n are constants, is likely to be good model for the relationship between s and t. **(1)**

c The regression line of $\log t$ on $\log s$ is given as $\log t = -0.9051 + 1.4437 \log s$. Determine the values of the constants a and n in the equation given in part **b**. **(2)**

← Sections 1.1, 1.2

(E) **2** The table shows some data collected on the pressure, in pascals, of some gases (P) and the temperature (t °C).

Pressure (P)	Temperature (t)
45	3.65
73	11.01
81	15.24
90	21.95
102	35.21
115	58.43

The data is coded using the changes of variable $x = P$ and $y = \log t$. The regression line of y on x is found to be $y = -0.2139 + 0.0172x$.

a Given that the data can be modelled by an equation of the form $t = ab^P$ where a and b are constants, find, correct to three significant figures, the values of a and b. **(3)**

b Explain why this model is not reliable for estimating the temperature when the pressure is 250 pascals. **(1)**

← Section 1.1

(E) **3** Over a period of time, researchers took 10 blood samples from one patient with a blood disease. For each sample, they measured the levels of serum magnesium, s mg/dl, in the blood and the corresponding level of the disease protein, d mg/dl. The results are shown in the table.

s	1.2	1.9	3.2	3.9	2.5	4.5	5.7	4.0	1.1	5.9
d	3.8	7.0	11.0	12.0	9.0	12.0	13.5	12.2	2.0	13.9

a Calculate the value of the product moment correlation coefficient between s and d for this sample. **(1)**

b Stating your hypotheses clearly, test, at the 1% significance level, whether there is a positive correlation between serum magnesium and disease protein levels in this patient. **(3)**

← Sections 1.2, 1.3

(E) **4** A sailor from Jacksonville believes there is a negative correlation between the daily mean air temperature and the daily mean windspeed. He collects data from 6 days in June 2015.

Temperature (°C)	Windspeed (knots)
24.8	4.9
23.3	2.6
22.7	7.6
24.7	4.5
24.5	4.3
25.1	4.0

By calculating the product moment correlation coefficient for this data, test at the 10% level of significance, whether there is evidence to support the sailor's claim. State your hypotheses clearly. **(4)**

← Sections 1.2, 1.3

5 Data about the number of miles done by a sample of one-year-old cars and their value is collected from a dealer. The dealer believes there is a negative correlation between the number of miles done and the value.

Number of miles	2000	3500	4200	6500	7800
Value (£, 000)	12	9.1	8.2	7.7	6.1

a Test at the 2.5% level of significance the dealer's claim. State your hypotheses clearly. **(4)**

b State the effect that changing the level of significance to 1% would have on the dealer's conclusion. **(1)**

← Sections 1.2, 1.3

6 At a college, there are 148 students studying either engineering, childcare or tourism. Of these students, 89 wear glasses and the others do not. There are 30 engineering students of whom 18 wear glasses. There are 68 childcare students, of whom 44 wear glasses.

A student is chosen at random.

Find the probability that this student:

a is studying tourism **(1)**

b does not wear glasses, given that the student is studying tourism. **(2)**

Amongst the engineering students, 80% are right-handed. Corresponding percentages for childcare and tourism students are 75% and 70% respectively.

A student is again chosen at random.

c Find the probability that this student is right-handed. **(2)**

d Given that this student is right-handed, find the probability that the student is studying engineering. **(2)**

← Sections 2.2, 2.4

(E) 7 A group of 100 people produced the following information relating to three attributes. The attributes were wearing glasses, being left-handed and having dark hair.

Glasses were worn by 36 people, 28 were left-handed and 36 had dark hair. There were 17 who wore glasses and were left-handed, 19 who wore glasses and had dark hair and 15 who were left-handed and had dark hair. Only 10 people wore glasses, were left-handed and had dark hair.

a Represent these data on a Venn diagram. **(3)**

A person was selected at random from this group.

Find the probability that this person:

b wore glasses but was not left-handed and did not have dark hair **(1)**

c did not wear glasses, was not left-handed and did not have dark hair **(1)**

d had only two of the attributes **(2)**

e wore glasses given they were left-handed and had dark hair. **(2)**

← Sections 2.3, 2.4

(E) **8** A survey of the reading habits of some students revealed that, on a regular basis, 25% read broadsheet newspapers, 45% read tabloid newspapers and 40% do not read newspapers at all.

a Find the proportion of students who read both broadsheet and tabloid newspapers. **(2)**

b Draw a Venn diagram to represent this information. **(3)**

A student is selected at random. Given that this student reads newspapers on a regular basis,

c find the probability that this student only reads broadsheet newspapers. **(2)**

← Sections 2.3, 2.4

(E) **9** A bag contains 3 blue counters and 5 red counters. One counter is drawn at random from the bag and not replaced. A second counter is then drawn.

a Draw a tree diagram to represent this situation. **(2)**

b Find the probability that:

i the second counter drawn is blue. **(2)**

ii both counters selected are blue, given that the second counter is blue. **(2)**

← Sections 2.4, 2.5

(E) **10** For the events A and B, $\mathrm{P}(A \cap B') = 0.34$, $\mathrm{P}(A' \cap B) = 0.13$ and $\mathrm{P}(A \cup B) = 0.62$.

a Draw a Venn diagram to illustrate the complete sample space for the events A and B. **(2)**

b Write down the values of $\mathrm{P}(A)$ and $\mathrm{P}(B)$. **(2)**

c Find $\mathrm{P}(A|B')$. **(2)**

d Determine whether or not A and B are independent. **(2)**

← Sections 2.1, 2.3, 2.4

(E/P) **11** Two events A and B are such that $\mathrm{P}(B) = 0.3$ and $\mathrm{P}(A \cap B) = 0.15$. If A and B are independent, find:

a $\mathrm{P}(A)$ **(1)**

b $\mathrm{P}(A' \cap B')$ **(1)**

A third event C has $\mathrm{P}(C) = 0.4$. Given that B and C are mutually exclusive and $\mathrm{P}(A \cap C) = 0.1$,

c Draw a Venn diagram to illustrate this situation. **(2)**

d Find:

i $\mathrm{P}(A|C)$ **(2)**

ii $\mathrm{P}(A \cap (B \cup C'))$ **(2)**

iii $\mathrm{P}(A|(B \cup C'))$ **(2)**

← Sections 2.1, 2.3, 2.4

(E/P) **12** The probability that Joanna oversleeps is 0.15. If she oversleeps, the probability that she is late to college is 0.75. If she gets up on time, the probability that she is late to college is 0.1.

a Find the probability that Joanna is late to college on any particular day. **(2)**

b Find the probability that Joanna overslept given that she is late to college. **(2)**

← Sections 2.4, 2.5

(E/P) **13** The measure of intelligence, IQ, of a group of students is assumed to be normally distributed with mean 100 and standard deviation 15.

a Find the probability that a student selected at random has an IQ less than 91. **(1)**

The probability that a randomly selected student as an IQ of at least $100 + k$ is 0.2090.

b Find, to the nearest integer, the value of k. **(2)**

← Sections 3.2, 3.3

E/P **14** The heights of a group of athletes are modelled by a normal distribution with mean 180 cm and standard deviation 5.2 cm. The weights of this group of athletes are modelled by a normal distribution with mean 85 kg and standard deviation 7.1 kg.

Find the probability that a randomly chosen athlete:

a is taller than 188 cm **(1)**

b weighs less than 97 kg. **(1)**

c Assuming that for these athletes height and weight are independent, find the probability that a randomly chosen athlete is taller than 188 cm and weighs more than 97 kg. **(2)**

d Comment on the assumption that height and weight are independent. **(1)**

←Section 3.2

E/P **15** From experience a high jumper knows that he can clear a height of at least 1.78 m once in five attempts. He also knows that he can clear a height of at least 1.65 m on seven out of 10 attempts.

Assuming that the heights cleared by the high jumper follow a normal distribution,

a find, to three decimal places, the mean and the standard deviation of the heights cleared by the high jumper **(3)**

b calculate the probability that the high jumper will clear a height of 1.74 m. **(1)**

← Sections 3.2, 3.5

E/P **16** A company makes dinner plates with an average diameter of 22 cm. The diameter, D cm, can be modelled using a normal distribution. Given that 32% of plates are less than 21.5 cm in diameter, find

a the standard deviation of the diameter **(2)**

b the proportion of plates with diameter between 21 cm and 22.5 cm. **(2)**

A plate can be used in a restaurant if the diameter is between 21 cm and 22.5 cm. A sample of 30 plates is taken.

c Find the probability that at least 10 of these plates can be used. **(2)**

← Section 3.5

E/P **17** For a particular type of plant 45% have white flowers and the remainder have coloured flowers. Gardenmania sells plants in batches of 12. A batch is selected at random.

Calculate the probability this batch contains:

a exactly 5 plants with white flowers **(1)**

b more plants with white flowers than coloured ones. **(2)**

Gardenmania takes a random sample of 10 batches of plants.

c Find the probability that exactly 3 of these batches contain more plants with white flowers than coloured ones. **(2)**

Due to an increasing demand for these plants by large companies, Gardenmania decides to sell them in batches of 150.

d Use a suitable approximation to calculate the probability that a batch of 150 plants contains more than 75 with white flowers. **(3)**

←Section 3.6

E/P **18** At a school fair, the probability of getting a winning ticket on the tombola is 0.48. A random sample of 80 tickets is taken.

a Find the probability that there are exactly 35 winning tickets, giving your answer to 5 decimal places. **(2)**

b Find the percentage error when using a normal approximation to calculate the probability that there are exactly 35 winning tickets. **(4)**

← Section 3.6

(E) **19** The time, in minutes, it takes Robert to complete the puzzle in his morning newspaper each day is normally distributed with mean 18 and standard deviation 3. After taking a holiday, Robert records the times taken to complete a random sample of 15 puzzles and he finds that the mean time is 16.5 minutes. You may assume that the holiday has not changed the standard deviation of times taken to complete the puzzle.

Stating your hypotheses clearly test, at the 5% level of significance, whether or not there has been a reduction in the mean time Robert takes to complete the puzzle. **(4)**

← Section 3.7

(E/P) **20** The length of adult male rattlesnakes, L metres, is normally distributed with standard deviation 0.4 metres.

Given that 5% of rattlesnakes have a length less than 1.7 metres,

a find the mean length of a rattlesnake. **(3)**

Ten rattlesnakes are chosen at random.

b Find the probability that at least 6 have a length greater than 2.3 metres. **(3)**

Adult female rattlesnakes are of length M metres, where M is normally distributed with standard deviation 0.3 metres.

Adult female rattlesnakes are thought to have a mean length of 1.9 metres. Given that a random sample of 20 female rattlesnakes is taken,

c find, at the 5% level of significance, the critical regions for a hypothesis test that the mean length is not equal to 1.9 metres. State your hypotheses clearly. **(4)**

d The mean length of the sample of 20 female rattlesnakes is found to be 2.09 metres. Comment on this observation in the light of your answer to part **c**. **(2)**

← Sections 3.5, 3.7

(E) **21** Daily mean temperature in Hurn is modelled as being normally distributed with a standard deviation of 2.3 °C.

A random sample of 20 recorded daily mean temperatures is taken in 2015.

Given that the mean of the sample is 11.1 °C, test at the 5% level of significance whether the mean of the daily mean temperatures is less than 12 °C. State your hypotheses clearly. **(4)**

← Section 3.7

Challenge

1 $P(A) = 0.7$ and $P(B) = 0.3$.

a Given that $P(A \cap B') = p$, find the range of possible values of p.

$P(C) = 0.5$ and $P(A \cap B \cap C) = 0.05$.

b Given that $P(A' \cap B \cap C) = q$, find the range of possible values of q. ← Sections 2.3, 2.4

2 A politician claims to have the support of 53% of constituents. A random sample of 300 constituents was taken, and is used to test the politician's claim at the 10% significance level.

a Use a suitable approximation to find the critical regions for this test. Round your answers to 2 decimal places.

It was found that 173 constituents said they supported the politician.

b Using your answer to part **a**, comment on the politician's claim. ← Sections 3.6, 3.7

4 Moments

Objectives

After completing this chapter you should be able to:

- Calculate the turning effect of a force applied to a rigid body → pages 71–72
- Calcualte the resultant moment of a set of forces acting on a rigid body → pages 73–76
- Solve problems involving uniform rods in equilibrium → pages 76–80
- Solve problems involving non-uniform rods → pages 80–83
- Solve problems involving rods on the point of tilting → pages 83–85

Prior knowledge check

1 Find the value of x in each of the following:

a

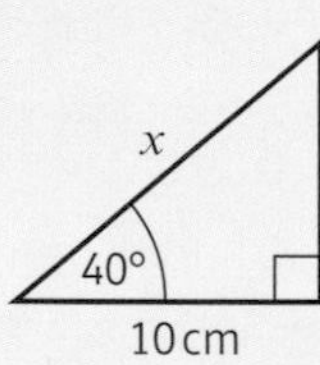

b

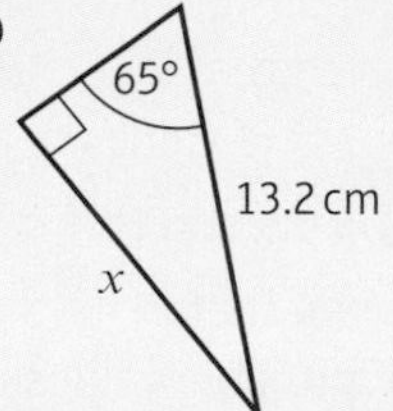

← GCSE Mathematics

2 Masses A and B rest on a light scale-pan supported by two strings, each with tension T.

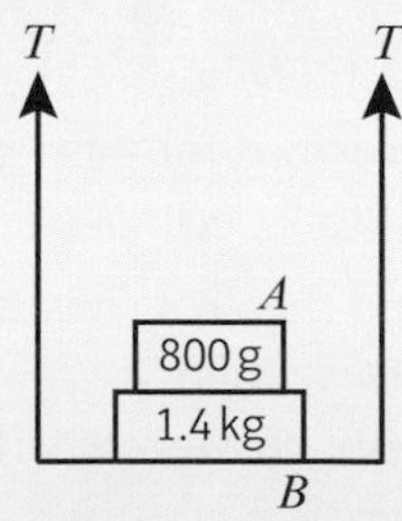

Find:

a the value of T

b the normal reaction of the scale-pan on mass B

c the normal reaction of mass B on mass A.

← Year 1, Chapter 10

Moments measure the turning effect of a force. Engineers use moments to work out how much load can be applied safely to a crane. → Mixed exercise Q12

4.1 Moments

So far you have looked mostly at situations involving particles. This means you can ignore rotational effects. In this chapter, you begin to model objects as **rigid bodies**. This allows you to consider the size of the object as well as where forces are applied.

The moment of a force measures the turning effect of the force on a rigid body.

It is the product of the magnitude of the force and the perpendicular distance from the axis of rotation.

- **Clockwise moment of F about $P = |\mathbf{F}| \times d$**

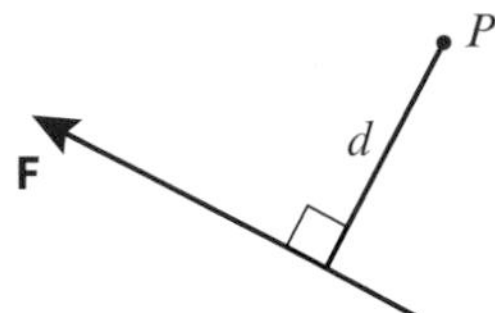

The moment of the force, **F**, is acting about the point P.

Watch out When you describe a moment, you need to give the direction of rotation.

In the diagram above, the distance given is perpendicular to the line of action of the force. When this is not the case, you need to use trigonometry to find the perpendicular distance.

- **Clockwise moment of F about $P = |\mathbf{F}| \times d\sin\theta$**

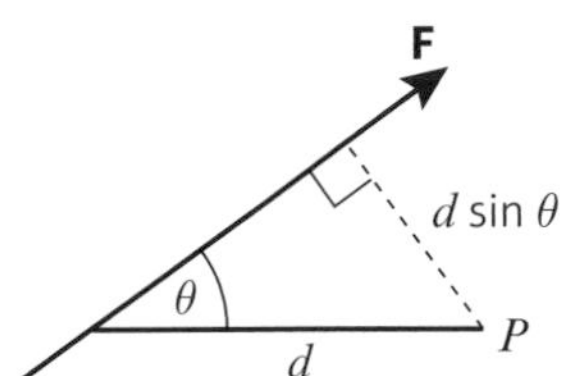

Notation A moment is a force multiplied by a distance, so its units are **newton metres (N m)** or **newton centimetres (N cm)**.

Online Explore the moment of a force acting about a point using GeoGebra.

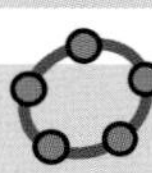

Example 1

Find the moment of each force about the point P.

a 6 N, P, 3 m

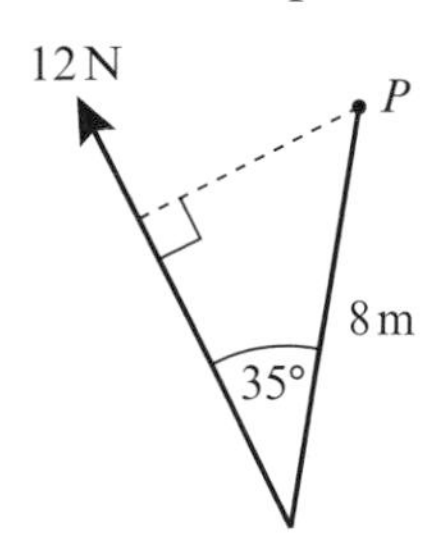

a Moment of the 6 N force about P
= magnitude of force × perpendicular distance
= 6 × 3 = 18 N m anticlockwise

The distance given on the diagram is the perpendicular distance, so you can substitute the given values directly into the formula.

Don't forget to include the direction of the rotation when you describe the moment of the force.

b Moment of the 12 N force about P
= magnitude of force × perpendicular distance
= $12 \times 8\sin 35°$ = 55.1 N m clockwise (3 s.f.)

This time you need to use the perpendicular distance $8\sin 35°$.

Example 2

The diagram shows two forces acting on a lamina. Find the moment of each of the forces about the point P.

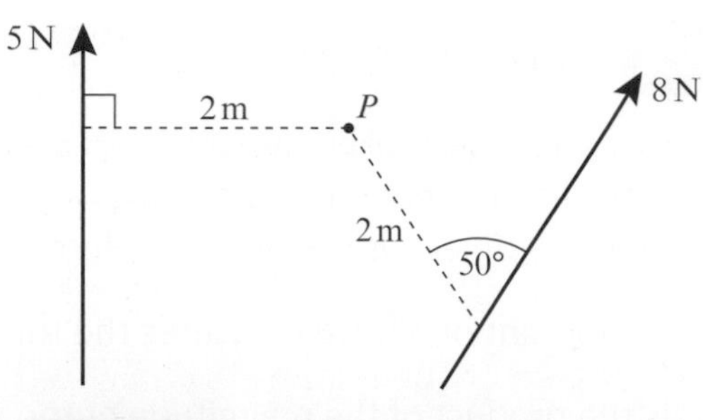

Moment of the 5 N force
= magnitude of force × perpendicular distance
= $5 \times 2 = 10$ N m clockwise

Moment of the 8 N force
= magnitude of force × perpendicular distance
= $8 \times 2\sin 50° = 12.3$ N m anticlockwise (3 s.f.)

The moments act in opposite directions.

Notation A lamina is a 2D object whose thickness can be ignored.

Exercise 4A

1 Calculate the moment about P of each of these forces acting on a lamina.

a b c d

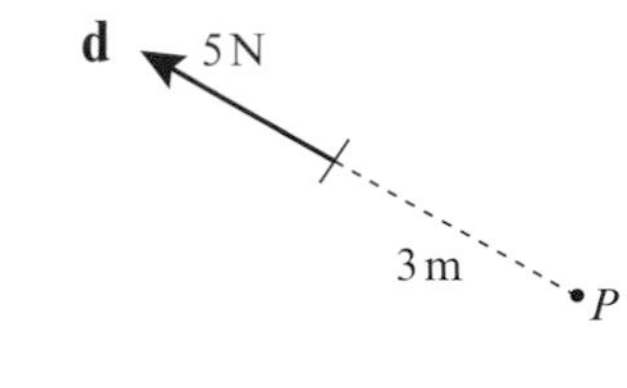

2 Calculate the moment about P of each of these forces acting on a lamina.

a b c d

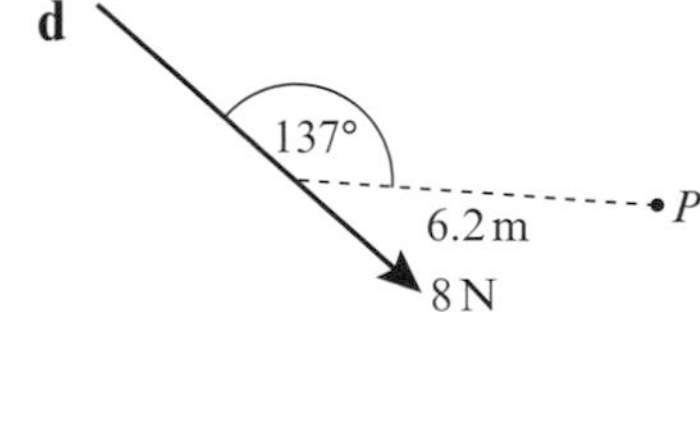

3 The diagram shows a sign hanging from a wooden beam. The sign has a mass of 4 kg.

a Calculate the moment of the weight of the mass:

i about P ii about Q

b Comment on any modelling assumptions you have made.

(P) 4 $ABCD$ is a rectangular lamina. A force of 12 N acts horizontally at B, as shown in the diagram. Find the moment of this force about:

a A b B c C d D

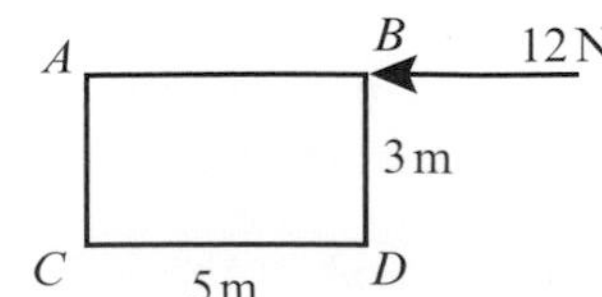

(P) 5 In the diagram, the force **F** produces a moment of 15 N m clockwise about the pivot P. Calculate the magnitude of **F**.

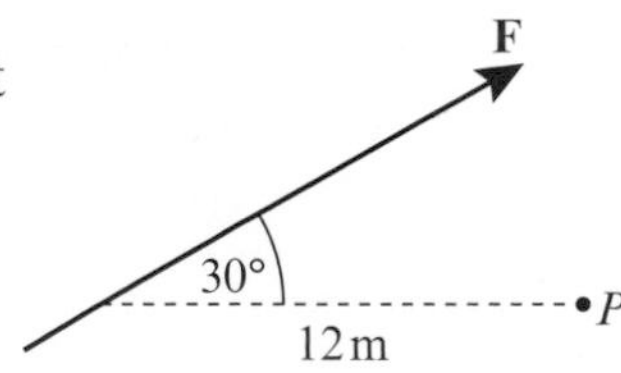

4.2 Resultant moments

When you have several **coplanar** forces acting on a body, you can determine the turning effect around a given point by choosing a positive direction (clockwise or anticlockwise) and then finding the sum of the moments produced by each force.

Notation **Coplanar forces** are forces that act in the same plane.

- **The sum of the moments acting on a body is called the resultant moment.**

Example 3

The diagram shows a set of forces acting on a light rod. Calculate the resultant moment about the point P.

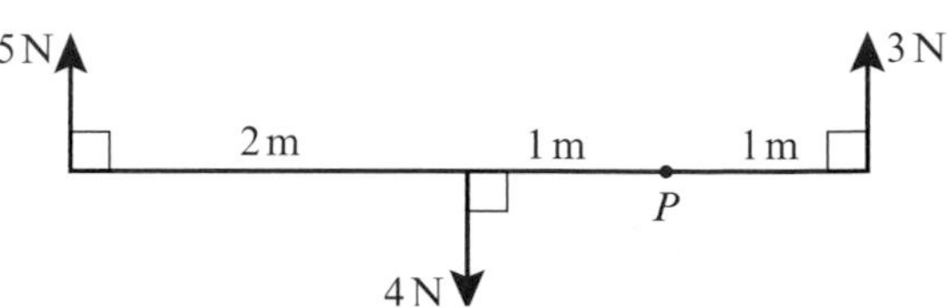

The moment of the 5 N force is
$5 \times (2 + 1) = 15$ N m clockwise.

The moment of the 4 N force is
$4 \times 1 = 4$ N m anticlockwise.

The moment of the 3 N force is
$3 \times 1 = 3$ N m anticlockwise.

Choosing clockwise as positive:

Resultant moment $= 15 + (-4) + (-3) = 8$ N m

$\therefore$ resultant moment is 8 N m clockwise.

Your positive direction is clockwise, so the anticlockwise moments are negative.

Problem-solving

You could also solve this problem by considering the clockwise and anticlockwise moments separately.
Sum of clockwise moments = 15 N m
Sum of anticlockwise moments = 3 + 4 = 7 N m
Resultant moment = 15 − 7 = 8 N m clockwise

Example 4

The diagram shows a set of forces acting on a light rod. Calculate the resultant moment about the point P.

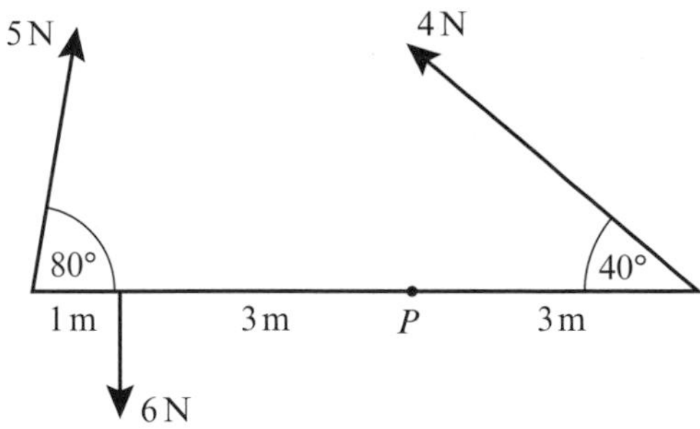

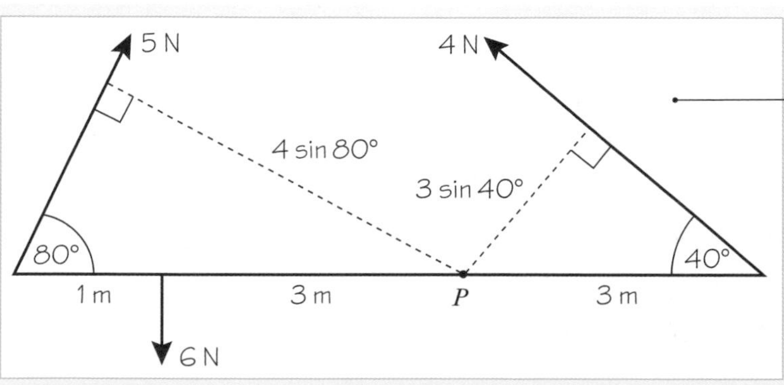

Draw a diagram to find the perpendicular distances.

Moment of 6 N force
= 6 × 3 = 18 N m anticlockwise

Moment of 5 N force
= 5 × 4 sin 80° = 19.696... N m clockwise

Moment of 4 N force
= 4 × 3 sin 40° = 7.713... N m anticlockwise

Find the moment of each force.

Take clockwise as the positive direction:

Choose a positive direction.

Resultant moment = (−18) + 19.696... + (−7.713...) = −6.017...

∴ resultant moment is 6.02 N m anticlockwise.

Add the unrounded values then round your answer to three significant figures.

Example 5

The diagram shows two forces acting on a lamina.
Calculate the resultant moment about the point P.

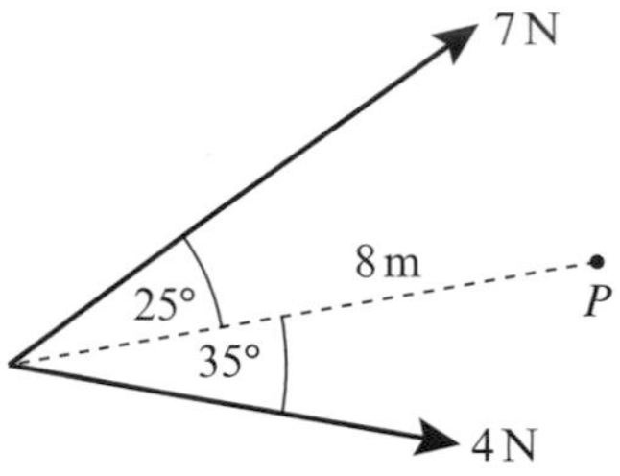

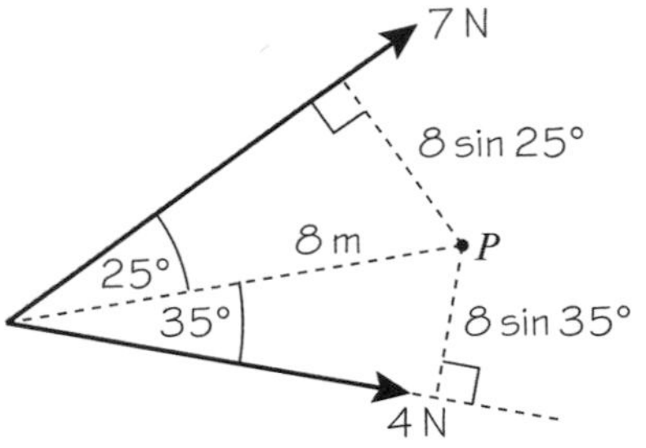

Draw a diagram and find the perpendicular distances.

Moment of 7 N force
= 7 × 8 sin 25° = 23.66... N m clockwise.

Moment of 4 N force
= 4 × 8 sin 35° = 18.35... N m anticlockwise.

Find the moments of both forces.

Take clockwise as the positive direction:

Choose a positive direction.

Resultant moment = 23.66... + (−18.35...) = 5.31...

∴ resultant moment is 5.31 N m clockwise.

Sum together your moments to find the resultant moment.

Exercise 4B

1 These diagrams show sets of forces acting on a light rod. In each case, calculate the resultant moment about P.

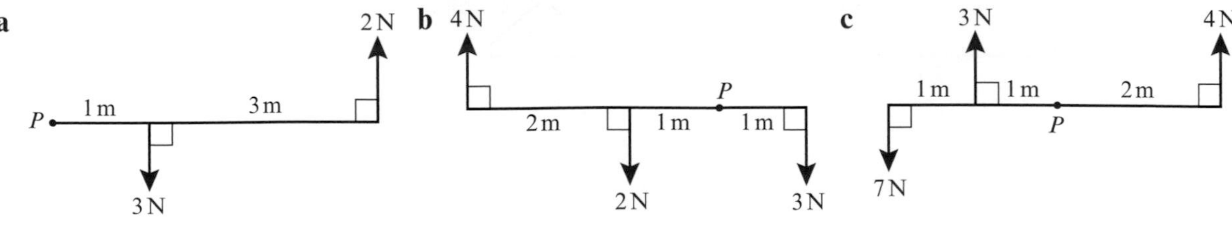

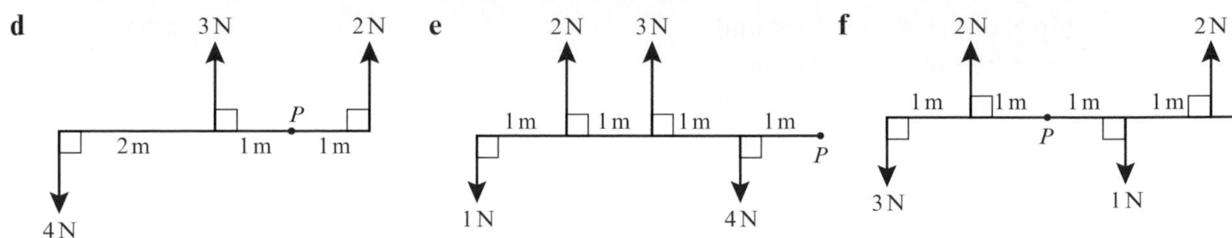

2 These diagrams show forces acting on a lamina. In each case, find the resultant moment about P.

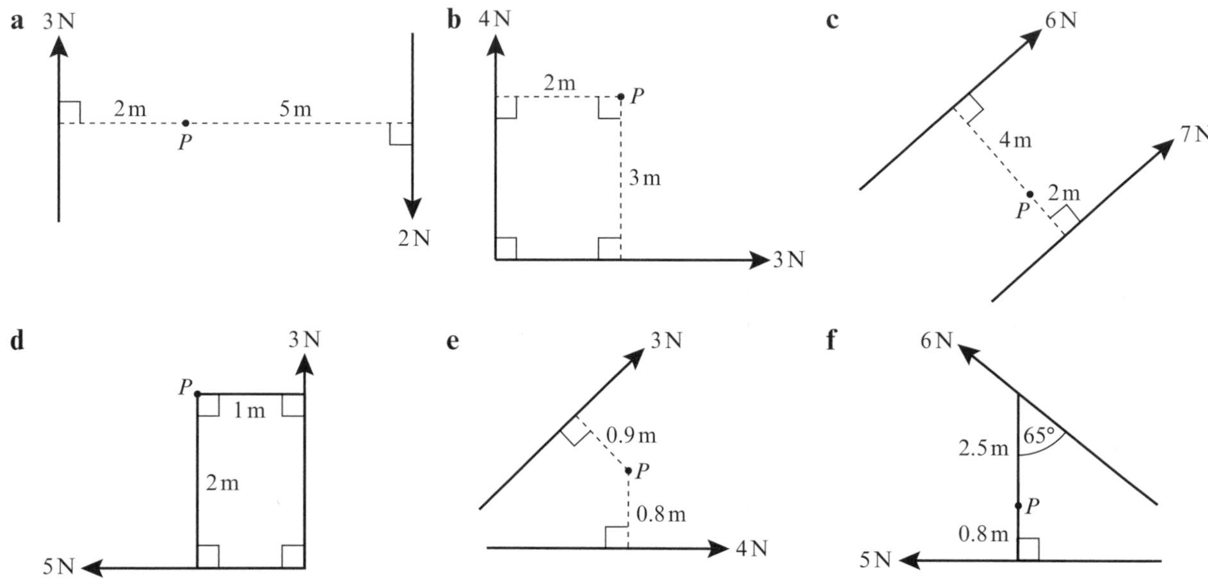

(P) **3** The diagram shows a set of forces acting on a light rod. The resultant moment about P is 17 N m clockwise. Find the length, d.

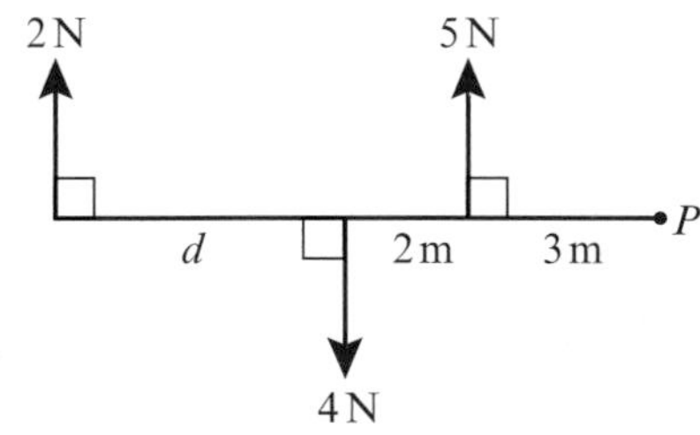

(E/P) **4** The diagram shows a set of forces acting on a light rod. The resultant moment about P is 12.8 N m clockwise. Find the value of x.

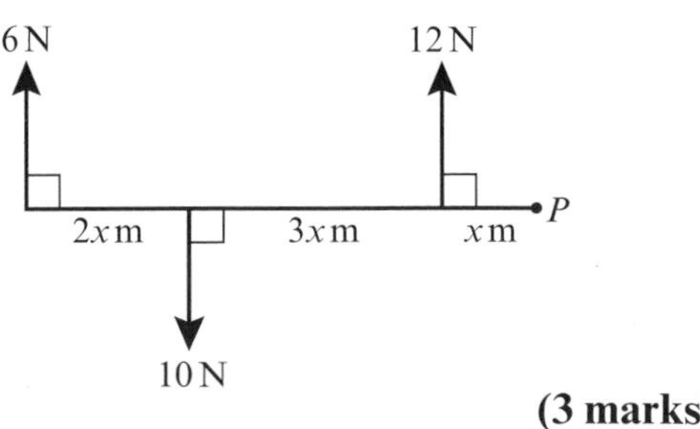

(3 marks)

(E/P) **5** A cruise ship is tethered to a dock and is being moved by three tugs. The cruise ship is modelled as a rectangular lamina *PQRS* fixed at *P* under the action of three coplanar forces. *A* is the midpoint of *PS* and *B* is the midpoint of *RS*.

Determine the direction of the rotation of the cruise ship and the magnitude of the resultant moment about *P*. **(5 marks)**

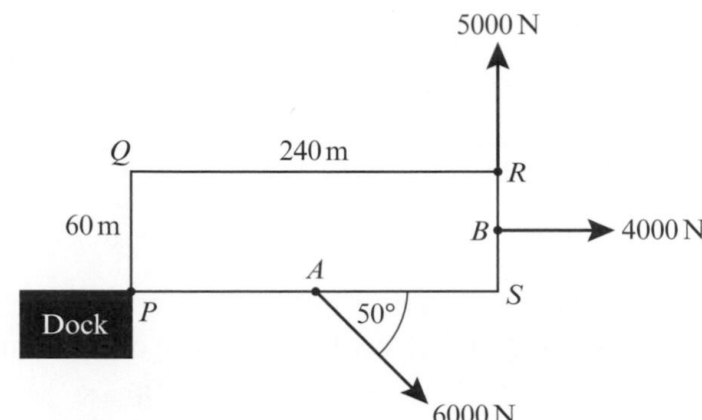

(E/P) **6** The diagram shows a drawbridge inclined at an angle of θ to the horizontal, where $0 < \theta < 90°$. The drawbridge is modelled as a uniform rod of weight 8000 N. A horizontal force of magnitude 6000 N is applied at the top of the drawbridge. Given that the drawbridge is rising, prove that $\tan\theta > \frac{2}{3}$ **(5 marks)**

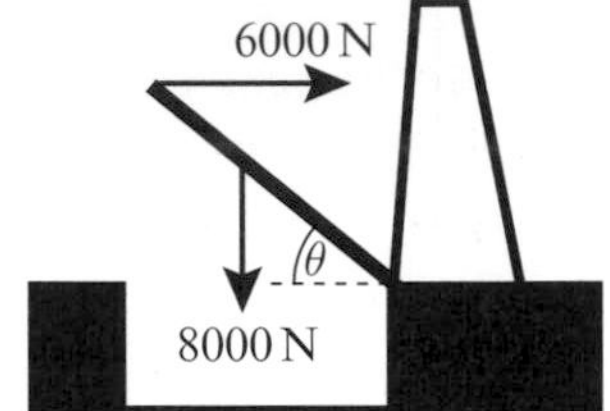

Hint The drawbridge is modelled as a uniform rod so its weight acts at its midpoint.

4.3 Equilibrium

- **When a rigid body is in equilibrium, the resultant force in any direction is 0 N and the resultant moment about any point is 0 N m.**

Hint If the resultant moment is zero then the sum of the clockwise moments equals the sum of the anticlockwise moments.

You can simplify many problems involving rigid bodies by choosing which point(s) to take moments about. When you take moments at a given point, you can ignore the rotational effect of any forces acting at that point.

Example 6

The diagram shows a uniform rod *AB*, of length 3 m and weight 20 N, resting horizontally on supports at *A* and *C*, where *AC* = 2 m.

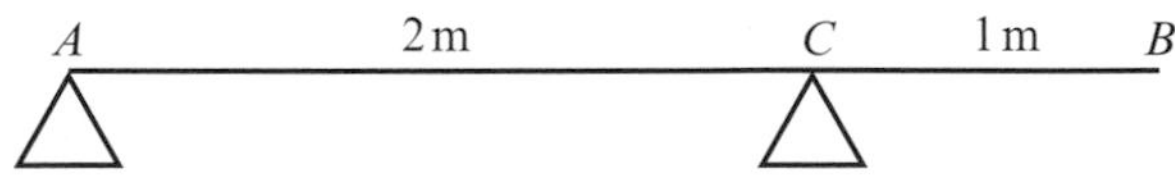

Calculate the magnitude of the reaction at each of the supports.

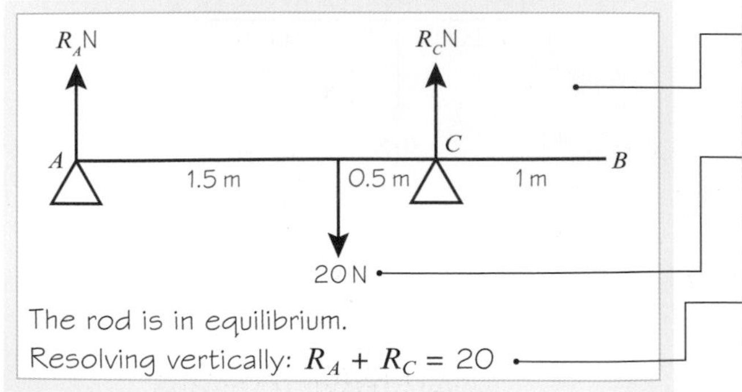

The rod is in equilibrium.

Resolving vertically: $R_A + R_C = 20$

Draw a diagram showing all the forces acting.

The weight of the rod acts at its centre of mass. You are told that this is a uniform rod, so the weight acts at the midpoint of the rod.

Total of forces acting upwards = total of forces acting downwards.

Considering the moments about point A:
$20 \times 1.5 = R_C \times (1.5 + 0.5)$
$30 = 2R_C$
$15 = R_C$
$R_A + 15 = 20$
$R_A = 5$
Therefore the reaction at A is 5 N and the reaction at C is 15 N.

Clockwise moment = anticlockwise moment

Problem-solving

Take moments about the point that makes the algebra as simple as possible. Taking moments about A results in an equation with just one unknown.

Substituting the value of R_C into the first equation.

Example 7

A uniform beam AB, of mass 40 kg and length 5 m, rests horizontally on supports at C and D, where $AC = DB = 1$ m. When a man of mass 80 kg stands on the beam at E the magnitude of the reaction at D is twice the magnitude of the reaction at C. By modelling the beam as a rod and the man as a particle, find the distance AE.

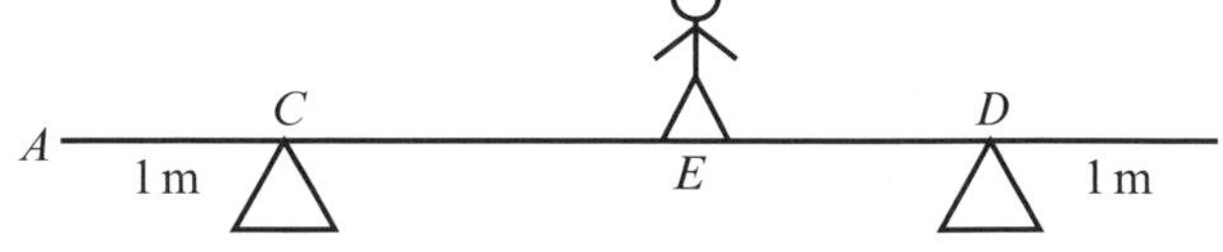

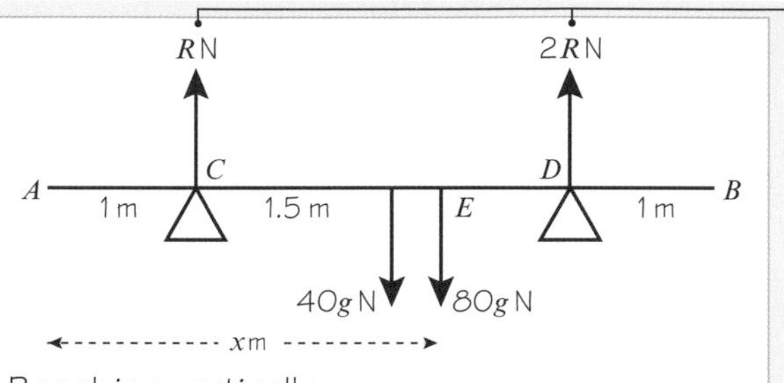

Draw a diagram showing the forces.

Because you are told a relationship between the reaction at C and the reaction at D you can use this on your diagram.

Resolving vertically:
$R + 2R = 40g + 80g$
$3R = 120g$
$R = 40g$

The rod is in equilibrium so there is no resultant force.

Let the distance AE be x m.
Taking moments about A:
$40g \times 2.5 + 80g \times x = 40g \times 1 + 80g \times 4$
$100g + 80g \times x = 360g$
$80g \times x = 260g$
$\Rightarrow \quad x = \frac{260g}{80g} = \frac{26}{8} = 3.25$
Distance $AE = 3.25$ m

Clockwise moment = anticlockwise moment

Problem-solving

How have you used the modelling assumptions in the question?

- Since the beam is a rod, you can ignore its width
- Since the man is a particle, his weight acts at the point E

Example 8

A uniform rod PQ is hinged at the point P, and is held in equilibrium at an angle of 50° to the horizontal by a force of magnitude **F** acting perpendicular to the rod at Q. Given that the rod has a length of 3 m and a mass of 8 kg, find the value of **F**.

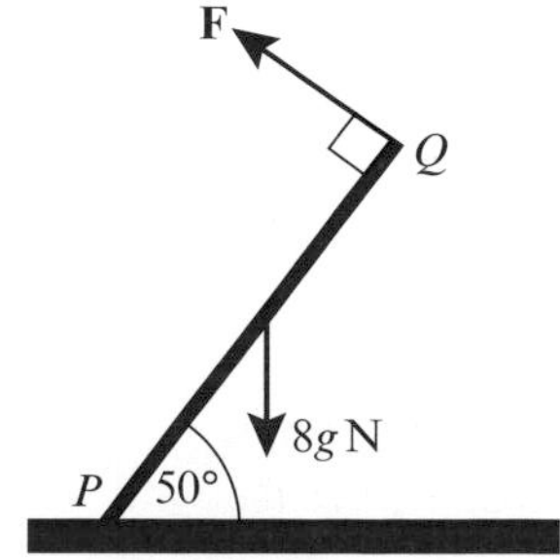

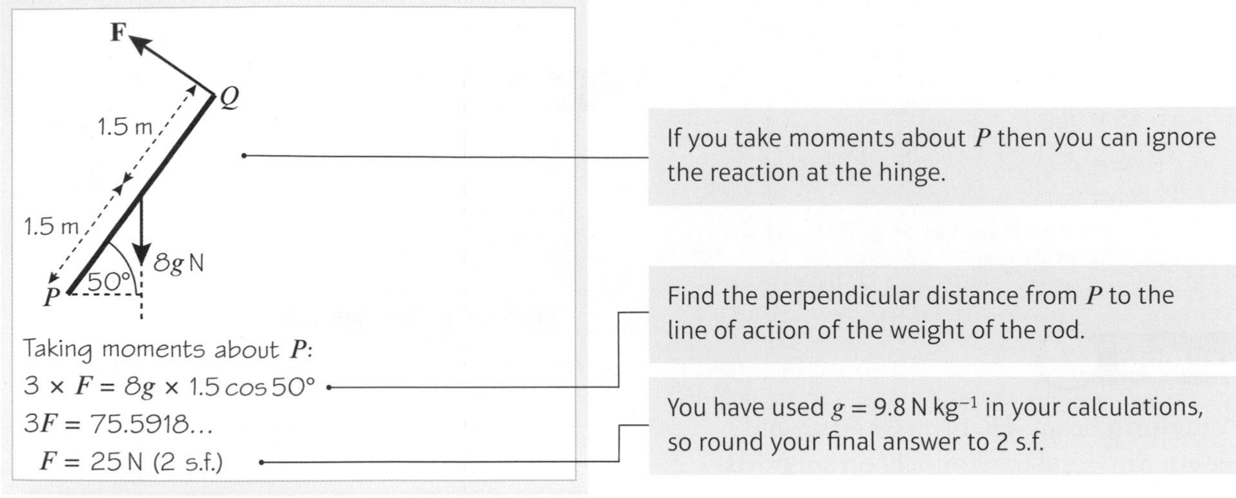

Taking moments about P:

$3 \times F = 8g \times 1.5 \cos 50°$

$3F = 75.5918...$

$F = 25$ N (2 s.f.)

If you take moments about P then you can ignore the reaction at the hinge.

Find the perpendicular distance from P to the line of action of the weight of the rod.

You have used $g = 9.8$ N kg^{-1} in your calculations, so round your final answer to 2 s.f.

Exercise 4C

1 AB is a uniform rod of length 5 m and weight 20 N. In these diagrams AB is resting in a horizontal position on supports at C and D. In each case, find the magnitudes of the reactions at C and D.

a A — 1 m — C — 3 m — D — 1 m — B

b A — 2 m — C — 2 m — D — 1 m — B

c A — 1.5 m — C — 2.5 m — D — 1 m — B

d A — 1.5 m — C — 2.7 m — D — 0.8 m — B

2 Each of these diagrams shows a light rod in equilibrium in a horizontal position under the action of a set of forces. Find the values of the unknown forces and distances.

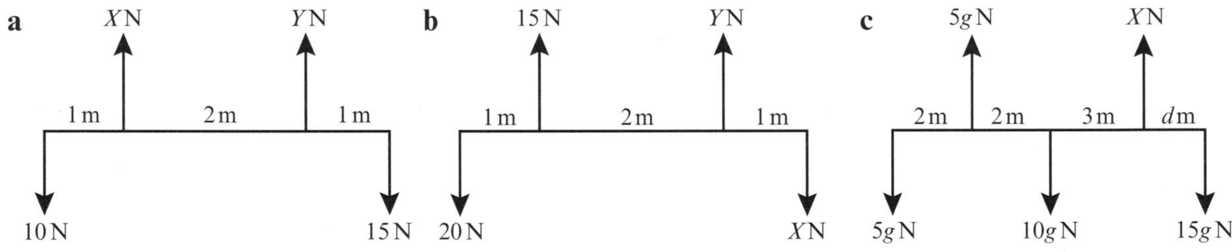

3 Jack and Jill are playing on a seesaw made from a uniform plank AB, of length 5 m pivoted at M, the midpoint of AB. Jack has mass 35 kg and Jill has mass 28 kg. Jill sits at A and Jack sits at a distance x m from B. The plank is in equilibrium. Find the value of x.

4 A uniform rod AB, of length 3 m and mass 12 kg, is pivoted at C, where $AC = 1$ m. A vertical force **F** applied at A maintains the rod in horizontal equilibrium. Calculate the magnitude of **F**.

5 A broom consists of a broomstick of length 130 cm and mass 5 kg and a broomhead of mass 5.5 kg attached at one end. By modelling the broomstick as a uniform rod and the broomhead as a particle, find where a support should be placed so that the broom will balance horizontally.

(P) **6** A uniform rod AB, of length 4 m and weight 20 N, is suspended horizontally by two vertical strings attached at A and at B. A particle of weight 10 N is attached to the rod at point C, where $AC = 1.5$ m.

a Find the magnitudes of the tensions in the two strings.

The particle is moved so that the rod remains in horizontal equilibrium with the tension in the string at B 1.5 times the tension in the string at A.

b Find the new distance of the particle from A.

(E/P) **7** 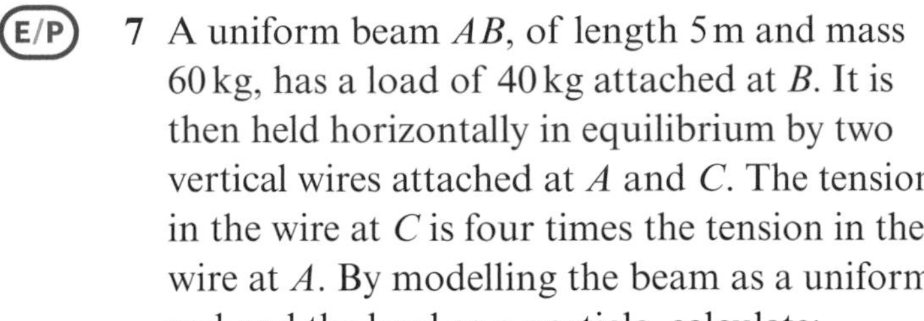
A uniform beam AB, of length 5 m and mass 60 kg, has a load of 40 kg attached at B. It is then held horizontally in equilibrium by two vertical wires attached at A and C. The tension in the wire at C is four times the tension in the wire at A. By modelling the beam as a uniform rod and the load as a particle, calculate:

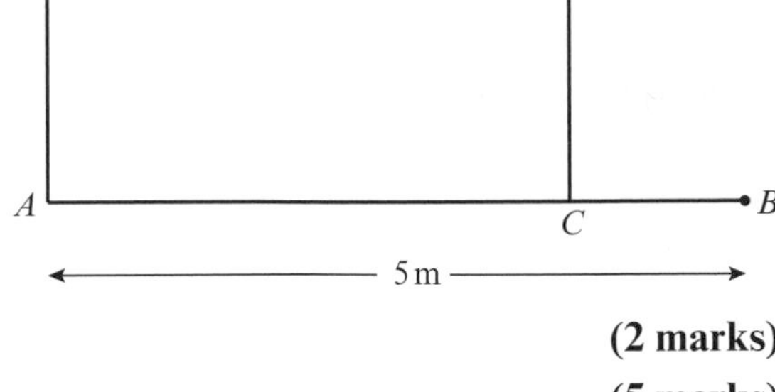

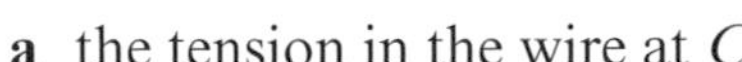
a the tension in the wire at C **(2 marks)**

b the distance CB. **(5 marks)**

(E) **8** A uniform plank AB has length 5 m and mass 15 kg. The plank is held in equilibrium horizontally by two smooth supports A and C as shown in the diagram, where $BC = 2$ m.

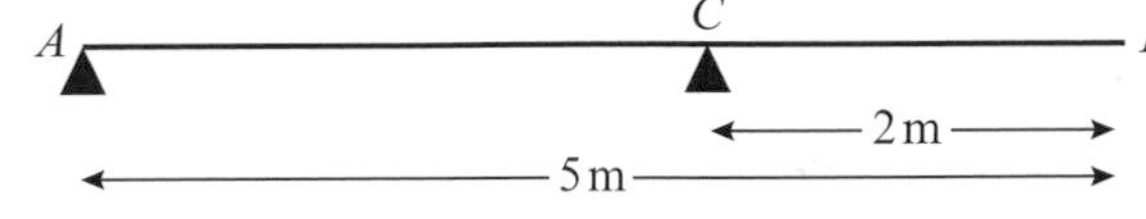

a Find the reaction on the plank at C. **(3 marks)**

A person of mass 45 kg stands on the plank at the point D and it remains in equilibrium. The reactions on the plank at A and C are now equal.

b Find the distance AD. **(7 marks)**

(E/P) **9** A uniform beam AB has weight W N and length 8 m. The beam is held in a horizontal position in equilibrium by two vertical light inextensible wires attached to the beam at the points A and C where $AC = 4.5$ m, as shown in the diagram. A particle of weight 30 N is attached to the beam at B.

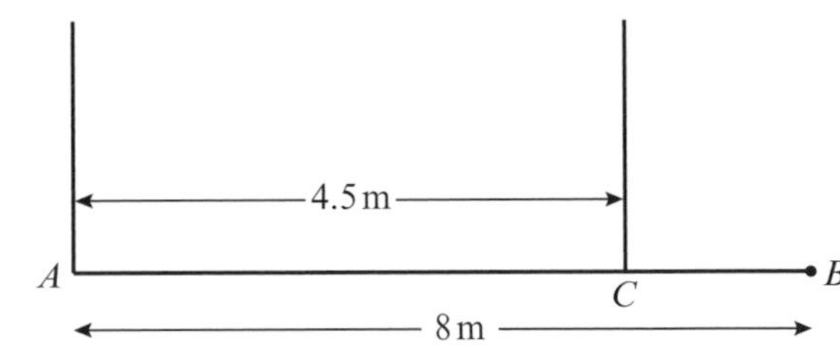

a Show that the tension in the wire attached to the beam at C is $\left(\frac{8}{9}W + \frac{160}{3}\right)$N. **(4 marks)**

b Find, in terms of W, the tension in the wire attached to the beam at A. **(3 marks)**

Given that the tension in the wire attached to the beam at C is twelve times the tension in the wire attached to the beam at A,

c find the value of W. **(3 marks)**

(E/P) **10** A metal lever of mass 5 kg and length 1.5 m is attached by a smooth hinge to a vertical wall. The lever is held at an angle of 30° to the vertical by a horizontal force of magnitude **F** N applied at the other end of the lever. By modelling the lever as a uniform rod, find the value of **F**. **(4 marks)**

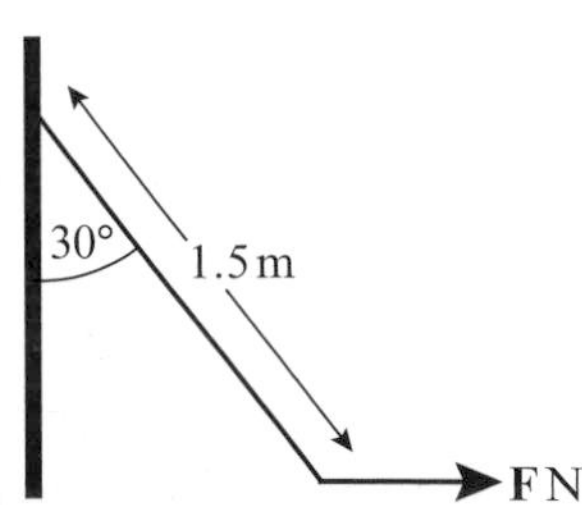

(E) **11** A uniform ladder, AB, is leaning against a smooth vertical wall on rough horizontal ground at an angle of 70° to the horizontal. The ladder has length 8 m, and is held in equilibrium by a frictional force of magnitude 60 N acting horizontally at B, as shown in the diagram.

a Write down the magnitude of the normal reaction of the wall on the ladder at A. **(1 mark)**

b Find the mass of the ladder. **(4 marks)**

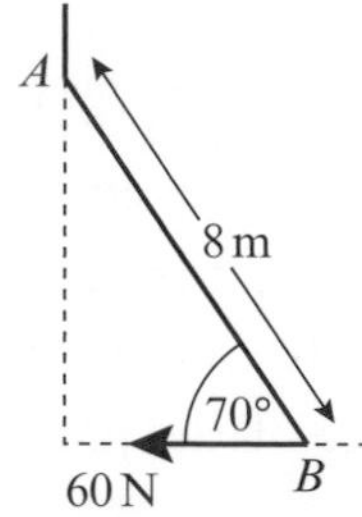

Problem-solving

In part **b** you can ignore the normal reaction at B by taking moments about that point.

Challenge

The diagram shows a kinetic sculpture made from hanging rods. The distances between the points marked on each rod are equal. Arrange 1 kg, 2 kg, 3 kg, 4 kg and 5 kg weights onto the marked squares, using each weight once, so that the sculpture hangs in equilibrium with the rods horizontal.

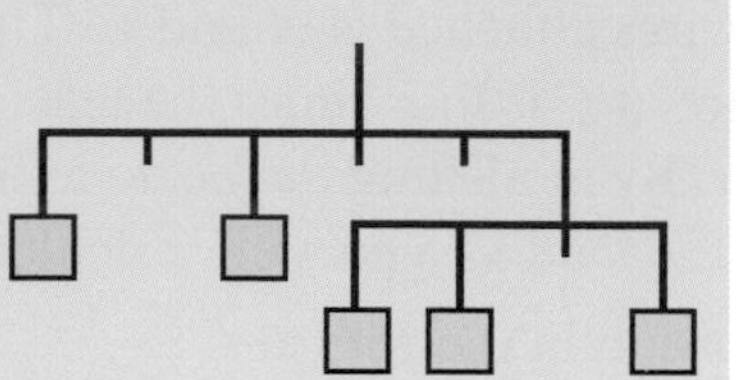

4.4 Centres of mass

So far you have only considered **uniform** rods, where the centre of mass is always at the midpoint. If a rod is **non-uniform** the centre of mass is not necessarily at the midpoint of the rod.

You might need to consider the moment due to the weight of a non-uniform rod, or find the position of its centre of mass.

Example 9

Sam and Tamsin are sitting on a non-uniform plank AB of mass 25 kg and length 4 m. The plank is pivoted at M, the midpoint of AB. The centre of mass of AB is at C where AC is 1.8 m. Sam has mass 35 kg.
Tamsin has mass 25 kg and sits at A.
Where must Sam sit for the plank to be horizontal?

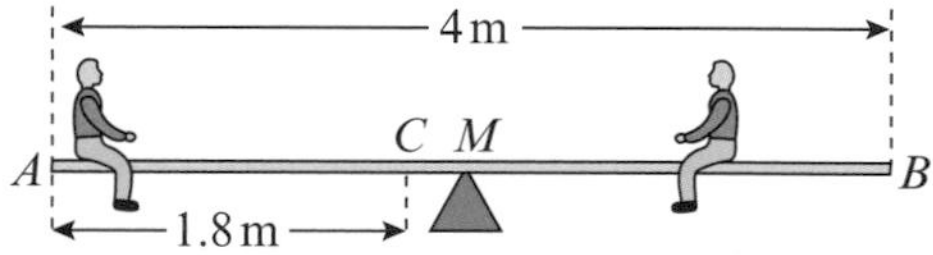

R N; A; 1.8 m; 0.2 m; 2 m; B; M; 25g N; 25g N; 35g N; x m

Model the plank as a rod and the children as paticles. Then draw a diagram to represent the situation.

Suppose that Sam sits at a point x m from A.

Taking moments about M:

$25g \times 2 + 25g \times 0.2 = 35g \times (x - 2)$

Take moments about M to eliminate the reaction at M from your calculations.

$50 + 5 = 35x - 70$

Divide both sides of the equation by g.

$35x = 125$

$x = 3.57$

Sam should sit 3.57 m from end A (or 0.43 m from end B).

Online Explore the moment acting about pirot M using GeoGebra.

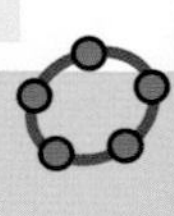

Example 10

A non-uniform rod AB is 3 m long and has weight 20 N. It is in a horizontal position resting on supports at points C and D, where $AC = 1$ m and $AD = 2.5$ m. The magnitude of the reaction at C is three times the magnitude of the reaction at D. Find the distance of the centre of mass of the rod from A.

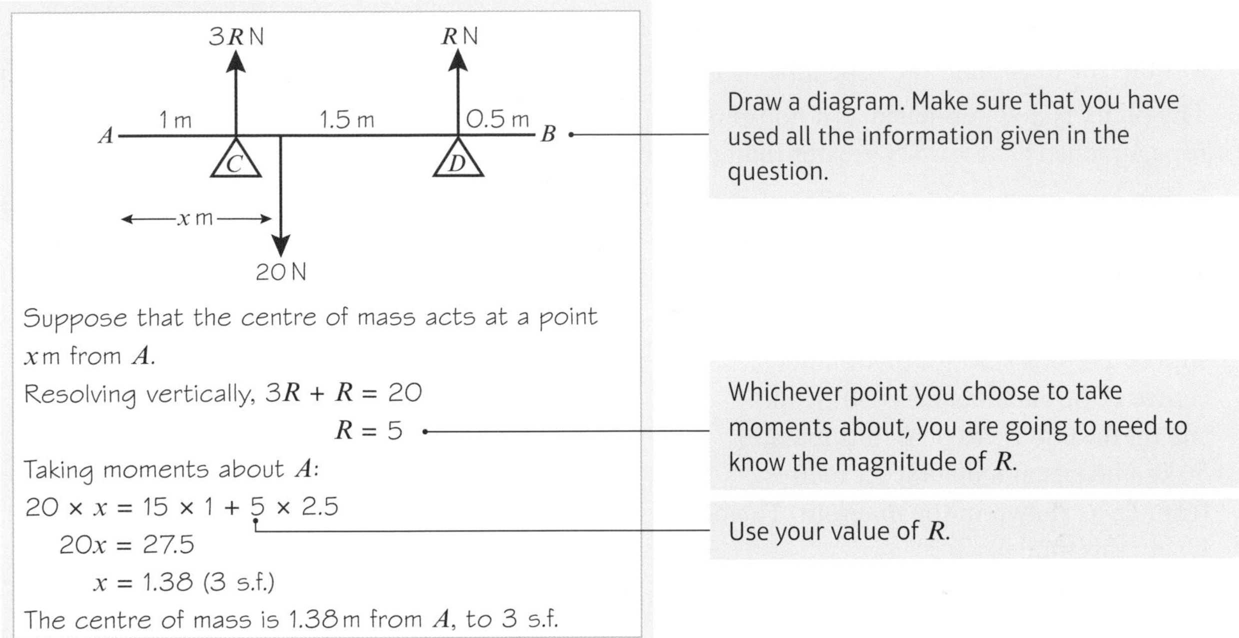

Suppose that the centre of mass acts at a point x m from A.

Resolving vertically, $3R + R = 20$

$R = 5$

Whichever point you choose to take moments about, you are going to need to know the magnitude of R.

Taking moments about A:

$20 \times x = 15 \times 1 + 5 \times 2.5$

$20x = 27.5$

$x = 1.38$ (3 s.f.)

Use your value of R.

The centre of mass is 1.38 m from A, to 3 s.f.

Exercise 4D

1 A non-uniform rod AB, of length 4 m and weight 6 N, rests horizontally on two supports, A and B. Given that the centre of mass of the rod is 2.4 m from the end A, find the reactions at the two supports.

2 A non-uniform bar AB of length 5 m is supported horizontally on supports, A and B. The reactions at these supports are $3g$ N and $7g$ N respectively.

a State the weight of the bar.

b Find the distance of the centre of mass of the bar from A.

3 A non-uniform plank AB, of length 4 m and weight 120 N, is pivoted at its midpoint. The plank is in equilibrium in a horizontal position with a child of weight 200 N sitting at A and a child of weight 300 N sitting at B. By modelling the plank as a rod and the two children as particles, find the distance of the centre of mass of the plank from A.

(P) 4 A non-uniform rod AB, of length 5 m and mass 15 kg, rests horizontally suspended from the ceiling by two vertical strings attached to C and D, where $AC = 1$ m and $AD = 3.5$ m.

a Given that the centre of mass is at E where $AE = 3$ m, find the magnitudes of the tensions in the strings.

When a particle of mass 9 kg is attached to the rod at F the tension in the string at D is twice the tension in the string at C.

b Find the distance AF.

E/P **5** A plank AB has mass 24 kg and length 4.8 m. A load of mass 15 kg is attached to the plank at the point C, where $AC = 1.4$ m. The loaded plank is held in equilibrium, with AB horizontal, by two vertical ropes, one attached at A and the other attached at B, as shown in the diagram. The plank is modelled as a uniform rod, the load as a particle and the ropes as light inextensible strings.

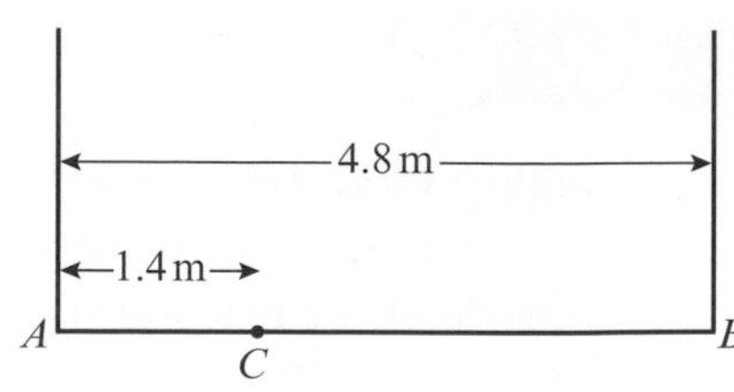

a Find the tension in the rope attached at B. **(4 marks)**

The plank is now modelled as a non-uniform rod. With the new model, the tension in the rope attached at A is 25 N greater than the tension in the rope attached at B.

b Find the distance of the centre of mass of the plank from A. **(6 marks)**

E **6** A seesaw in a playground consists of a beam AB of length 10 m which is supported by a smooth pivot at its centre C. Sophia has mass 30 kg and sits on the end A. Roshan has mass 50 kg and sits at a distance x metres from C, as shown in the diagram. The beam is initially modelled as a uniform rod. Using this model,

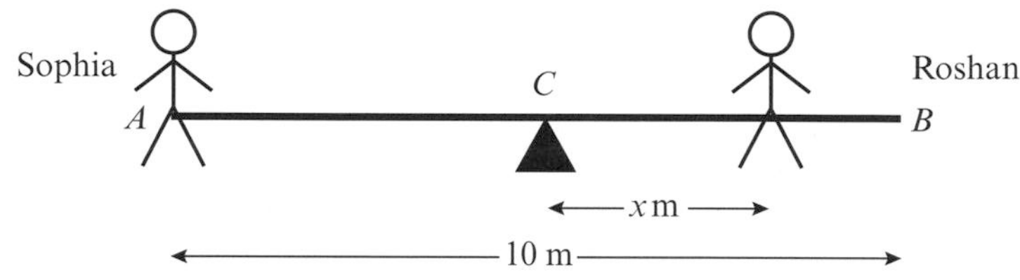

a find the value of x for which the seesaw can rest in equilibrium in a horizontal position. **(3 marks)**

b State what is implied by the modelling assumption that the beam is uniform. **(1 mark)**

Roshan finds he must sit at a distance 4 m from C for the seesaw to rest horizontally in equilibrium. The beam is now modelled as a non-uniform rod of mass 25 kg. Using this model,

c find the distance of the centre of mass of the beam from C. **(4 marks)**

E/P **7** A non-uniform rod AB, of length 25 m and weight 80 N, rests horizontally in equilibrium on supports C and D as shown in the diagram. The centre of mass of the rod is 10 m from A.

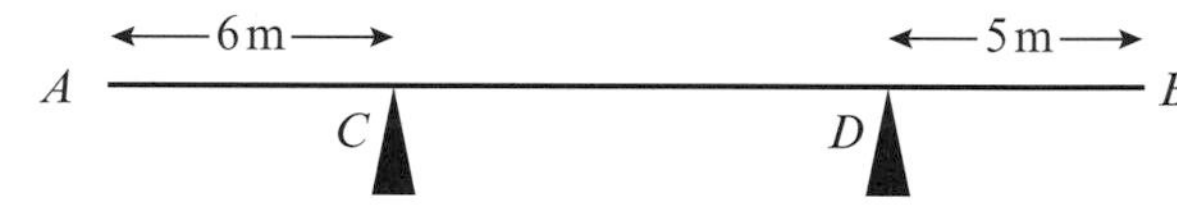

A particle of weight W newtons is attached to the rod at a point E, where E is x metres from A.

The rod remains in equilibrium and the magnitude of the reaction at C is five times the magnitude of the reaction at D.

Show that $W = \dfrac{400}{25 - 3x}$ **(5 marks)**

E/P **8** A non-uniform rod of weight 100 N and length 4 m is freely hinged to a vertical wall, and held in place by a cable attached at an angle of 30° to the end of the rod. Given that the tension in the cable is 80 N and that the rod is held in horizontal equilibrium, find the distance of the centre of mass of the rod from the wall. **(8 marks)**

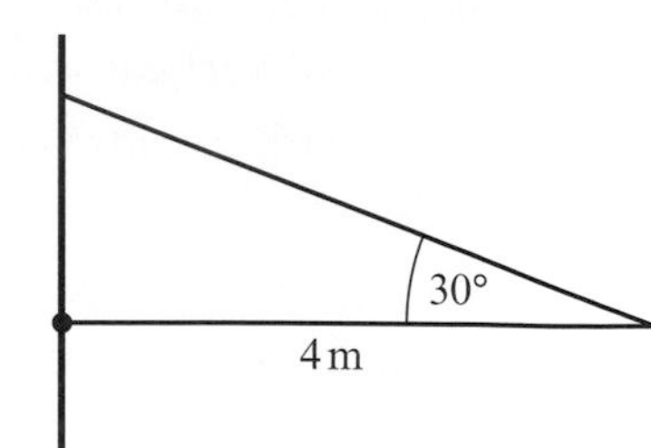

Challenge

A non-uniform beam of weight 120 N and length 5 m is smoothly pivoted at a point M and is held at an angle of 40° by a cable attached at point N. Given that the tension in the cable is 30 N and it makes an angle of 80° with the beam, find the distance of the centre of mass of the beam from M.

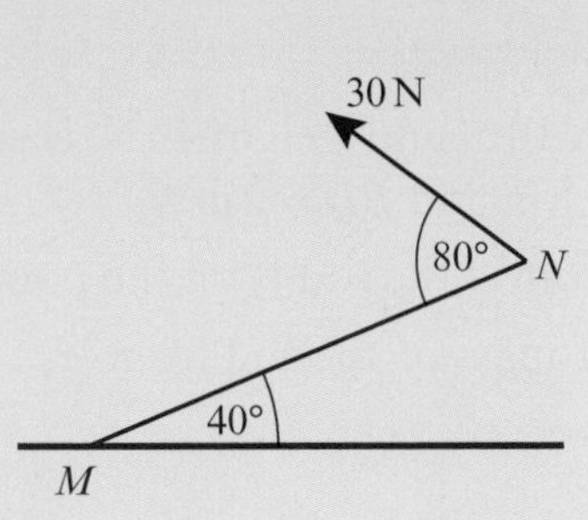

4.5 Tilting

You need to be able to answer questions involving rods that are on the point of tilting.

- **When a rigid body is on the point of tilting about a pivot, the reaction at any other support (or the tension in any other wire or string) is zero.**

Example 11

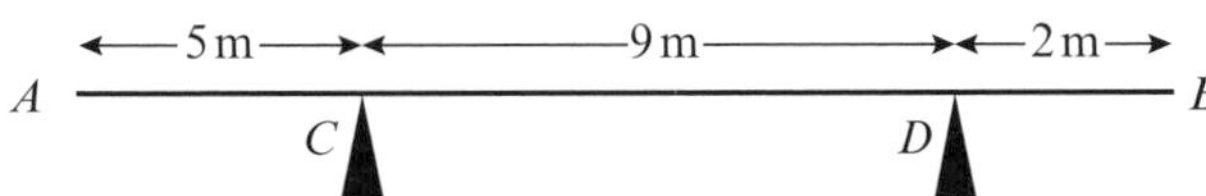

A uniform beam AB, of mass 45 kg and length 16 m, rests horizontally on supports C and D where $AC = 5$ m and $CD = 9$ m.

When a child stands at A, the beam is on the point of tilting about C.
Find the mass of the child.

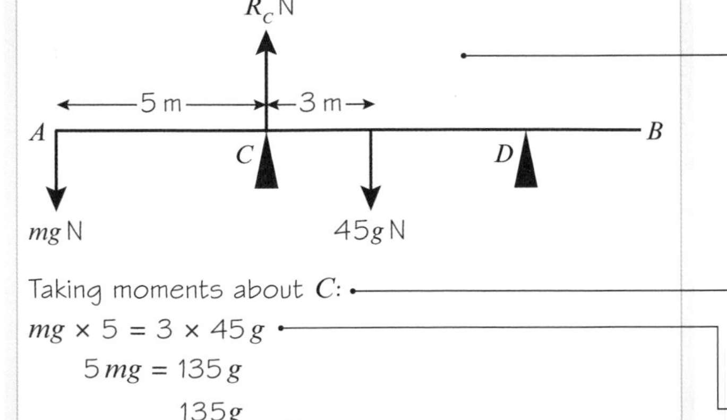

Draw a diagram showing the forces. Remember, as the beam is 'on the point of tilting', C is a pivot and the reaction force at D is zero.

Taking moments about C:

Take moments about C. This means you can ignore the reaction R_C.

$mg \times 5 = 3 \times 45g$

Anticlockwise moment = Clockwise moment

$5mg = 135g$

$$m = \frac{135g}{5g} = 27$$

The mass of the child is 27 kg.

Online See the point at which the beam starts to tilt due to the weight of the child and explore the problem with different forces and distances using GeoGebra.

Example 12

A non-uniform rod AB, of length 10 m and weight 40 N, is suspended from a pair of light cables attached to C and D where $AC = 3$ m and $BD = 2$ m.

When a weight of 25 N is hung from A the rod is on the point of rotating.

Find the distance of the centre of mass of the rod from A.

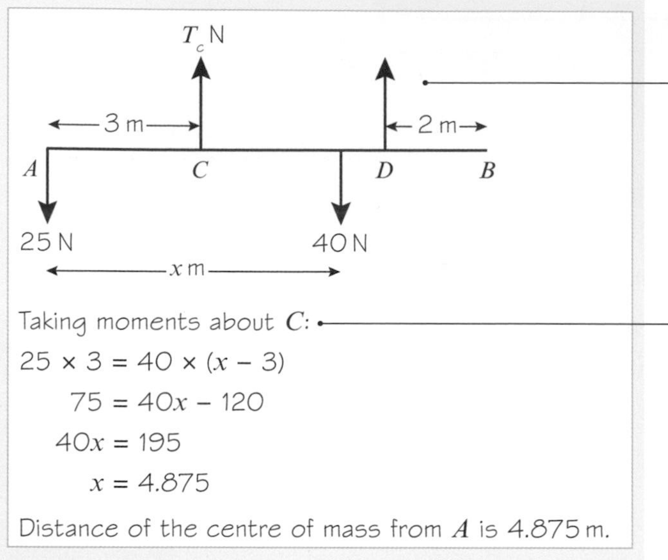

Taking moments about C:

$25 \times 3 = 40 \times (x - 3)$

$75 = 40x - 120$

$40x = 195$

$x = 4.875$

Distance of the centre of mass from A is 4.875 m.

Draw a diagram showing the forces about C. As the rod is 'on the point of rotating', C is a pivot and the tension force at D is zero. As the rod is non-uniform you don't know the distance from A to the centre of mass, so you can label it as x.

You don't require the tension force, T_C, so take moments about C.

Problem-solving

The distance from C to the centre of mass is $(x - 3)$ m. Equate the clockwise and anticlockwise moments then solve the equation to find the value of x.

Exercise 4E

1 A uniform rod AB has length 4 m and mass 8 kg. It is resting in a horizontal position on supports at points C and D where $AC = 1$ m and $AD = 2.5$ m. A particle of mass m kg is placed at point E where $AE = 3.3$ m. Given that the rod is about to tilt about D, calculate the value of m.

2 A uniform bar AB, of length 6 m and weight 40 N, is resting in a horizontal position on supports at points C and D where $AC = 2$ m and $AD = 5$ m. When a particle of weight 30 N is attached to the bar at point E the bar is on the point of tilting about C. Calculate the distance AE.

3 A plank AB, of mass 12 kg and length 3 m, is in equilibrium in a horizontal position resting on supports at C and D where $AC = 0.7$ m and $DB = 1.1$ m. A boy of mass 32 kg stands on the plank at point E. The plank is about to tilt about D. By modelling the plank as a uniform rod and the boy as a particle, calculate the distance AE.

(P) **4** A uniform rod AB has length 5 m and weight 20 N. The rod is resting on supports at points C and D where $AC = 2$ m and $BD = 1$ m.

a Find the magnitudes of the reactions at C and D.

A particle of weight 12 N is placed on the rod at point A.

b Show that this causes the rod to tilt about C.

A second particle of weight 100 N is placed on the rod at E to hold it in equilibrium.

c Find the minimum and maximum possible distances of E from A.

E 5 A uniform plank of mass 100 kg and length 10 m rests horizontally on two smooth supports, A and B, as shown in the diagram. A man of mass 80 kg starts walking from one end of the plank, A, to the other end.

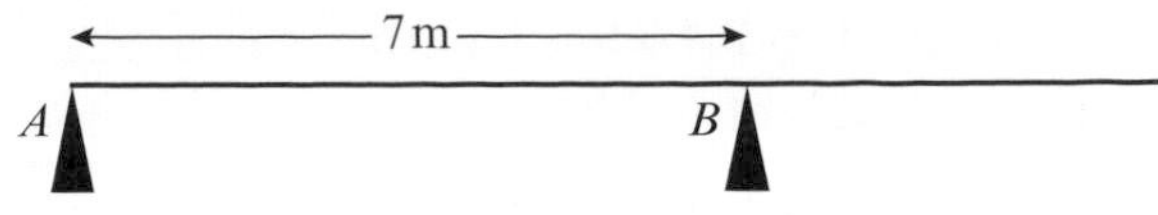

Find the distance he can walk past B before the plank starts to tip. **(4 marks)**

E/P 6 A non-uniform beam PQ, of mass m and length $8a$, hangs horizontally in equilibrium from two wires at M and N, where $PM = a$ and $QN = 2a$, as shown in the diagram. The centre of mass of the beam is at the point O. A particle of mass $\frac{3}{4}m$ is placed on the beam at Q and the beam is on the point of tipping about N.

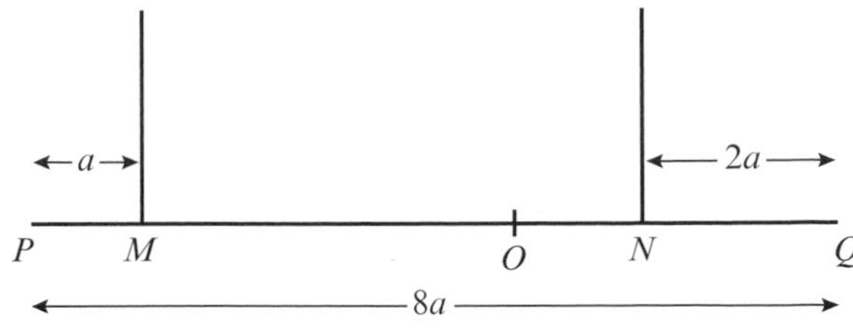

a Show that $ON = \frac{3}{2}a$. **(3 marks)**

The particle is removed and replaced at the midpoint of the beam and the beam remains in equilibrium.

b Find the magnitude of the tension in the wire attached at point N in terms of m. **(5 marks)**

E/P 7 A uniform beam AB, of weight W and length 14 m, hangs in equilibrium in a horizontal position from two vertical cables attached at points C and D where $AC = 4$ m and $BD = 6$ m.

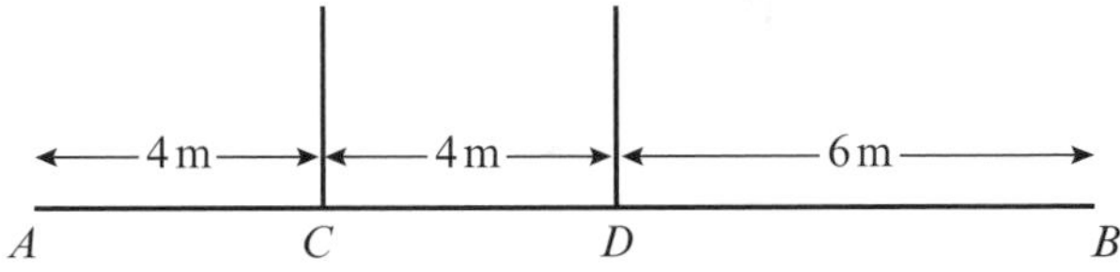

A weight of 180 N is hung from A and the beam is about to tilt. The weight is removed and a different weight, V, is hung from B and the beam does not tilt. Find the maximum value of V. **(6 marks)**

Mixed exercise 4

E 1 A plank AE, of length 6 m and weight 100 N, rests in a horizontal position on supports at B and D, where $AB = 1$ m and $DE = 1.5$ m. A child of weight 145 N stands at C, the midpoint of AE, as shown in the diagram. The child is modelled as a particle and the plank as a uniform rod. The child and the plank are in equilibrium. Calculate:

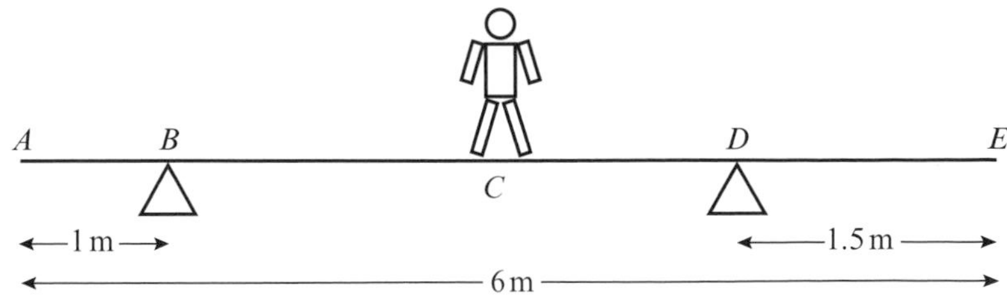

a the magnitude of the force exerted by the support on the plank at B **(3 marks)**

b the magnitude of the force exerted by the support on the plank at D. **(2 marks)**

The child now stands at a different point F on the plank. The plank is in equilibrium and on the point of tilting about D.

c Calculate the distance DF. **(4 marks)**

(E/P) **2** A uniform rod AB has length 4 m and weight 150 N. The rod rests in equilibrium in a horizontal position, smoothly supported at points C and D, where $AC = 1$ m and $AD = 2.5$ m as shown in the diagram. A particle of weight W N is attached to the rod at a point E where $AE = x$ metres. The rod remains in equilibrium and the magnitude of the reaction at C is now equal to the magnitude of the reaction at D.

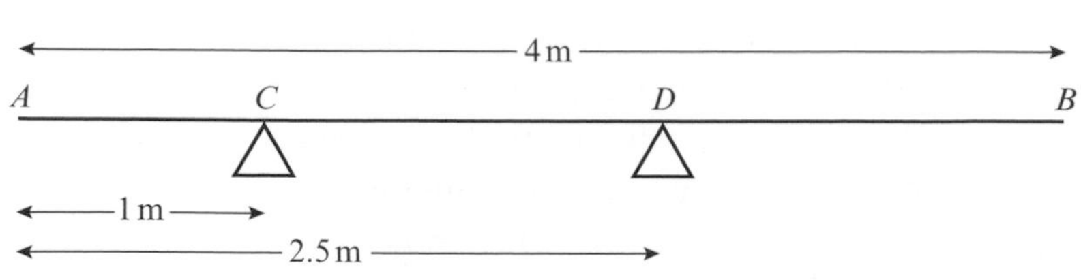

a Show that $W = \dfrac{150}{7 - 4x}$ **(6 marks)**

b Hence deduce the range of possible values of x. **(3 marks)**

(E) **3** A uniform plank AB has mass 40 kg and length 4 m. It is supported in a horizontal position by two smooth pivots. One pivot is at the end A and the other is at the point C where $AC = 3$ m, as shown in the diagram. A man of mass 80 kg stands on the plank which remains in equilibrium. The magnitude of the reaction at A is twice the magnitude of the reaction at C. The magnitude of the reaction at C is R N. The plank is modelled as a rod and the man is modelled as a particle.

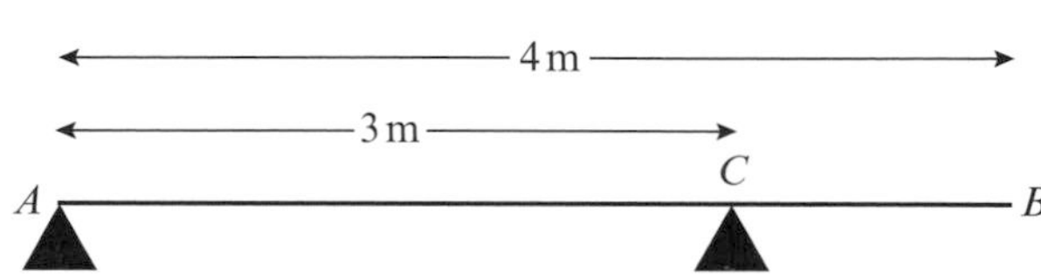

a Find the value of R. **(2 marks)**

b Find the distance of the man from A. **(3 marks)**

c State how you have used the modelling assumption that:

i the plank is uniform

ii the plank is a rod

iii the man is a particle. **(3 marks)**

(E/P) **4** A non-uniform rod AB has length 4 m and weight 150 N. The rod rests horizontally in equilibrium on two smooth supports C and D, where $AC = 1$ m and $DB = 0.5$ m, as shown in the diagram. The centre of mass of AB is x metres from A. A particle of weight W N is placed on the rod at A. The rod remains in equilibrium and the magnitude of the reaction of C on the rod is 100 N.

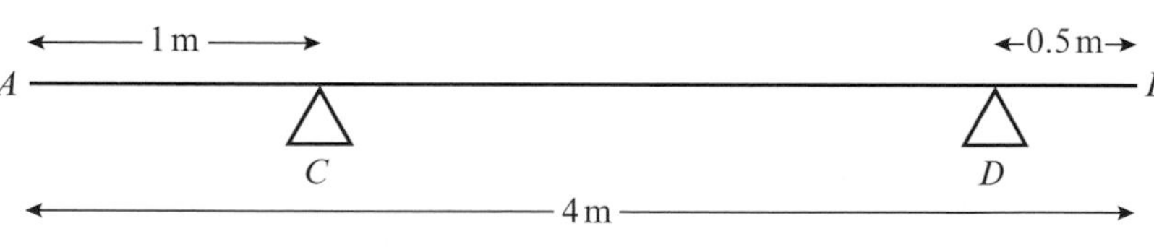

a Show that $550 + 7W = 300x$. **(4 marks)**

The particle is now removed from A and placed on the rod at B. The rod remains in equilibrium and the reaction of C on the rod now has magnitude 52 N.

b Obtain another equation connecting W and x. **(4 marks)**

c Calculate the value of x and the value of W. **(3 marks)**

(E) 5 A lever consists of a uniform steel rod AB, of weight 100 N and length 2 m, which rests on a small smooth pivot at a point C. A load of weight 1700 N is suspended from the end B of the rod by a rope. The lever is held in equilibrium in a horizontal position by a vertical force applied at the end A, as shown in the diagram. The rope is modelled as a light string.

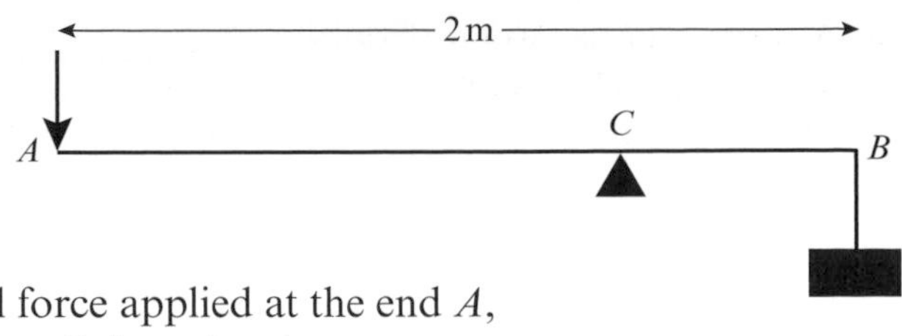

a Given that $BC = 0.25$ m find the magnitude of the force applied at A. **(4 marks)**

The position of the pivot is changed so that the rod remains in equilibrium when the force at A has magnitude 150 N.

b Find, to the nearest centimetre, the new distance of the pivot from B. **(4 marks)**

(E) 6 A plank AB has length 4 m. It lies on a horizontal platform, with the end A lying on the platform and the end B projecting over the edge, as shown in the diagram. The edge of the platform is at the point C.

Jack and Jill are experimenting with the plank. Jack has mass 48 kg and Jill has mass 36 kg. They discover that if Jack stands at B and Jill stands at A and $BC = 1.8$ m, the plank is in equilibrium and on the point of tilting about C.

a By modelling the plank as a uniform rod, and Jack and Jill as particles, find the mass of the plank. **(4 marks)**

They now alter the position of the plank in relation to the platform so that, when Jill stands at B and Jack stands at A, the plank is again in equilibrium and on the point of tilting about C.

b Find the distance BC in this position. **(4 marks)**

(E) 7 A plank of wood AB has mass 12 kg and length 5 m. It rests in a horizontal position on two smooth supports. One support is at the end A. The other is at the point C, 0.5 m from B, as shown in the diagram. A girl of mass 30 kg stands at B with the plank in equilibrium.

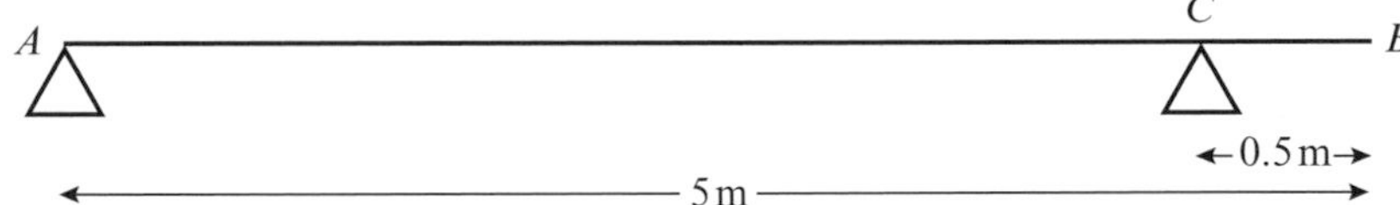

a By modelling the plank as a uniform rod and the girl as a particle, find the reaction on the plank at A. **(4 marks)**

The girl gets off the plank. A boulder of mass m kg is placed on the plank at A and a man of mass 93 kg stands on the plank at B. The plank remains in equilibrium and is on the point of tilting about C.

b By modelling the plank again as a uniform rod, and the man and the boulder as particles, find the value of m. **(5 marks)**

(E/P) 8 A plank AB has mass 50 kg and length 4 m. A load of mass 25 kg is attached to the plank at B. The loaded plank is held in equilibrium, with AB horizontal, by two vertical ropes attached at A and C, as shown in the diagram. The plank is modelled as a uniform rod and the load as a particle. Given that the tension in the rope at C is four times the tension in the rope at A, calculate the distance CB. **(7 marks)**

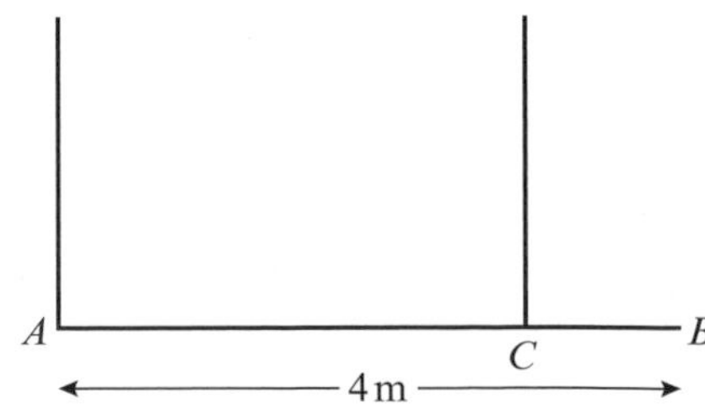

(E) **9** A beam AB has weight 200 N and length 5 m. The beam rests in equilibrium in a horizontal position on two smooth supports.

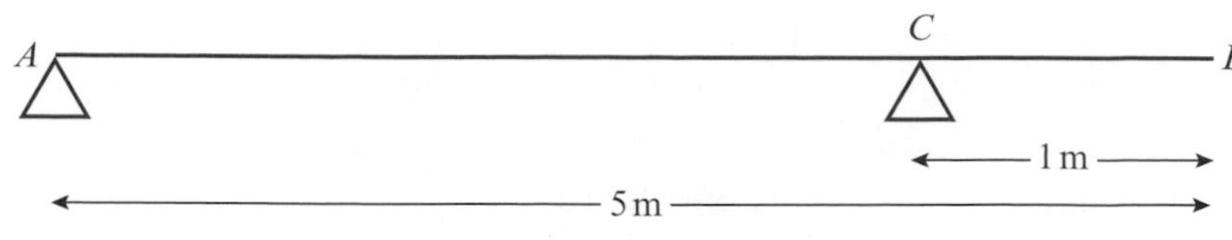

One support is at end A and the other is at a point C on the beam, where $BC = 1$ m, as shown in the diagram. The beam is modelled as a uniform rod.

a Find the reaction on the beam at C. **(4 marks)**

A woman of weight 500 N stands on the beam at the point D. The beam remains in equilibrium. The reactions on the beam at A and C are now equal.

b Find the distance AD. **(5 marks)**

(E/P) **10** A non-uniform plank MN of length 7 m is attached to a pivot at M and is held in a horizontal position by a force of 50 N applied at N at an angle of 125° to the plank as shown in the diagram. The centre of mass of the plank is at the point P. Given that the plank is in equilibrium and has a mass of 6 kg, find the distance MP. **(4 marks)**

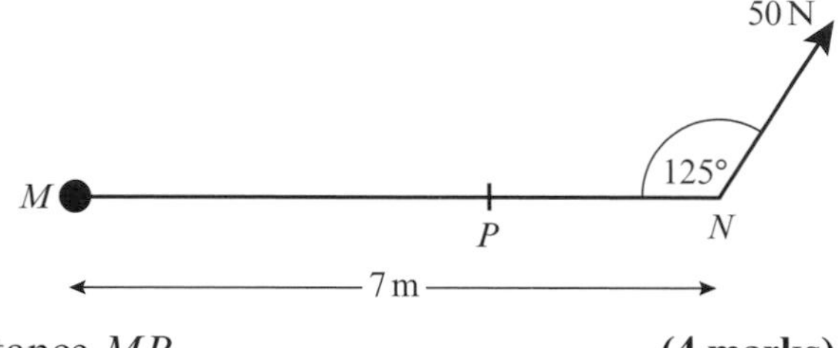

(E/P) **11** A ladder, AB, is leaning against a smooth vertical wall and on rough horizontal ground at an angle of 55° to the horizontal. The ladder has length 10 m and mass 20 kg. A man of mass 80 kg is standing at the point C on the ladder. Given that the magnitude of the frictional force at A is 200 N, find the distance AC. Model the ladder as a uniform rod and the man as a particle. **(5 marks)**

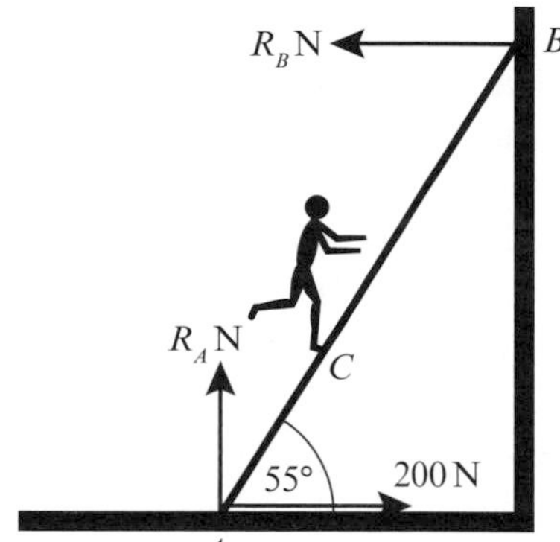

(E/P) **12** The beam of a crane is modelled as a uniform rod AB, of length 30 m and weight 4000 kg, resting in horizontal equilibrium. The beam is supported by a tower at C, where $AC = 10$ m. A counterbalance mass of weight 3000 kg is placed at A and a load of mass M is placed a variable distance x m from the supporting tower, where $x \geqslant 5$.

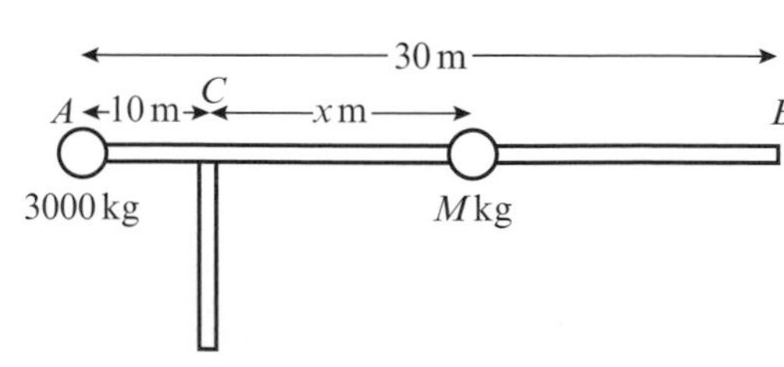

a Find an expression for M in terms of x. **(4 marks)**

b Hence determine the maximum and minimum loads that can be lifted by the crane. **(2 marks)**

c Criticise this model in relation to the beam. **(1 mark)**

Challenge

1 A non-uniform beam AB, of mass 10 kg and length 8 m, is pivoted at a point A. A particle of mass 2 kg is attached to the beam at a point D which is 1 m from B. The beam is held in equilibrium at an angle of 35° to the horizontal by a rope attached at point B. Given that the tension in the rope is 50 N and it makes an angle of 70° to the beam, find the distance of the centre of mass of the beam from A.

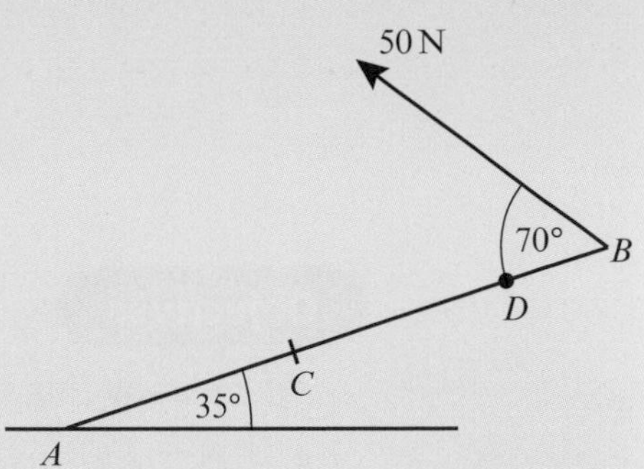

2 A builder is attempting to tip over a refrigerator. The refrigerator is modelled as a rectangular lamina of weight 1200 N. The centre of mass of the lamina is at the point of intersection of the diagonals of the rectangle, as shown in the diagram.

Given that the refrigerator is pivoting at vertex P and that the base of the refrigerator makes an angle of 20° to the floor, find the minimum force needed to tip the refrigerator if the force is applied:

a horizontally at A

b vertically at B.

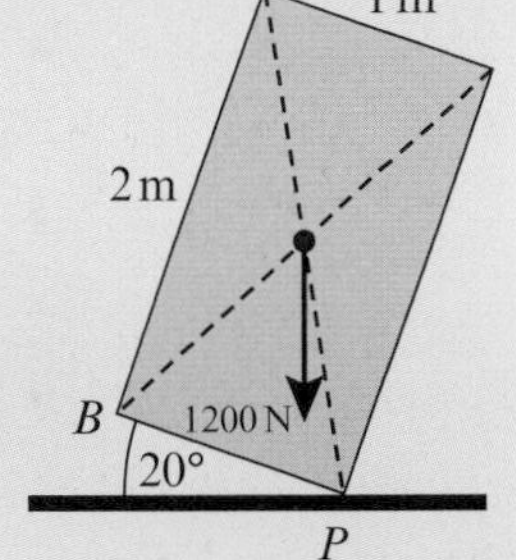

Summary of key points

1 Moment of **F** about $P = |\mathbf{F}| \times d$ clockwise

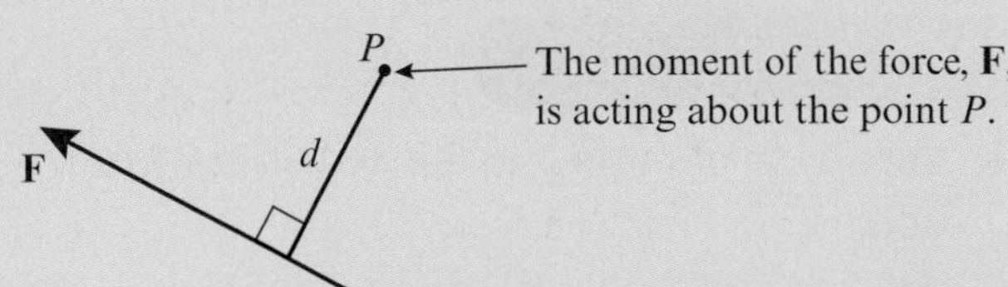

2 Moment of **F** about $P = |\mathbf{F}| \times d\sin\theta$ clockwise

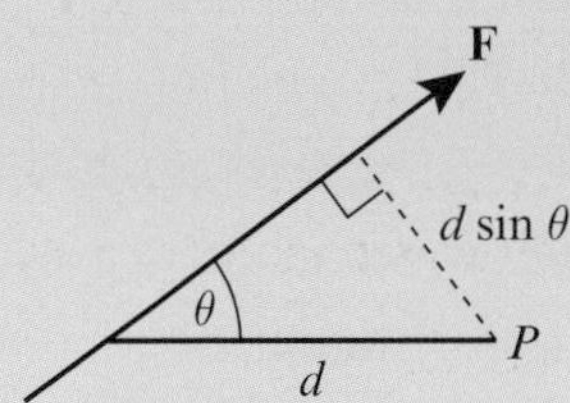

3 The sum of the moments acting on a body is called the resultant moment.

4 When a rigid body is in equilibrium the resultant force in any direction is 0 N and the resultant moment about any point is 0 N m.

5 When a rigid body is on the point of tilting about a pivot, the reaction at any other support (or the tension in any other wire or string) is zero.

5 Forces and friction

Objectives

After completing this chapter you should be able to:

- Resolve forces into components → pages 91–96
- Use the triangle law to find a resultant force → pages 93–96
- Solve problems involving smooth or rough inclined planes → pages 96–99
- Understand friction and the coefficient of friction. → pages 100–103
- Use $F \leqslant \mu R$ → pages 100–103

Prior knowledge check

1 A particle of mass 5 kg is acted on by two forces:

$\mathbf{F}_1 = (8\mathbf{i} + 2\mathbf{j})$ N and $\mathbf{F}_2 = (-3\mathbf{i} + 8\mathbf{j})$ N.

Find the acceleration of the particle in the form $(p\mathbf{i} + q\mathbf{j})$ m s^{-2}.

← Year 1, Chapter 10

2 In the diagram below, calculate

a the length of the hypotenuse

b the size of α.

Give your answers correct to 2 d.p.

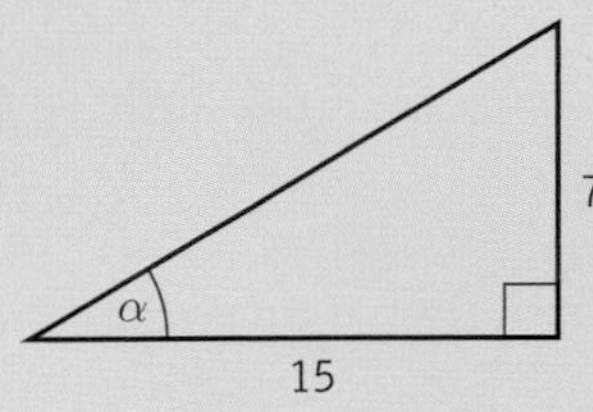

← GCSE Mathematics

A car's braking distance is determined by its speed and the frictional force between the car's wheels and the road. In wet or icy conditions, friction is reduced so the braking distance is increased. → Exercise 5C Q9

5.1 Resolving forces

- **If a force is applied at an angle to the direction of motion you can resolve it to find the component of the force that acts in the direction of motion.**

This book is being dragged along the table by means of a force of magnitude F. The book is moving horizontally, and the angle between the force and the direction of motion is θ.

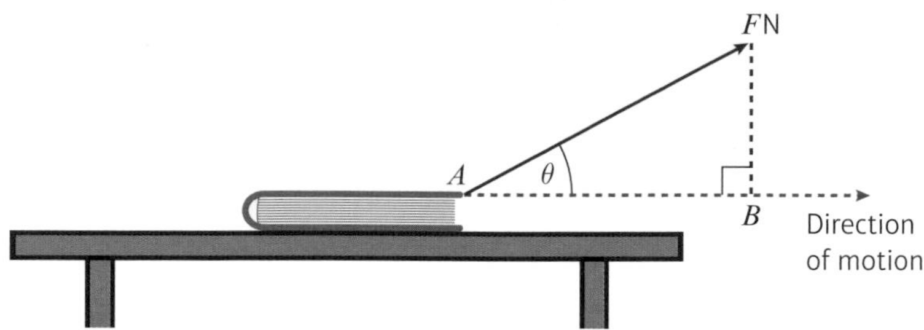

The effect of the force in the direction of motion is the length of the line AB. This is called the **component of the force in the direction of motion**. Using the rule for a right-angled triangle $\cos\theta = \frac{\text{adjacent}}{\text{hypotenuse}}$, you can see that the length of AB is $F \times \cos\theta$. Finding this value is called **resolving** the force in the direction of motion.

- **The component of a force of magnitude F in a certain direction is $F\cos\theta$, where θ is the size of the angle between the force and the direction.**

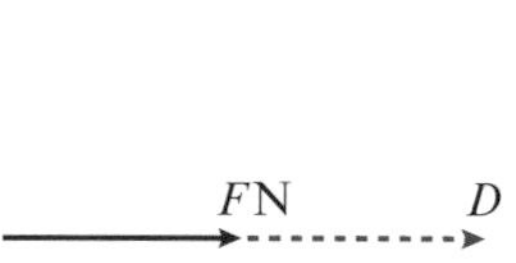

If F acts in the direction D, then the component of F in that direction is $F\cos 0° = F \times 1 = F$

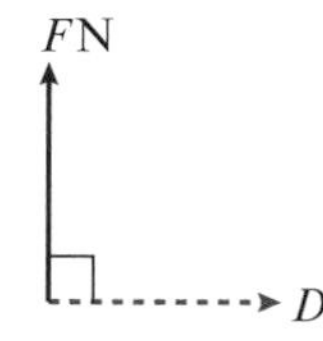

If F acts at the right angles to D, then the component of F in that direction is $F\cos 90° = F \times 0 = 0$

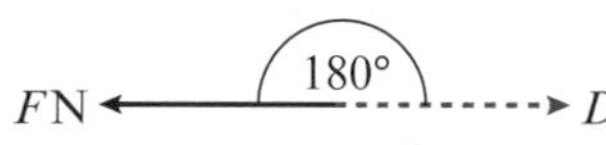

If F acts in the opposite direction to D, then the component of F in that direction is $F\cos 180° = F \times -1 = -F$

Example 1

Find the component of each force in **i** the x-direction **ii** the y-direction

iii Hence write each force in the form $p\mathbf{i} + q\mathbf{j}$ where $\mathbf{i}$ and $\mathbf{j}$ are the unit vectors in the x and y directions respectively.

a

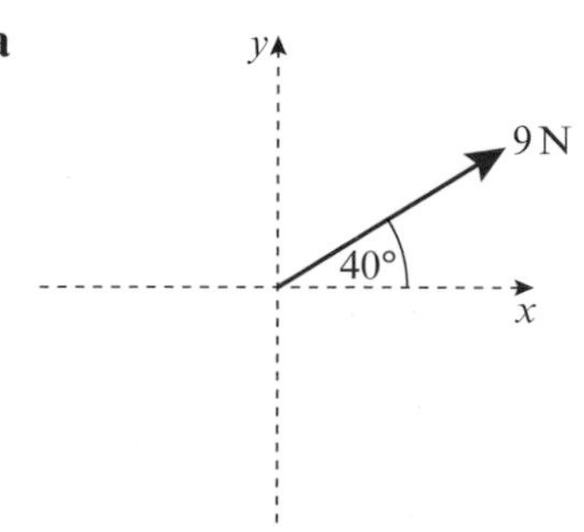

b

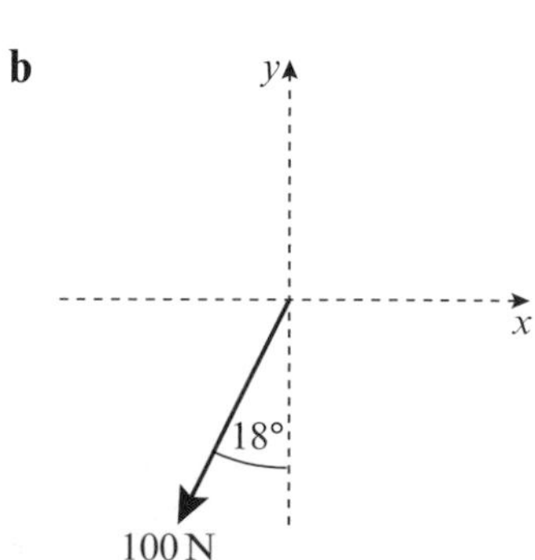

a i $\theta = 40°$

Component in x-direction $= F\cos\theta$
$= 9 \times \cos 40°$
$= 6.89$ N (3 s.f.)

Give your answers correct to three significant figures.

ii $\theta = 90° - 40°$
$= 50°$

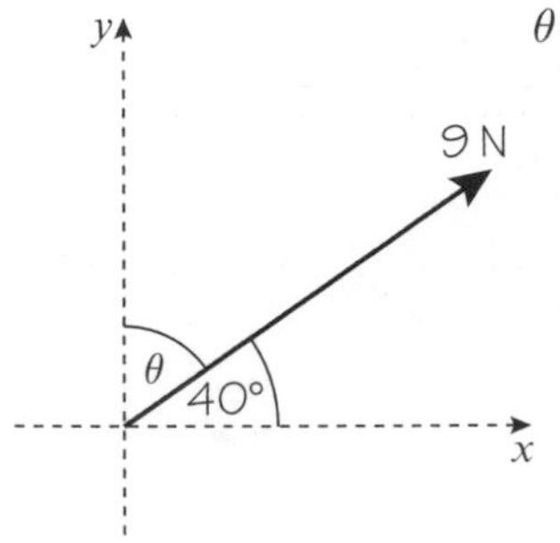

Make sure you find the angle between the force and the direction you are resolving in.

Component in y-direction $= F\cos\theta$
$= 9 \times \cos 50°$
$= 5.79$ N (3 s.f.)

You could also use $F\sin 40°$ as $\sin 40° = \cos(90° - 40°) = \cos 50°$

iii $(6.89\mathbf{i} + 5.79\mathbf{j})$ N

b i $\theta = 90° + 18°$
$= 108°$

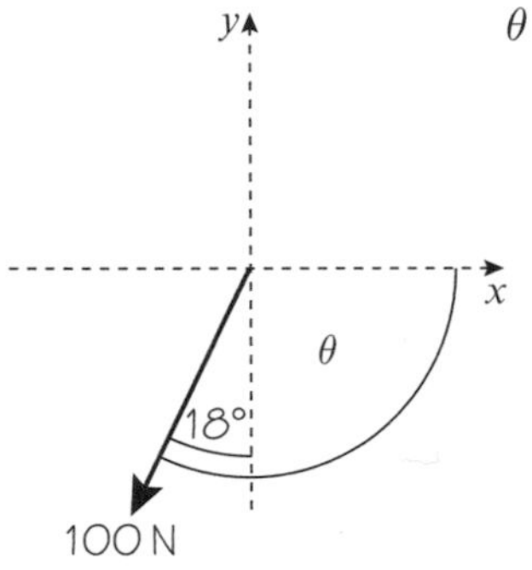

Component in x-direction $= F\cos\theta$
$= 100 \times \cos 108°$
$= -30.9$ N (3 s.f.)

You get a negative answer because you are resolving in the positive x-direction. You could also resolve in the negative x-direction using $\theta = 90° - 18° = 72°$, then change the sign of your answer from positive to negative:

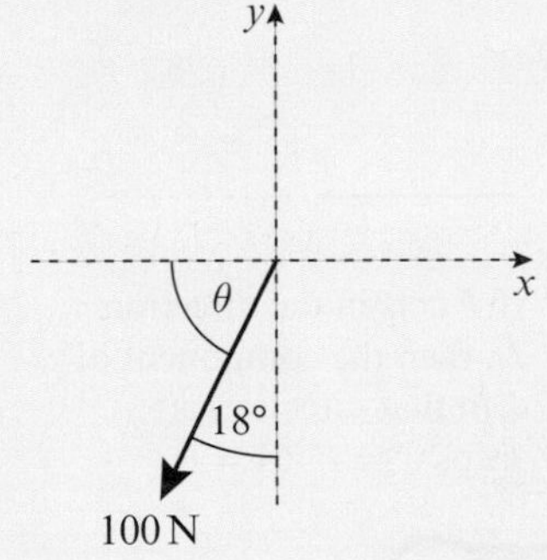

ii 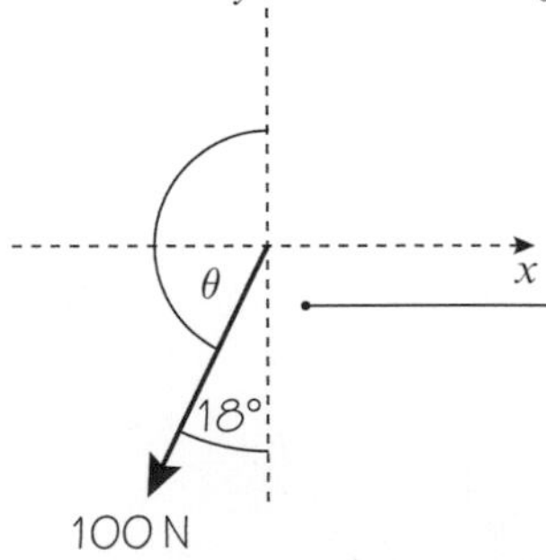 $\theta = 180° - 18°$
$= 162°$

You could use $\theta = 18°$ then change the sign of your answer from positive to negative: $-100\cos 18° = -95.1$ N (3 s.f.).

You can measure θ in either the clockwise or the anticlockwise direction since $\cos\theta = \cos(360° - \theta)$.

Component in y-direction $= F\cos\theta$
$= 100 \times \cos 162°$
$= -95.1$ N (3 s.f.)

iii $(-30.9\mathbf{i} - 95.1\mathbf{j})$ N

Example 2

A box of mass 8 kg lies on a smooth horizontal floor.
A force of 10 N is applied at an angle of 30° causing the box to accelerate horizontally along the floor.

a Work out the acceleration of the box.

b Calculate the normal reaction between the box and the floor.

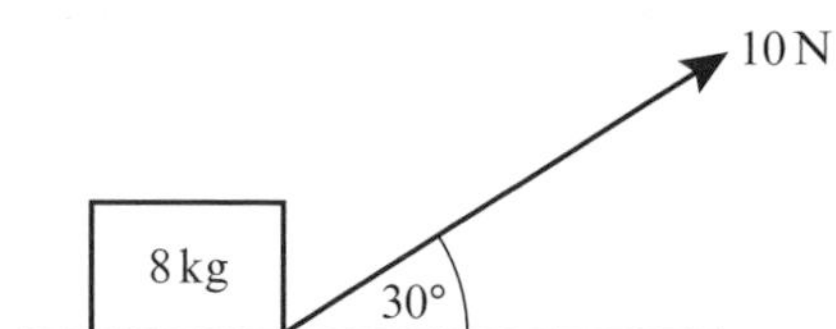

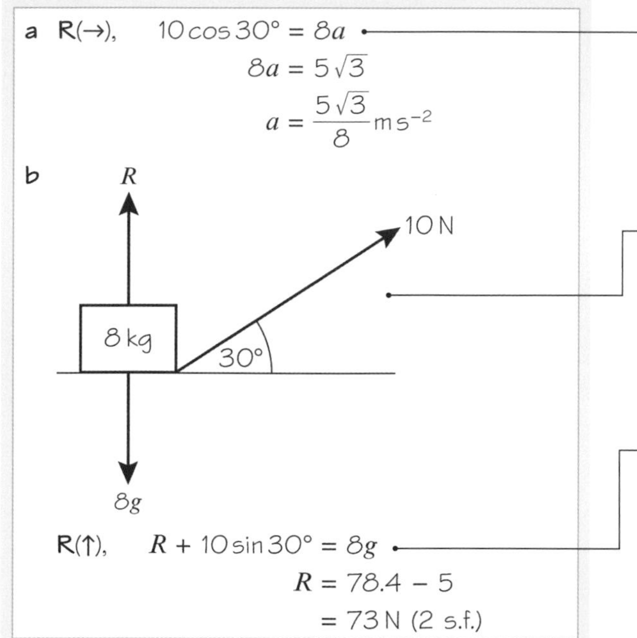

Resolve the force horizontally and write an equation of motion for the box. ← Year 1, Chapter 10

Add the weight of the box and the normal reaction to the force diagram.

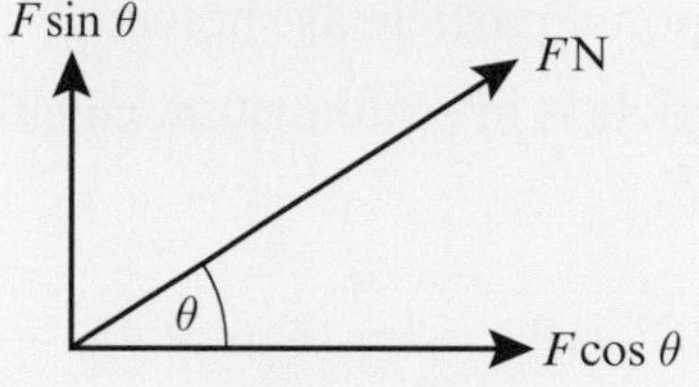

The component in the y-direction is $F\cos(90° - \theta) = F\sin\theta$

You can use the **triangle law** of vector addition to find the resultant of two forces acting at an angle without resolving them into components.

Example 3

Two forces P and Q act on a particle as shown. P has a magnitude of 10 N and Q has a magnitude of 8 N. Work out the magnitude and direction of the resultant force.

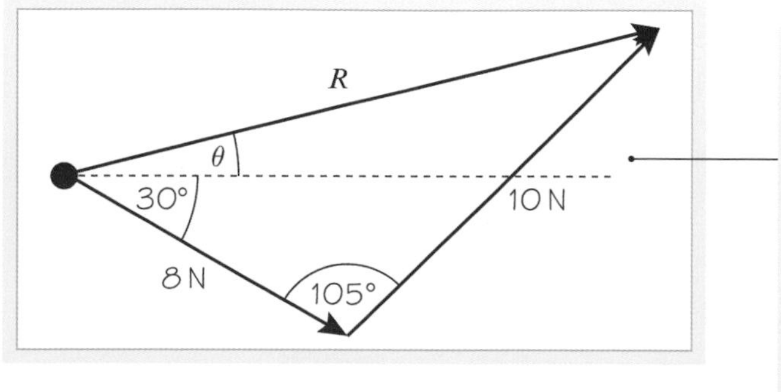

Use the triangle law for vector addition. The resultant force is the third side of a triangle formed by forces P and Q. You might need to use geometry to work out missing angles in the triangle:

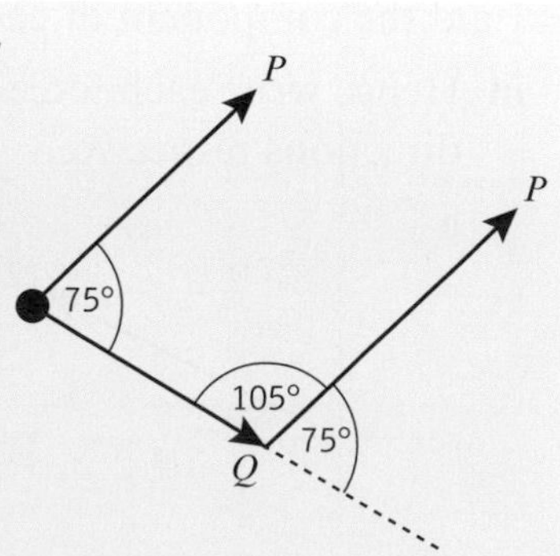

$R^2 = 8^2 + 10^2 - 2 \times 8 \times 10\cos 105°$
$= 164 - 160\cos 105° = 205.411...$
$R = 14.332... = 14.3\text{ N (3 s.f.)}$

Use the cosine rule to calculate the magnitude of R.

$$\frac{\sin(\theta + 30°)}{10} = \frac{\sin 105°}{14.332...}$$

Use the sine rule to work out θ.

$$\sin(\theta + 30°) = \frac{10\sin 105°}{14.332...} = 0.673...$$

Remember to use unrounded values in your calculations then round your final answer.

$\theta + 30° = 42.373...$
$\theta = 12.4°\text{ (3 s.f.)}$

The resultant force R has a magnitude of 14.3 N and acts at an angle of 12.4° above the horizontal.

Use your diagram to check that your answer makes sense.

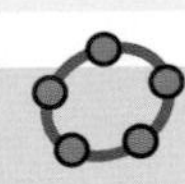

Online Explore the resultant of two forces using GeoGebra.

Example 4

Three forces act upon a particle as shown.

Given that the particle is in equilibrium, calculate the magnitude of P.

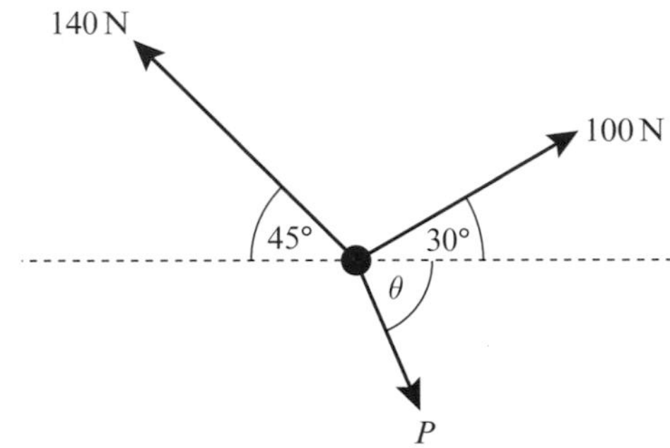

$R(\rightarrow)$, $100\cos 30° + P\cos\theta = 140\cos 45°$
$P\cos\theta = 12.392...$ (1)
$R(\uparrow)$, $100\sin 30° + 140\sin 45° = P\sin\theta$
$P\sin\theta = 148.994...$ (2)

Resolve horizontally and vertically. You can solve these two equations simultaneously by dividing to eliminate P.

$$\frac{P\sin\theta}{P\cos\theta} = \frac{148.994...}{12.392...}$$

$\tan\theta = 12.023...$
$\theta = 85.245...°$
$P\cos 85.2454...° = 12.392...$
$P = 150\text{ N (3 s.f.)}$

Problem-solving

You could also solve this problem by drawing a **triangle of forces**.
The particle is in equilibrium, so the three forces will form a closed triangle:

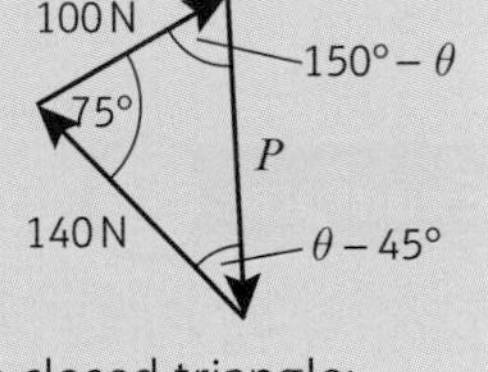

Exercise 5A

1 Find the component of each force in **i** the x-direction **ii** the y-direction

iii Hence write each force in the form $p\mathbf{i} + q\mathbf{j}$ where $\mathbf{i}$ and $\mathbf{j}$ are the unit vectors in the x and y directions respectively.

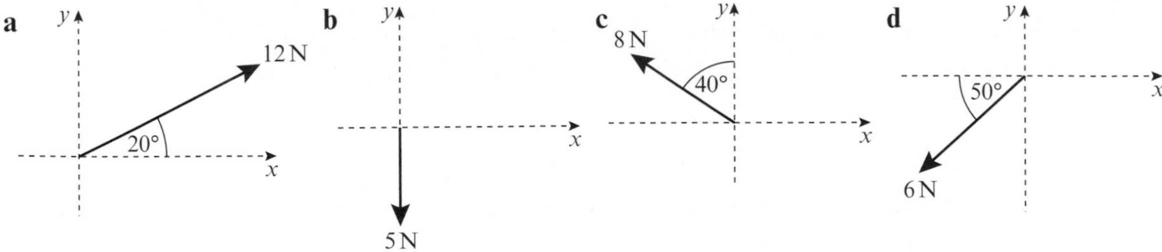

2 For each of the following systems of forces, find the sum of the components in **i** the x-direction, **ii** the y-direction.

a

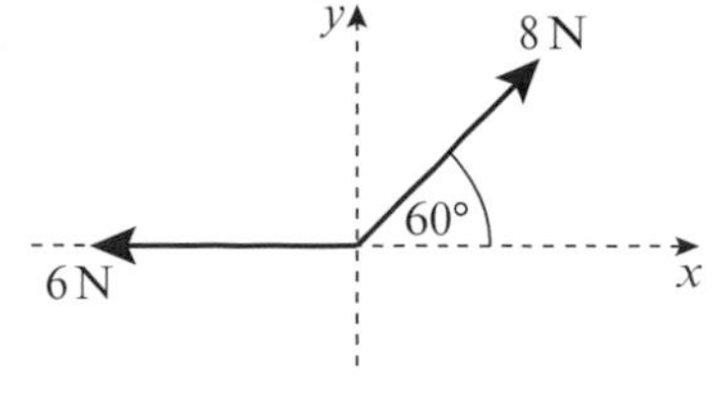

b

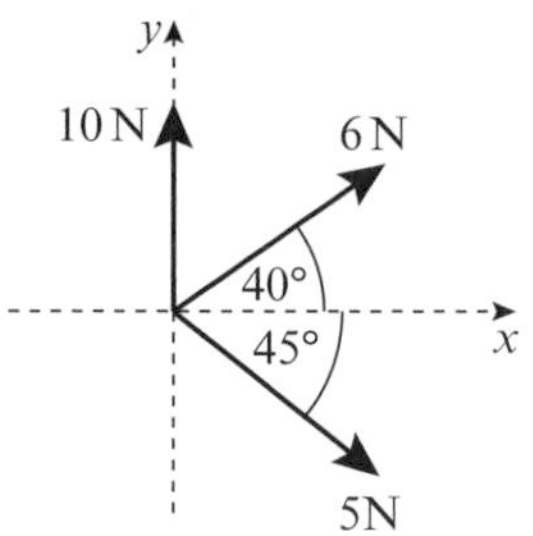

c

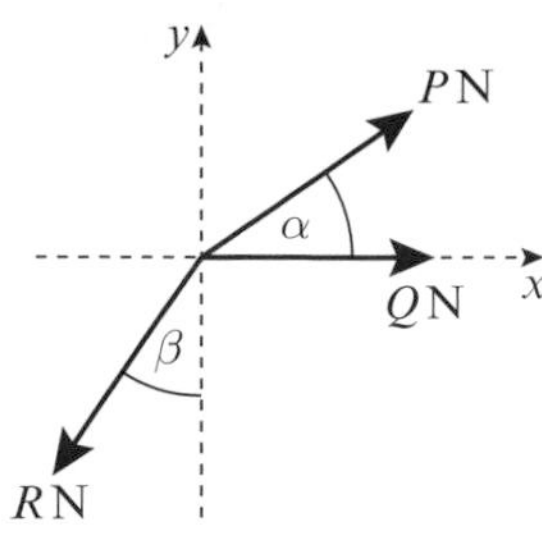

3 Find the magnitude and direction of the resultant force acting on each of the particles shown below.

a

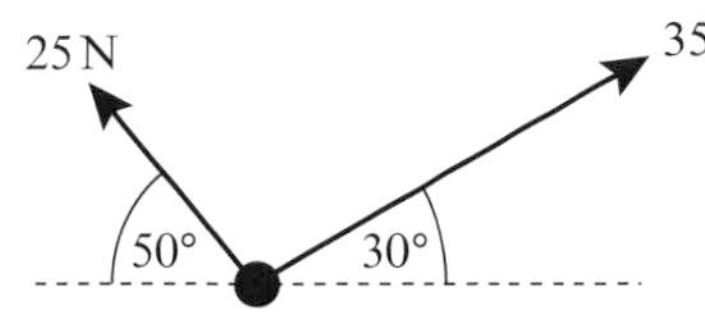

b

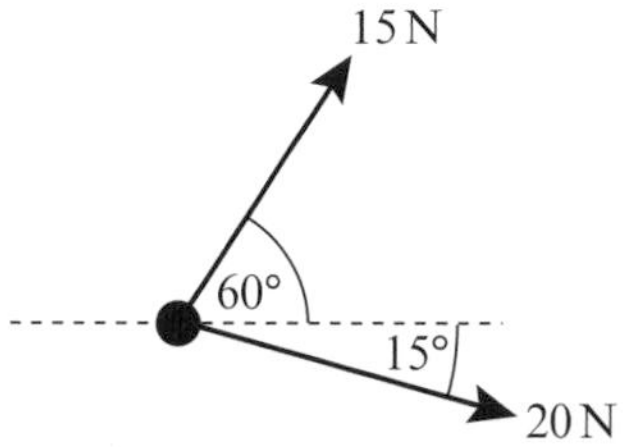

c

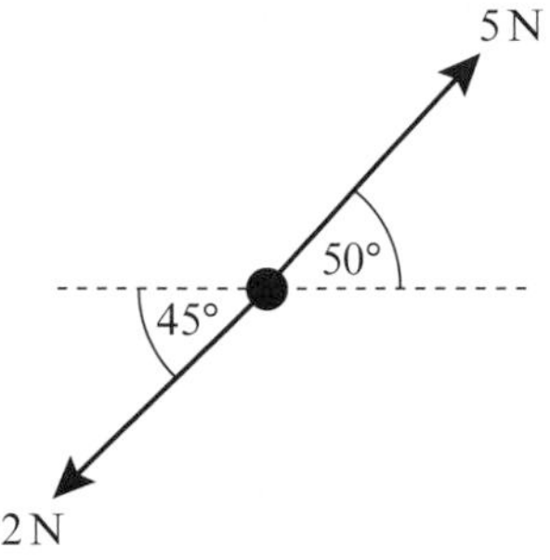

Ⓟ **4** Three forces act upon a particle as shown in the diagrams below.

Given that the particle is in equilibrium, calculate the magnitude of B and the value of θ.

a

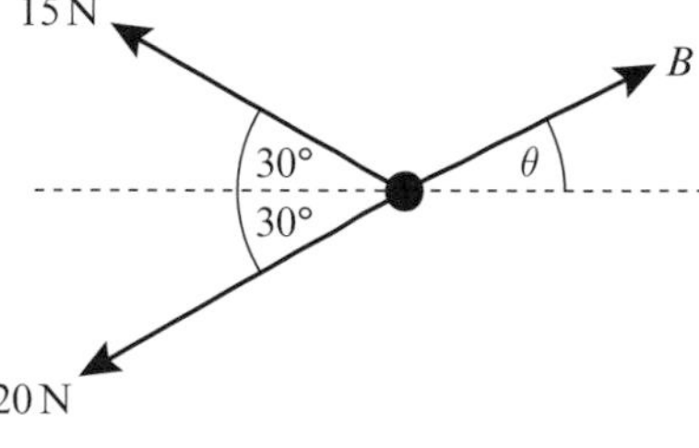

b

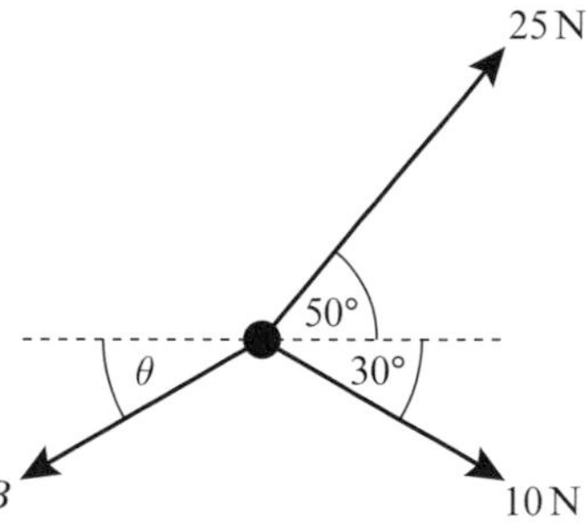

c

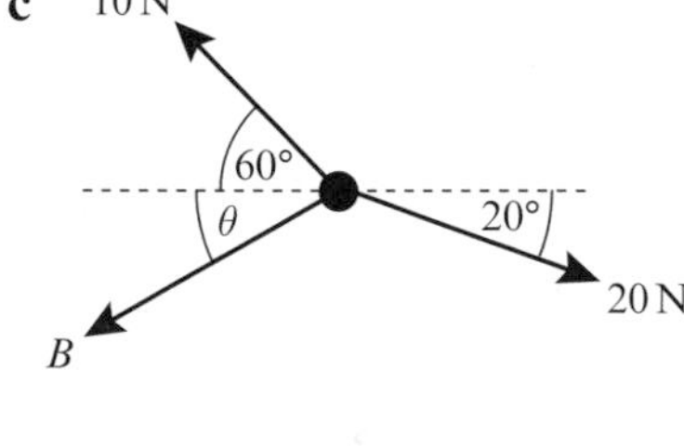

5 A box of mass 5 kg lies on a smooth horizontal floor. The box is pulled by a force of 2 N applied at an angle of 30° to the horizontal, causing the box to accelerate horizontally along the floor.

a Work out the acceleration of the box.

b Work out the normal reaction of the box with the floor.

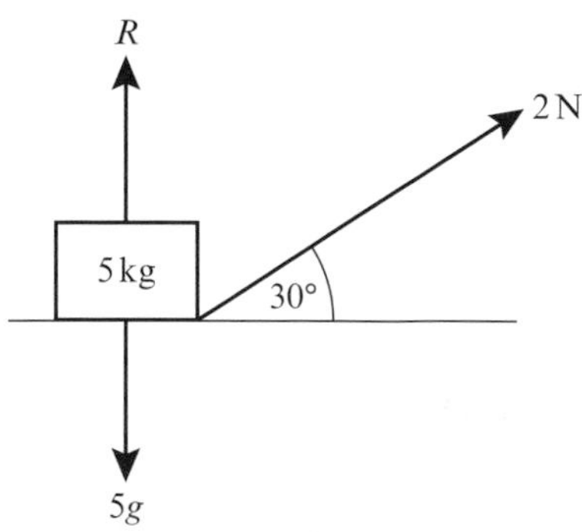

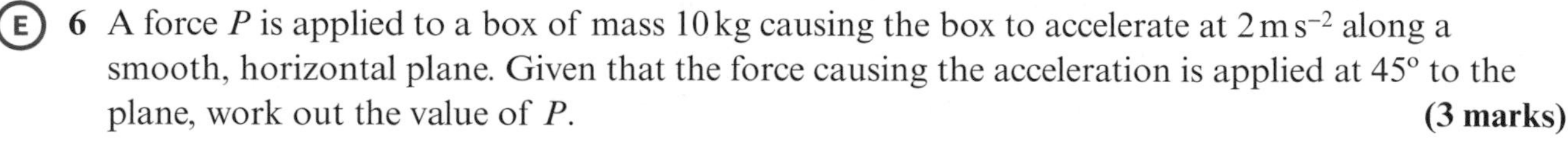

Ⓔ **6** A force P is applied to a box of mass 10 kg causing the box to accelerate at $2\,\text{m}\,\text{s}^{-2}$ along a smooth, horizontal plane. Given that the force causing the acceleration is applied at 45° to the plane, work out the value of P. **(3 marks)**

Ⓔ **7** A force of 20 N is applied to a box of mass m kg causing the box to accelerate at $0.5\,\text{m}\,\text{s}^{-2}$ along a smooth, horizontal plane. Given that the force causing the acceleration is applied at 25° to the plane, work out the value of m. **(3 marks)**

E/P **8** A parachutist of mass 80 kg is attached to a canopy by two lines, each with tension T. The parachutist is falling with constant velocity, and experiences a resistance to motion due to air resistance equal to one quarter of her weight. Show that the tension in each line, T, is $20\sqrt{3}\,g$ N.

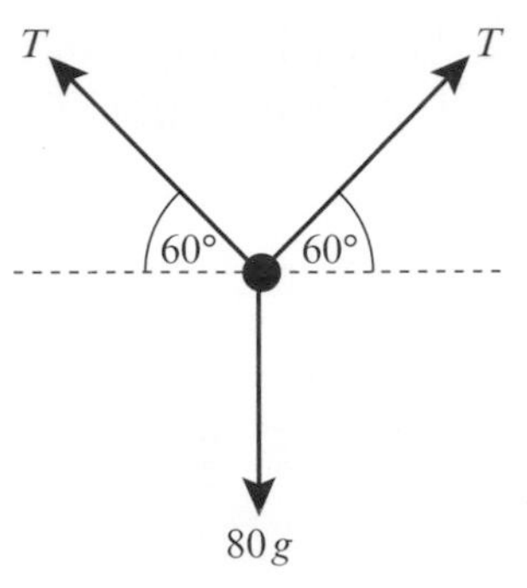

(3 marks)

E/P **9** A system of forces act upon a particle as shown in the diagram. The resultant force on the particle is $(2\sqrt{3}\mathbf{i} + 2\mathbf{j})$ N. Calculate the magnitudes of $\mathbf{F}_1$ and $\mathbf{F}_2$.

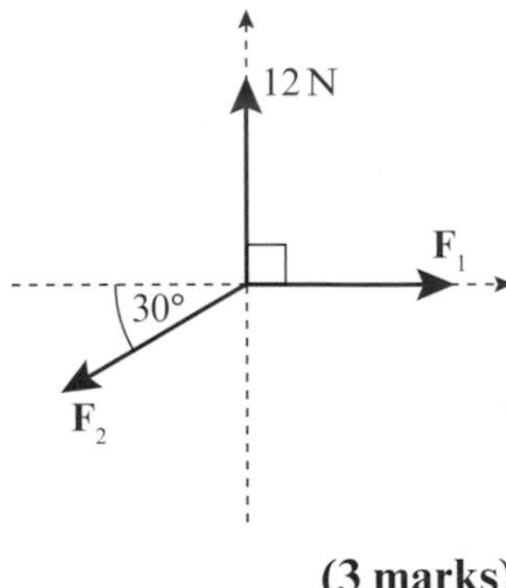

(3 marks)

Challenge

Two forces act upon a particle as shown in the diagram. The resultant force on the particle is $(3\mathbf{i} + 5\mathbf{j})$ N. Calculate the magnitudes of $\mathbf{F}_1$ and $\mathbf{F}_2$.

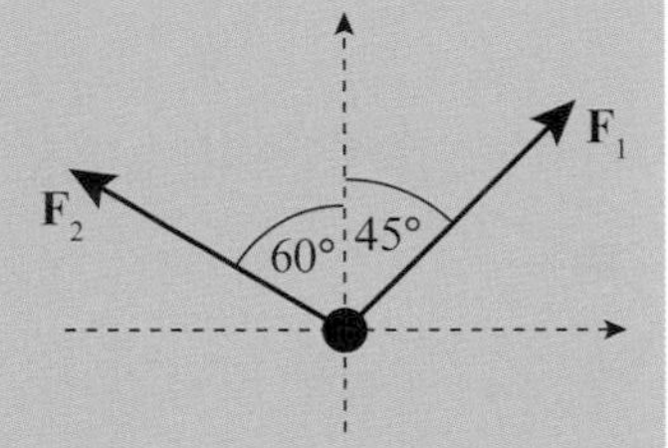

5.2 Inclined planes

Force diagrams may be used to model situations involving objects on inclined planes.

- **To solve problems involving inclined planes, it is usually easier to resolve parallel to and at right angles to the plane.**

Example 5

A block of mass 10 kg slides down a smooth slope angled at 15° to the horizontal.

a Draw a force diagram to show all the forces acting on the block.

b Calculate the magnitude of the normal reaction of the slope on the block.

c Find the acceleration of the block.

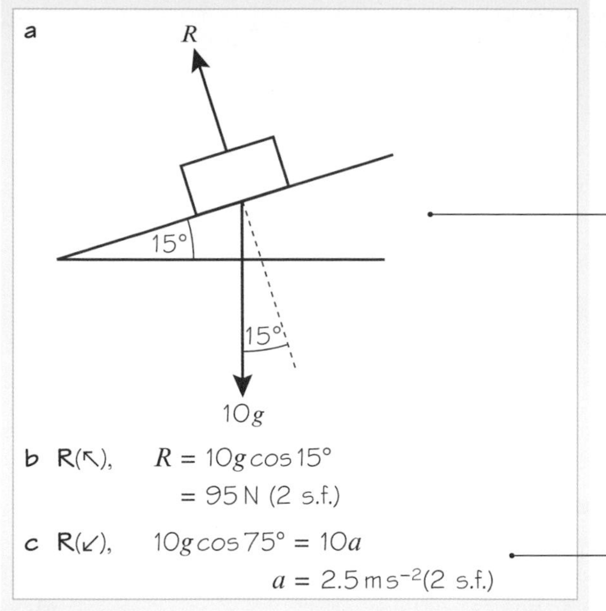

Watch out The normal reaction force acts at right angles to the plane, not vertically.

Your working will be easier if you resolve at right angles to the plane. The weight of the block acts at an angle of 15° to this direction.

Notation The diagonal arrows, **R**(↖) and **R**(↙), show that you are resolving down the slope and perpendicular to the slope. You can also use **R**(//) to show resolution parallel to the slope and **R**(⊥) to show resolution perpendicular to the slope.

b R(↖), $R = 10g\cos 15°$
$= 95\text{ N (2 s.f.)}$

c R(↙), $10g\cos 75° = 10a$
$a = 2.5\text{ m s}^{-2}\text{(2 s.f.)}$

Resolve down the slope and use $F = ma$.

Example 6

A particle of mass m is pushed up a smooth slope by a force of magnitude $5g$ N acting at an angle of 60° to the slope, causing the particle to accelerate up the slope at 0.5 m s^{-2}. Show that the mass of the particle is $\left(\frac{5g}{1+g}\right)$kg.

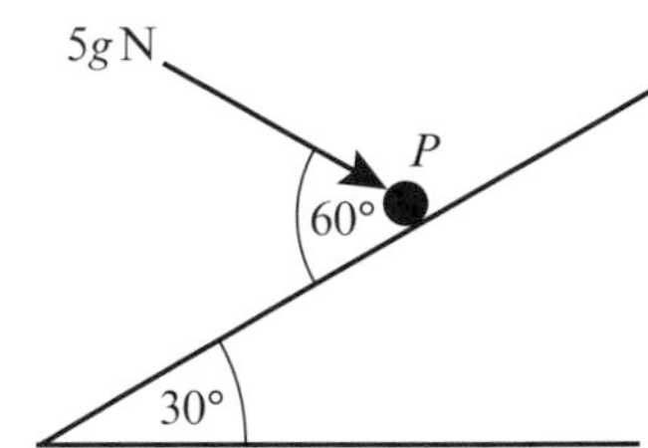

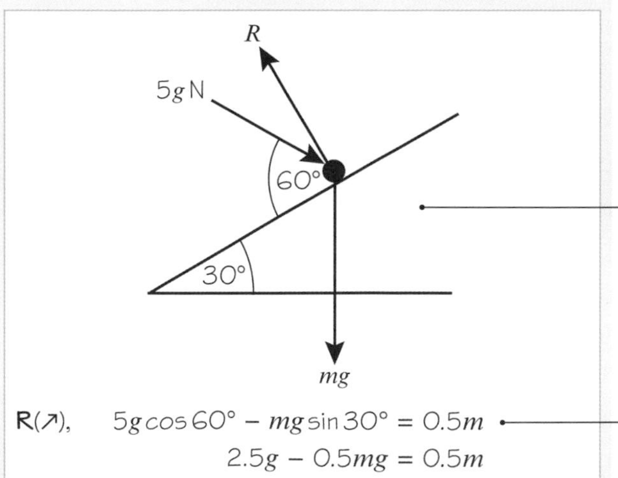

Draw a diagram to show all the forces acting on the particle.

R(↗), $5g\cos 60° - mg\sin 30° = 0.5m$
$2.5g - 0.5mg = 0.5m$
$2.5g = 0.5m + 0.5mg$
$5g = m + mg$
$5g = m(1 + g)$

$m = \left(\frac{5g}{1+g}\right)$ kg as required

Resolve up the slope, in the direction of the acceleration, and write an equation of motion for the particle.

You need to find the mass of the particle in terms of g, so you don't need to use $g = 9.8\text{ m s}^{-2}$ in your working.

Example 7

A particle P of mass 2 kg is moving on a smooth slope and is being acted on by a force of 4 N that acts parallel to the slope as shown.

The slope is inclined at an angle α to the horizontal, where $\tan\alpha = \frac{3}{4}$. Work out the acceleration of the particle.

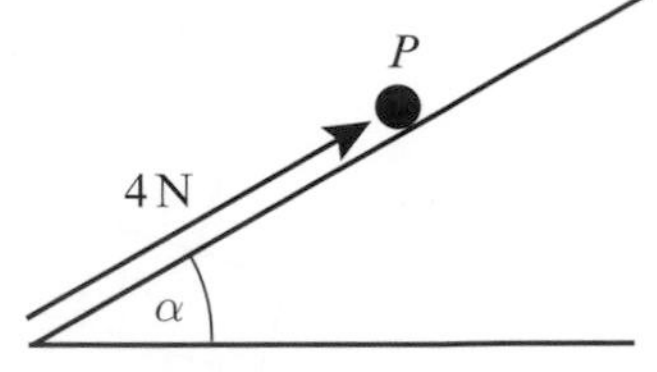

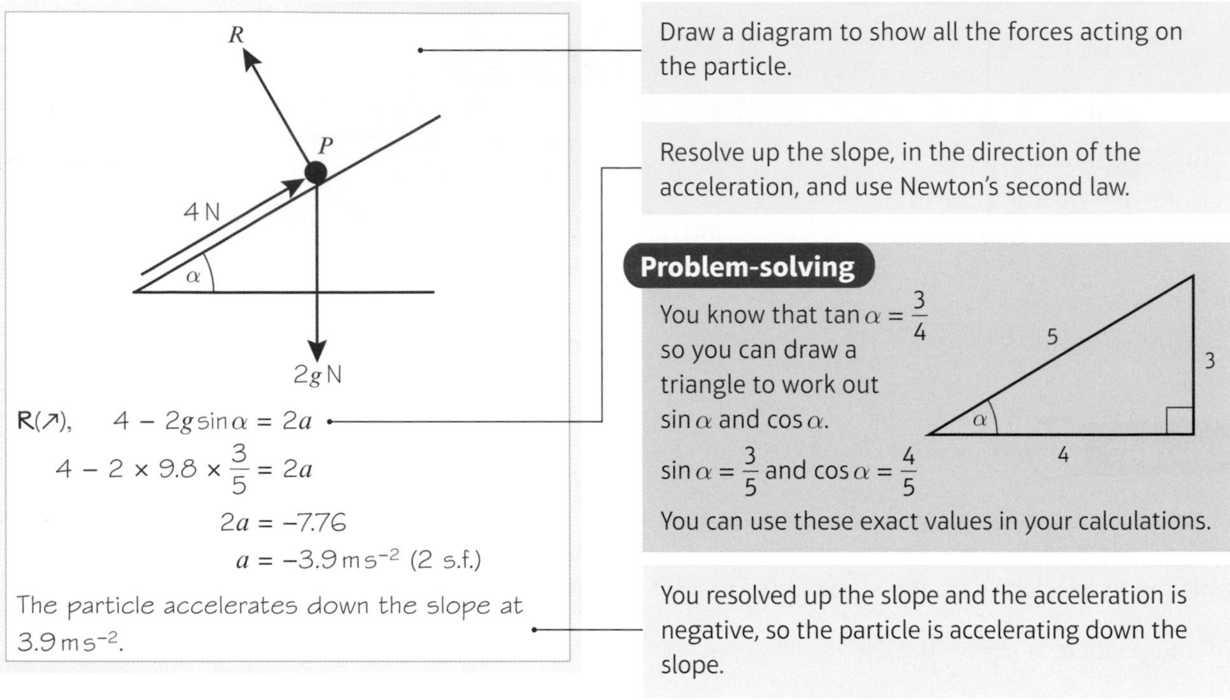

Exercise 5B

1 A particle of mass 3 kg slides down a smooth slope that is inclined at 20° to the horizontal.

a Draw a force diagram to represent all the forces acting on the particle.

b Work out the normal reaction between the particle and the plane.

c Find the acceleration of the particle.

2 A force of 50 N is pulling a particle of mass 5 kg up a smooth plane that is inclined at 30° to the horizontal. Given that the force acts parallel to the plane,

a draw a force diagram to represent all the forces acting on the particle

b work out the normal reaction between the particle and the plane

c find the acceleration of the particle.

3 A particle of mass 0.5 kg is held at rest on a smooth slope that is inclined at an angle α to the horizontal. The particle is released. Given that $\tan\alpha = \frac{3}{4}$, calculate:

a the normal reaction between the particle and the plane

b the acceleration of the particle.

E 4 A force of 30 N is pulling a particle of mass 6 kg up a rough slope that is inclined at 15° to the horizontal. The force acts in the direction of motion of the particle and the particle experiences a constant resistance due to friction.

a Draw a force diagram to represent all the forces acting on the particle. **(4 marks)**

Given that the particle is moving with constant speed,

b calculate the magnitude of the resistance due to friction. **(5 marks)**

E 5 A particle of mass m kg is sliding down a smooth slope that is angled at 30° to the horizontal. The normal reaction between the plane and the particle is 5 N.

a Calculate the mass m of the particle. **(3 marks)**

b Calculate the acceleration of the particle. **(3 marks)**

E/P 6 A force of 30 N acts horizontally on a particle of mass 5 kg that rests on a smooth slope that is inclined at 30° to the horizontal as shown in the diagram.
Find the acceleration of the particle.

(4 marks)

E/P 7 A particle of mass 3 kg is moving on a rough slope that is inclined at 40° to the horizontal. A force of 6 N acts vertically upon the particle. Given that the particle is moving at a constant velocity, calculate the value of F, the constant resistance due to friction.

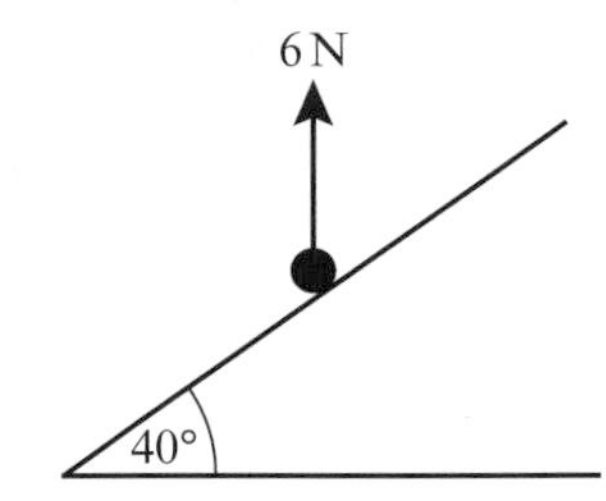

(4 marks)

E/P 8 A particle of mass m kg is pulled up a rough slope by a force of 26 N that acts at an angle of 45° to the slope. The particle experiences a constant frictional force of magnitude 12 N.

Given that $\tan\alpha = \frac{1}{\sqrt{3}}$ and that the acceleration of the particle is 1 m s^{-2},

show that $m = 1.08$ kg (3 s.f.).

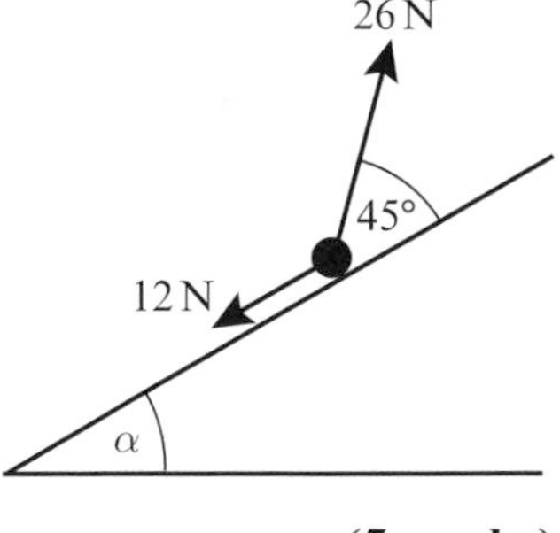

(5 marks)

Challenge

A particle is sliding down a smooth slope inclined at an angle θ to the horizontal, where $0 < \theta < 30°$. The angle of inclination of the slope is increased by 60°, and the magnitude of the acceleration of the particle increases from a to $4a$.

a Show that $\tan\theta = \frac{\sqrt{3}}{7}$

b Hence find θ, giving your answer to 3 significant figures.

5.3 Friction

Friction is a force which opposes motion between two rough surfaces. It occurs when the two surfaces are moving relative to one another, or when there is a **tendency** for them to move relative to one another.

This block is stationary. There is no horizontal force being applied, so there is no tendency for the block to move. There is no frictional force acting on the block.

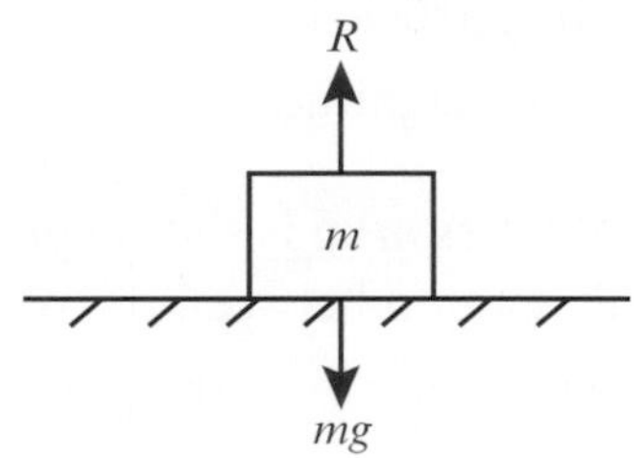

This block is also stationary. There is a horizontal force being applied which is not sufficient to move the block. There is a tendency for the block to move, but it doesn't because the force of friction is equal and opposite to the force being applied.

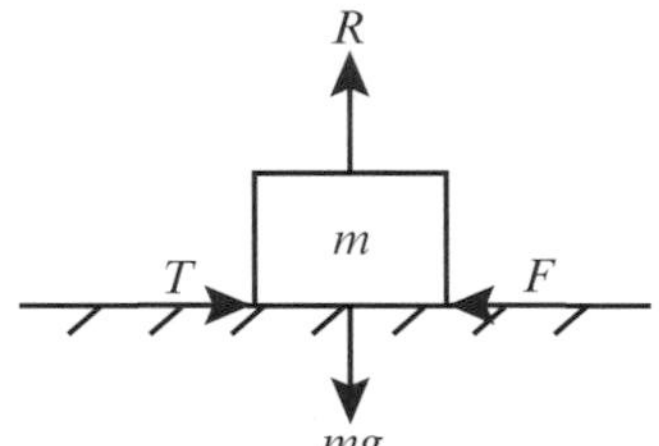

As the applied force increases, the force of friction increases to prevent the block from moving. If the magnitude of the applied force exceeds a certain **maximum** or **limiting value**, the block will move. While the block moves, the force of friction will remain constant at its maximum value.

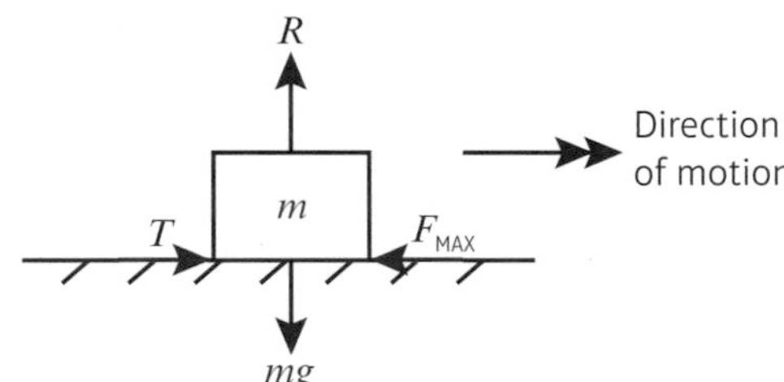

The limiting value of the friction depends on two things:

- the normal reaction R between the two surfaces in contact
- the roughness of the two surfaces in contact.

You can measure roughness using the **coefficient of friction**, which is represented by the letter μ (pronounced *myoo*). The rougher the two surfaces, the larger the value of μ. For smooth surfaces there is no friction and $\mu = 0$.

- **The maximum or limiting value of the friction between two surfaces, F_{MAX}, is given by**

$$F_{MAX} = \mu R$$

where μ is the coefficient of friction and R is the normal reaction between the two surfaces.

Example 8

A particle of mass 5 kg is pulled along a rough horizontal surface by a horizontal force of magnitude 20 N. The coefficient of friction between the particle and the floor is 0.2. Calculate:

a the magnitude of frictional force

b the acceleration of the particle.

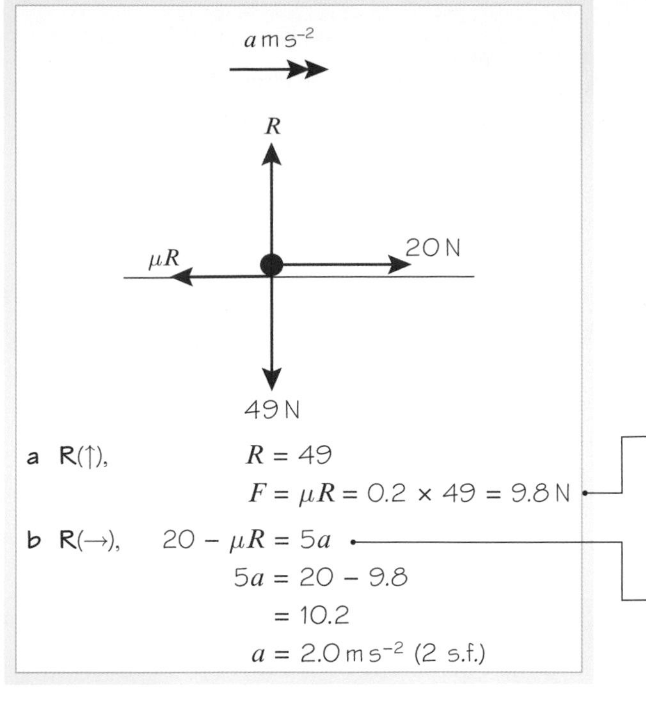

a R(↑), $R = 49$

$F = \mu R = 0.2 \times 49 = 9.8\text{ N}$

b R(→), $20 - \mu R = 5a$

$5a = 20 - 9.8$

$= 10.2$

$a = 2.0\text{ m s}^{-2}$ (2 s.f.)

For a moving particle $F = F_{\text{MAX}}$ so you can use $F = \mu R$ to find the magnitude of the frictional force.

Write an equation of motion for the particle, resolving horizontally. Note that the frictional force always acts in a direction so as to **oppose** the motion of the particle.

Example 9

A block of mass 5 kg lies on rough horizontal ground. The coefficient of friction between the block and the ground is 0.4. A horizontal force P is applied to the block. Find the magnitude of the frictional force acting on the block and the acceleration of the block when the magnitude of P is

a 10 N **b** 19.6 N **c** 30 N.

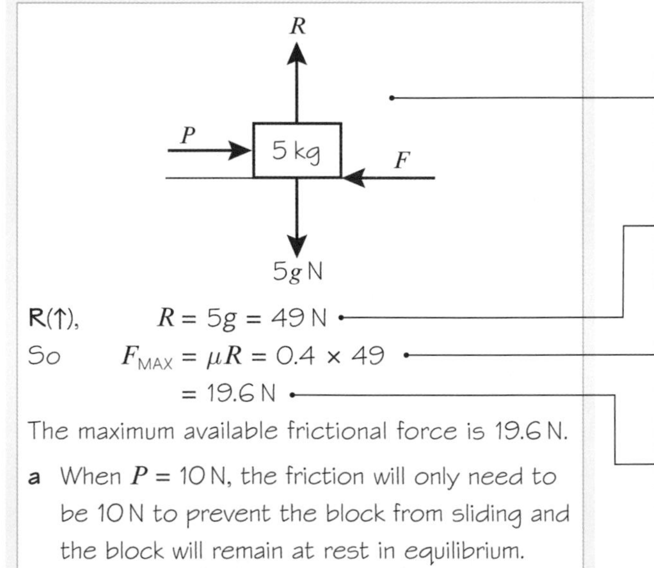

First draw a diagram showing all the forces acting on the block.

R(↑), $R = 5g = 49\text{ N}$

So $F_{\text{MAX}} = \mu R = 0.4 \times 49$

$= 19.6\text{ N}$

The maximum available frictional force is 19.6 N.

a When $P = 10\text{ N}$, the friction will only need to be 10 N to prevent the block from sliding and the block will remain at rest in equilibrium.

The normal reaction will equal the weight as the force P has no vertical component and there is no vertical acceleration.

You then need to calculate what the maximum possible frictional force is in this situation.

Do not round this value as you will need to use it in your calculations.

b When $P = 19.6$ N, the friction will need to be at its maximum value of 19.6 N to prevent the block from sliding, and the block will remain at rest in **limiting equilibrium**.

c When $P = 30$ N, the friction will be unable to prevent the block from sliding, and it will remain at its maximum value of 19.6 N. The block will accelerate from rest along the plane in the direction of P with acceleration a, where

$30 - 19.6 = 5a$

$a = 2.1\text{ m s}^{-2}$ (2 s.f.)

Watch out An object in limiting equilibrium can either be at rest or moving with constant velocity.

Example 10

A particle of mass 2 kg is sliding down a rough slope that is inclined at 30° to the horizontal. Given that the acceleration of the particle is 1 m s^{-2}, find the coefficient of friction, μ, between the particle and the slope.

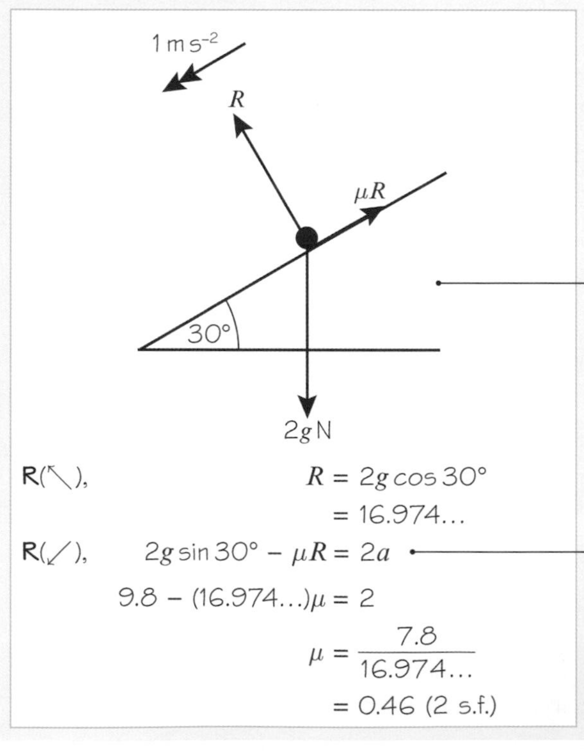

Online Explore this problem with different masses, slopes and frictional coefficients using GeoGebra.

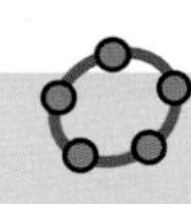

Draw a diagram showing the weight, the frictional force and the normal reaction.

R(↖), $R = 2g\cos 30°$

$= 16.974...$

R(↙), $2g\sin 30° - \mu R = 2a$

$9.8 - (16.974...)\mu = 2$

$\mu = \frac{7.8}{16.974...}$

$= 0.46$ (2 s.f.)

Use $F = ma$ to write an equation of motion for the particle

Watch out Make sure you use the normal reaction, not the weight, when substituting into $F_{MAX} = \mu R$.

Exercise 5C

1 Each of the following diagrams shows a body of mass 5 kg lying initially at rest on rough horizontal ground. The coefficient of friction between the body and the ground is $\frac{1}{7}$. In each diagram R is the normal reaction of the ground on the body and F is the frictional force exerted on the body. Any other forces applied to the body are as shown on the diagrams.

Hint The forces acting on the body can affect the magnitude of the normal reaction. In part **d** the normal reaction is $(5g + 14)$ N, so $F_{MAX} = \mu(5g + 14)$ N.

In each case

i find the magnitude of F,

ii state whether the body will remain at rest or accelerate from rest along the ground,

iii find, when appropriate, the magnitude of this acceleration.

a

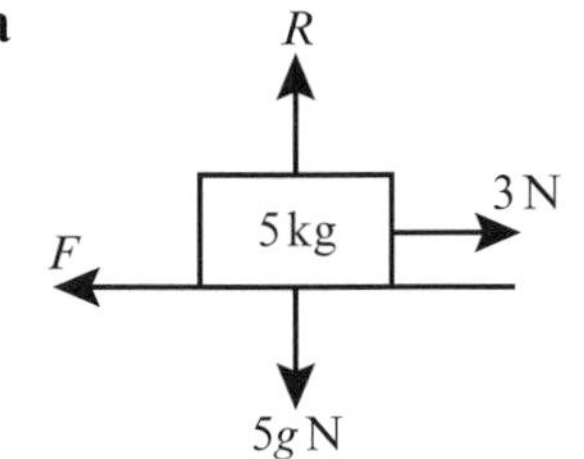

b

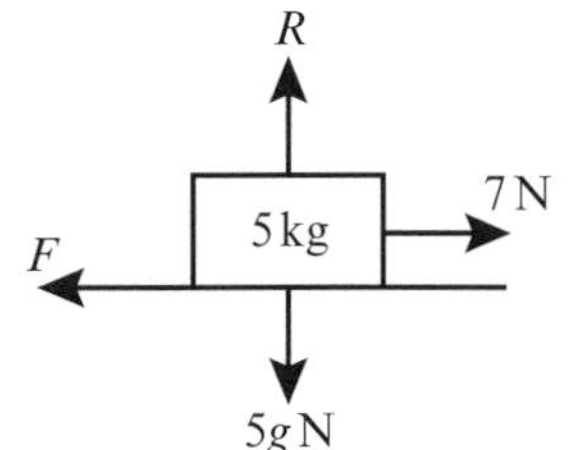

c

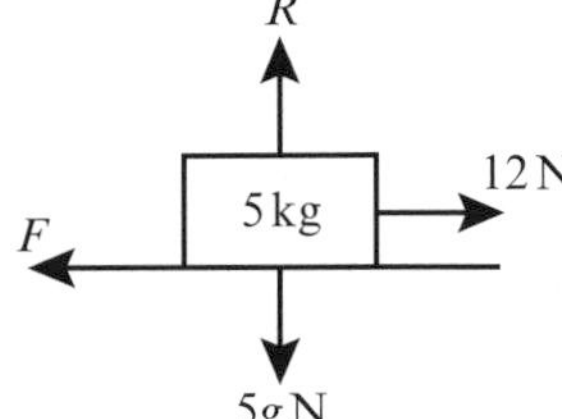

d

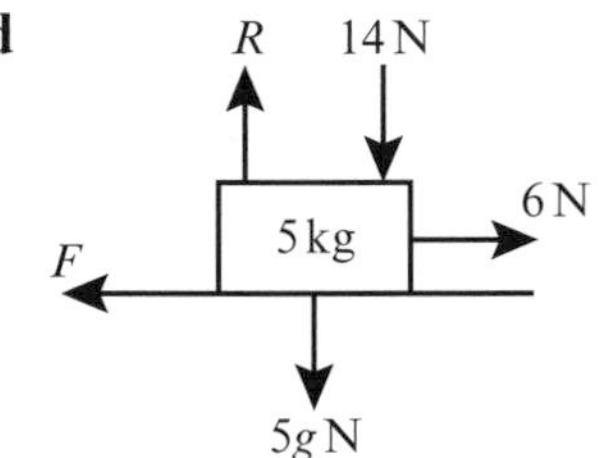

e

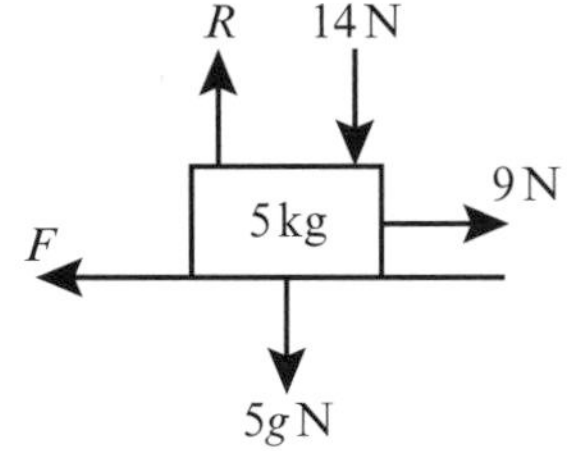

f

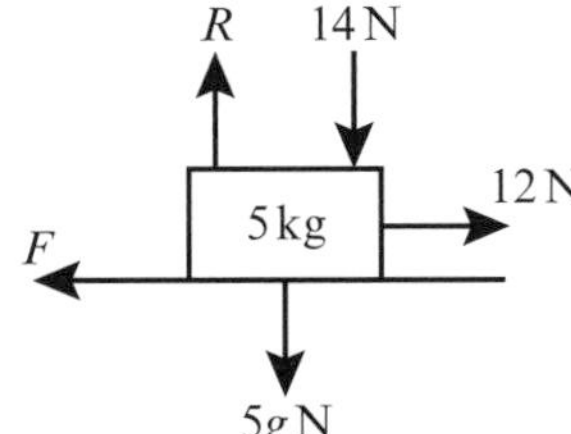

g

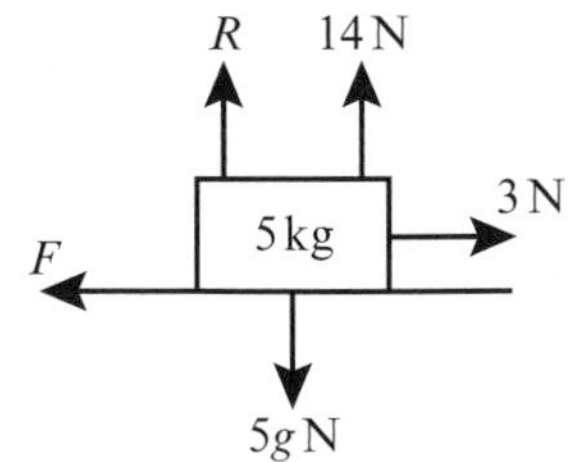

h

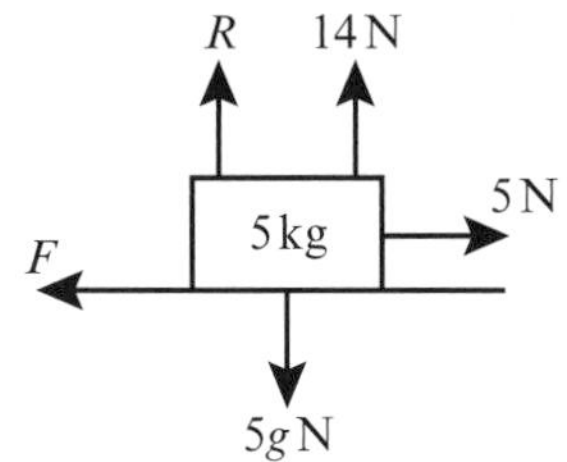

i

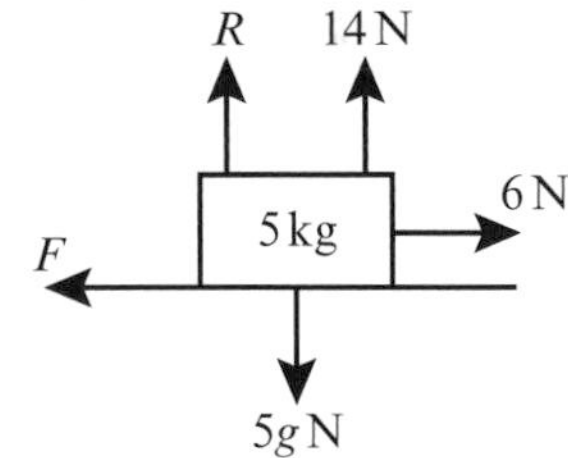

j

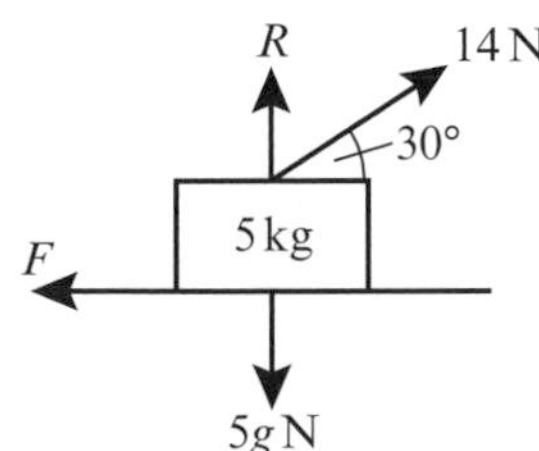

k

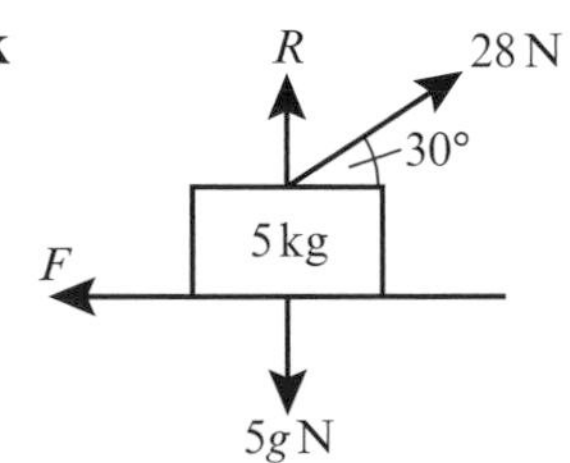

l

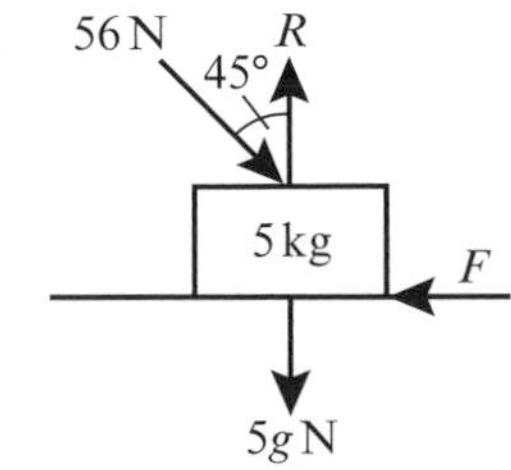

2 In each of the following diagrams, the forces shown cause the body of mass 10 kg to accelerate as shown along the rough horizontal plane. R is the normal reaction and F is the frictional force. Find the normal reaction and the coefficient of friction in each case.

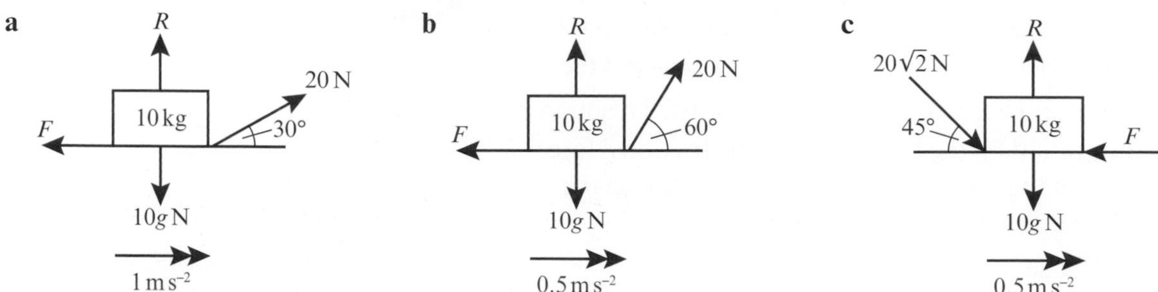

(E) **3** A particle of mass 0.5 kg is sliding down a rough slope that is angled at 15° to the horizontal. The acceleration of the particle is 0.25 m s^{-2}. Calculate the coefficient of friction between the particle and the slope. **(3 marks)**

(E) **4** A particle of mass 2 kg is sliding down a rough slope that is angled at 20° to the horizontal. A force of magnitude P acts parallel to the slope and is attempting to pull the particle up the slope. The acceleration of the particle is 0.2 m s^{-2} down the slope and the coefficient of friction between the particle and the slope is 0.3. Find the value of P. **(4 marks)**

5 A particle of mass 5 kg is being pushed up a rough slope that is angled at 30° to the horizontal by a horizontal force P. Given that the coefficient of friction is 0.2 and the acceleration of the particle is 2 m s^{-2} calculate the value of P.

(E/P) **6** A sled of mass 10 kg is being pulled along a rough horizontal plane by a force P that acts at an angle of 45° to the horizontal. The coefficient of friction between the sled and the plane is 0.1. Given that the sled accelerates at 0.3 m s^{-2}, find the value of P. **(7 marks)**

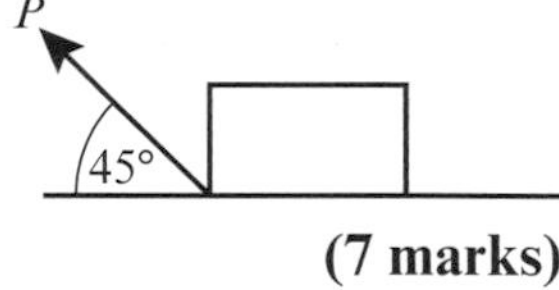

(E/P) **7** A train of mass m kg is travelling at 20 m s^{-1} when it applies its brakes, causing the wheels to lock up. The train decelerates at a constant rate, coming to a complete stop in 30 seconds.

a By modelling the train as a particle, show that the coefficient of friction between the railway track and the wheels of the train is $\mu = \frac{2}{3g}$. **(6 marks)**

> **Problem-solving**
>
> Use the formulae for constant acceleration.
>
> ← Year 1, Chapter 9

The train is no longer modelled as a particle, so that the effects of air resistance can be taken into account.

b State, with a reason, whether the coefficient of friction between the track and the wheels will increase or decrease in this revised model. **(2 marks)**

> **Challenge**
>
> A particle of mass m kg is sliding down a rough slope that is angled at α to the horizontal. The coefficient of friction between the particle and the slope is μ. Show that the acceleration of the particle is independent of its mass.

Mixed exercise 5

1 A box of mass 3 kg lies on a smooth horizontal floor. A force of 3 N is applied at an angle of 60° to the horizontal, causing the box to accelerate horizontally along the floor.

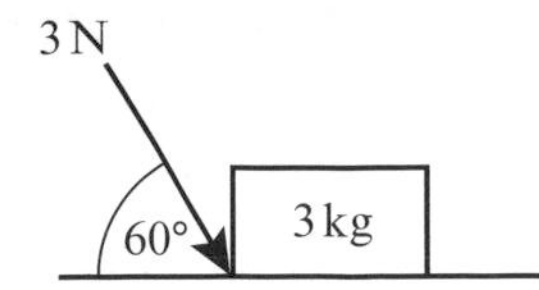

a Find the magnitude of the normal reaction of the floor on the box.

b Find the acceleration of the box.

(P) 2 A system of forces acts upon a particle as shown in the diagram. The resultant force on the particle is $(3\mathbf{i} + 2\mathbf{j})$ N. Calculate the magnitudes $\mathbf{F}_1$ and $\mathbf{F}_2$.

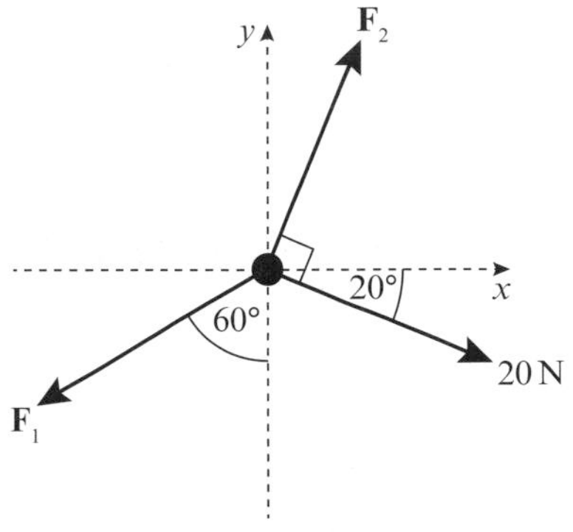

3 A force of 20 N is pulling a particle of mass 2 kg up a rough slope that is inclined at 45° to the horizontal. The force acts parallel to the slope, and the resistance due to friction is constant and has magnitude 4 N.

a Draw a force diagram to represent all the forces acting on the particle.

b Work out the normal reaction between the particle and the plane.

c Show that the acceleration of the particle is $1.1\ \text{m s}^{-2}$ (2 s.f.).

(E) 4 A particle of mass 5 kg sits on a smooth slope that is inclined at 10° to the horizontal. A force of 20 N acts on the particle at an angle of 20° to the plane, as shown in the diagram. Find the acceleration of the particle.

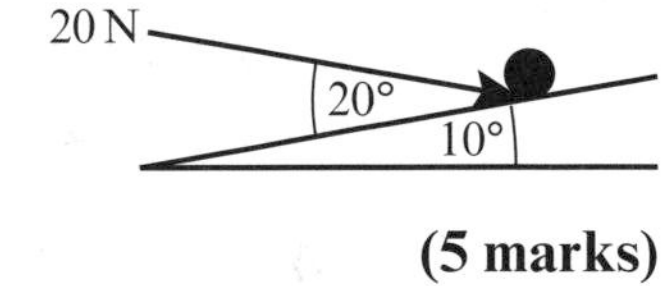

(5 marks)

(E/P) 5 A box is being pushed and pulled across a rough surface by constant forces as shown in the diagram. The box is moving at a constant speed. By modelling the box as a particle, show that the magnitude of the resistance due to friction F is $25(3\sqrt{2} + 2\sqrt{3})$ N. **(4 marks)**

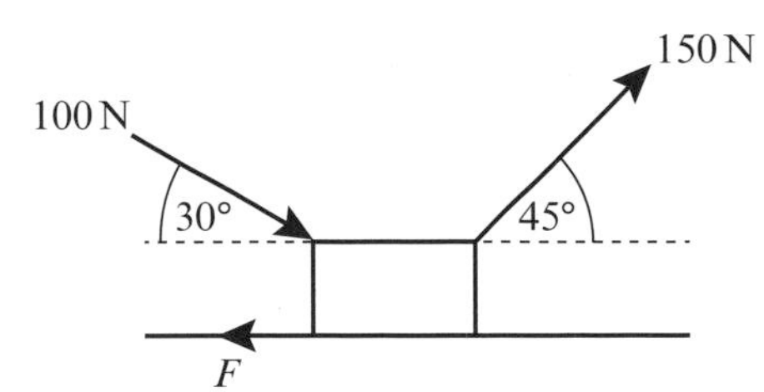

(E/P) 6 A trailer of mass 20 kg sits at rest on a rough horizontal plane. A force of 20 N acts on the trailer at an angle of 30° above the horizontal. Given that the trailer is in limiting equilibrium, work out the value of the coefficient of friction. **(6 marks)**

(E/P) 7 A particle of mass 2 kg is moving down a rough plane that is inclined at α to the horizontal, where $\tan \alpha = \frac{3}{4}$. A force of P N acts horizontally upon the particle towards the plane. Given that the coefficient of friction is 0.3 and that the particle is moving at a constant velocity, calculate the value of P. **(7 marks)**

(E/P) **8** A particle of mass 0.5 kg is being pulled up a rough slope that is angled at 30° to the horizontal by a force of 5 N. The force acts at an angle of 30° above the slope. Given that the coefficient of friction is 0.1, calculate the acceleration of the particle. **(7 marks)**

(E/P) **9** A car of mass 2150 kg is travelling down a rough road that is inclined at 10° to the horizontal. The engine of the car applies a constant driving force of magnitude 700 N, which acts in the direction of travel of the car. Any friction between the road and the tyres is initially ignored, and air resistance is modelled as a single constant force of magnitude F N that acts to oppose the motion of the car.

a Given that the car is travelling in a straight line at a constant speed of 22 m s^{-1}, find the magnitude of F. **(3 marks)**

The driver brakes suddenly. In the subsequent motion the car continues to travel in a straight line, and the tyres skid along the road, bringing the car to a standstill after 40 m. The driving force is removed, and the force due to air resistance is modelled as remaining constant.

b Find the coefficient of friction between the tyres and the road. **(7 marks)**

c Criticise this model with relation to

i the frictional forces acting on the car

ii the motion of the car. **(2 marks)**

Challenge

A boat of mass 400 kg is being pulled up a rough slipway at a constant speed of 5 m s^{-1} by a winch. The slipway is modelled as a plane inclined at an angle of 15° to the horizontal, and the boat is modelled as a particle. The coefficient of friction between the boat and the slipway is 0.2.

At the point when the boat is 8 m from the water-line, as measure along the line of greatest slope of the slipway, the winch cable snaps. Show that the boat will slide back down into the water, and calculate the total time from the winch cable breaking to the boat reaching the water-line.

Summary of key points

1. If a force is applied at an angle to the direction of motion, you can resolve it to find the component of the force that acts in the direction of motion.
2. The component of a force of magnitude F in a certain direction is $F\cos\theta$, where θ is the size of the angle between the force and the direction.
3. To solve problems involving inclined planes, it is usually easier to resolve parallel to and at right angles to the plane.
4. The maximum or limiting value of the friction between two surfaces, F_{MAX}, is given by $\mathbf{F_{MAX} = \mu R}$ where μ is the coefficient of friction and R is the normal reaction between the two surfaces.

Projectiles

6

Objectives

After completing this chapter you should be able to:

- Model motion under gravity for an object projected horizontally → pages 108–111
- Resolve velocity into components → pages 111–113
- Solve problems involving particles projected at an angle → pages 113–120
- Derive the formulae for time of flight, range and greatest height, and the equation of the path of a projectile → pages 120–125

A particle moving in a vertical plane under the action of gravity is sometimes called a **projectile**. You can use projectile motion to model the flight of a basketball. → Exercise 6C Q16

Prior knowledge check

1 A small ball is projected vertically upwards from a point P with speed 15 m s^{-1}. The ball is modelled as a particle moving freely under gravity. Find:

a the maximum height of the ball

b the time taken for the ball to return to P. ← Year 1, Chapter 9

2 Write expressions for x and y in terms of v and θ. ← GCSE Mathematics

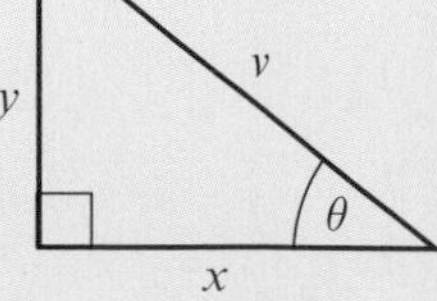

3 **a** Given $\sin\theta = \frac{5}{13}$ find

i $\cos\theta$ **ii** $\tan\theta$

b Given $\tan\theta = \frac{8}{15}$ find

i $\sin\theta$ **ii** $\cos\theta$ ← Pure Year 1, Chapter 10

6.1 Horizontal projection

You can model the motion of a projectile as a particle being acted on by a single force, gravity. In this model you ignore the effects of air resistance and any rotational movement on the particle.

You can analyse the motion of a projectile by considering its horizontal and vertical motion separately. Because gravity acts vertically downwards, there is **no force** acting on the particle in the horizontal direction.

- **The horizontal motion of a projectile is modelled as having constant velocity ($a = 0$). You can use the formula $s = vt$.**

The force due to gravity is modelled as being constant, so the vertical acceleration is constant.

- **The vertical motion of a projectile is modelled as having constant acceleration due to gravity ($a = g$).**

Use $g = 9.8\text{ m s}^{-2}$ unless the question specifies a different value.

Links You can use the constant acceleration formulae for the vertical motion of a projectile:

$v = u + at$

$s = \left(\frac{u+v}{2}\right)t$

$v^2 = u^2 + 2as$

$s = ut + \frac{1}{2}at^2$

$s = vt - \frac{1}{2}at^2$

← Year 1, Chapter 9

Example 1

A particle is projected horizontally at 25 m s^{-1} from a point 78.4 metres above a horizontal surface. Find:

a the time taken by the particle to reach the surface

b the horizontal distance travelled in that time.

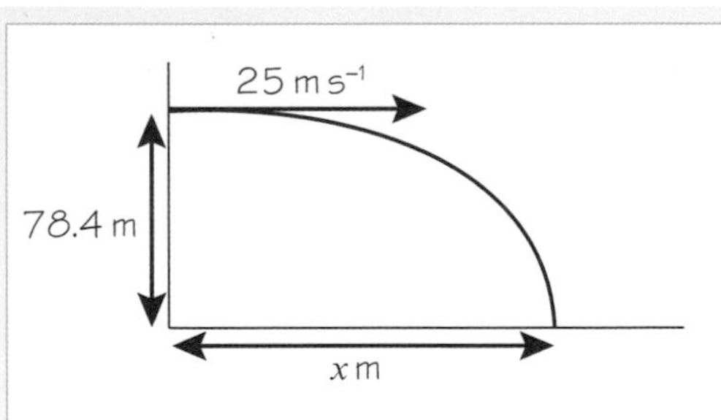

First draw a diagram showing all the information given in the question.

Notation u_x is the initial horizontal velocity. u_y is the initial vertical velocity.

Projected horizontally, R(→), $u_x = 25$

Taking the downwards direction as positive, R(↓), $u_y = 0$

The particle is projected horizontally so $u_y = 0$.

a R(↓), $u = 0$, $s = 78.4$, $a = 9.8$, $t = ?$

Write the values of u, s, a and t for the vertical motion.

$s = ut + \frac{1}{2}at^2$

$78.4 = 0 + \frac{1}{2} \times 9.8 \times t^2$

$78.4 = 4.9t^2$

$\frac{78.4}{4.9} = t^2$

$t^2 = 16$ so $t = 4\text{ s}$

Watch out The sign of g (positive or negative) depends on which direction is chosen as positive. Positive direction downwards: $g = 9.8\text{ m s}^{-2}$ Positive direction upwards: $g = -9.8\text{ m s}^{-2}$

The time taken must be positive so choose the positive square root.

b R(→), $u = 25$, $s = x$, $t = 4$

$s = vt$

$x = 25 \times 4$ so $x = 100\text{ m}$

Your answer to part **a** tells you the time taken for the particle to hit the surface. The horizontal motion has constant velocity so you can use: distance = speed × time.

Example 2

A particle is projected horizontally with a velocity of 15 m s^{-1}. Find:

a the horizontal and vertical components of the displacement of the particle from the point of projection after 3 seconds

b the distance of the particle from the point of projection after 3 seconds.

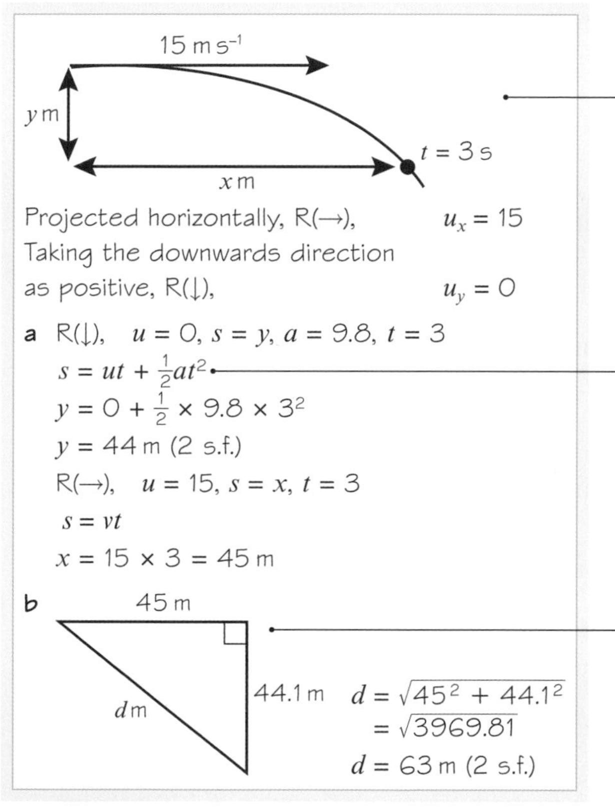

Draw a diagram based on the information in the question.

Use $s = ut + \frac{1}{2}at^2$ to find the vertical distance. This is the same distance as the particle would travel in 3 seconds if it was dropped and fell under the action of gravity.

The distance travelled is the magnitude of the displacement vector. Sketch a right-angled triangle showing the components and use Pythagoras' Theorem.

Example 3

A particle is projected horizontally with a speed of $U\text{ m s}^{-1}$ from a point 122.5 m above a horizontal plane. The particle hits the plane at a point which is at a horizontal distance of 90 m away from the starting point. Find the initial speed of the particle.

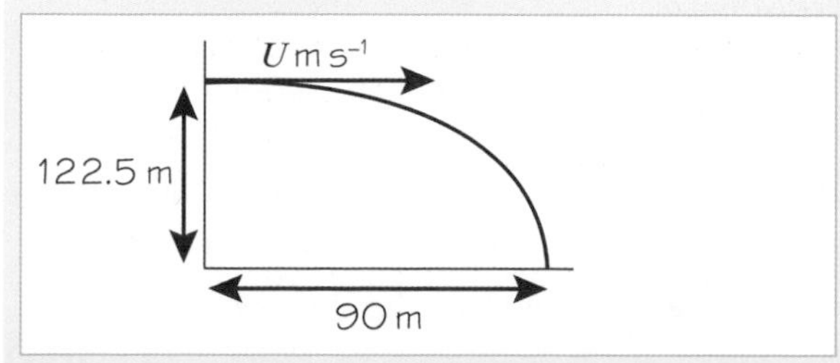

Projected horizontally, R(→), $u_x = U$
Taking the downwards direction as positive, R(↓), $u_y = 0$
R(↓), $u = 0$, $s = 122.5$, $a = 9.8$, $t = ?$

$s = ut + \frac{1}{2}at^2$

$122.5 = 0 + \frac{1}{2} \times 9.8 \times t^2$

$122.5 = 4.9t^2$

$t^2 = 25$ so $t = 5$ s

R(→), $v = U$, $s = 90$, $t = 5$

$s = vt$

$90 = U \times 5$

$U = 90 \div 5$ so $U = 18\,\text{m s}^{-1}$

Problem-solving

Many projectile problems can be solved by first using the **vertical motion** to find the total time taken.

Substitute $t = 5$ into the equation for horizontal motion to find U.

Exercise 6A

1 A particle is projected horizontally at $20\,\text{m s}^{-1}$ from a point h metres above horizontal ground. It lands on the ground 5 seconds later. Find:

a the value of h

b the horizontal distance travelled between the time the particle is projected and the time it hits the ground.

2 A particle is projected horizontally with a velocity of $18\,\text{m s}^{-1}$. Find:

a the horizontal and vertical components of the displacement of the particle from the point of projection after 2 seconds

b the distance of the particle from the point of projection after 2 seconds.

3 A particle is projected horizontally with a speed of $U\,\text{m s}^{-1}$ from a point 160 m above a horizontal plane. The particle hits the plane at a point which is at a horizontal distance of 95 m away from the point of projection. Find the initial speed of the particle.

4 A particle is projected horizontally from a point A which is 16 m above horizontal ground. The projectile strikes the ground at a point B which is at a horizontal distance of 140 m from A. Find the speed of projection of the particle.

(P) 5 A particle is projected horizontally with velocity $20\,\text{m s}^{-1}$ along a flat smooth table-top from a point 2 m from the table edge. The particle then leaves the table-top which is at a height of 1.2 m from the floor. Work out the total time taken for the particle to travel from the point of projection until it lands on the floor.

(E) 6 A darts player throws darts at a dart board which hangs vertically. The motion of a dart is modelled as that of a particle moving freely under gravity. The darts move in a vertical plane which is perpendicular to the plane of the dart board. A dart is thrown horizontally with an initial velocity of $14\,\text{m s}^{-1}$. It hits the board at a point which is 9 cm below the level from which it was thrown.

Find the horizontal distance from the point where the dart was thrown to the dart board.

(4 marks)

7 A particle of mass 2.5 kg is projected along a horizontal rough surface with a velocity of $5\,\text{m s}^{-1}$. After travelling a distance of 2 m the ball leaves the rough surface as a projectile and lands on the ground which is 1.2 m vertically below. Given that the total time taken for the ball to travel from the initial point of projection to the point when it lands is 1.0 seconds, find:

a the time for which the particle is in contact with the surface **(4 marks)**

b the coefficient of friction between the particle and the surface **(6 marks)**

c the horizontal distance travelled from the point of projection to the point where the particle hits the ground. **(3 marks)**

6.2 Horizontal and vertical components

Suppose a particle is projected with initial velocity U, at an angle α above the horizontal. The angle α is called the **angle of projection**.

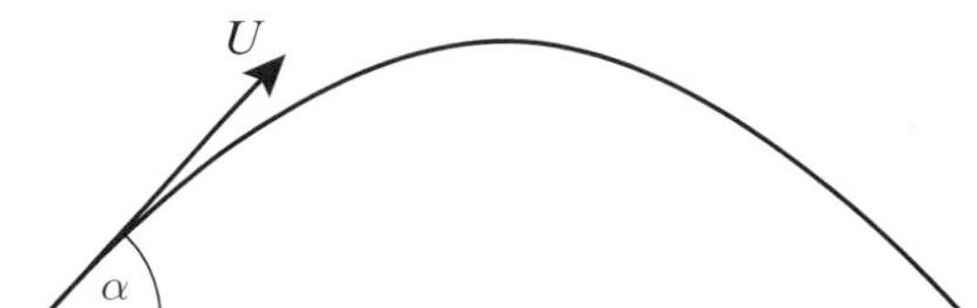

You can **resolve** the velocity into **components** that act horizontally and vertically:

Links This is the same technique as you use to resolve forces into components. ← Section 5.1

$\cos\alpha = \frac{u_x}{U}$ so $u_x = U\cos\alpha$

$\sin\alpha = \frac{u_y}{U}$ so $u_y = U\sin\alpha$

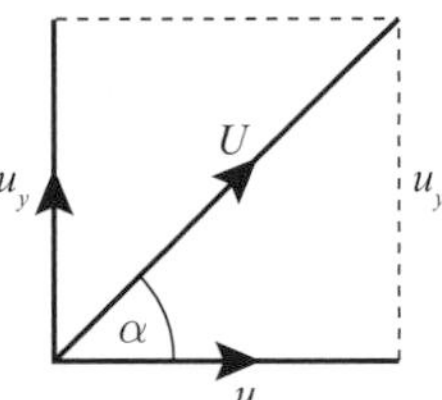

- **When a particle is projected with initial velocity U, at an angle α above the horizontal:**
  - **The horizontal component of the initial velocity is $U\cos\alpha$**
  - **The vertical component of the initial velocity is $U\sin\alpha$**

Example 4

A particle is projected from a point on a horizontal plane with an initial velocity of $40\,\text{m s}^{-1}$ at an angle α above the horizontal, where $\tan\alpha = \frac{3}{4}$.

a Find the horizontal and vertical components of the initial velocity.

Given that the vectors **i** and **j** are unit vectors acting in a vertical plane, horizontally and vertically respectively,

b express the initial velocity as a vector in terms of **i** and **j**.

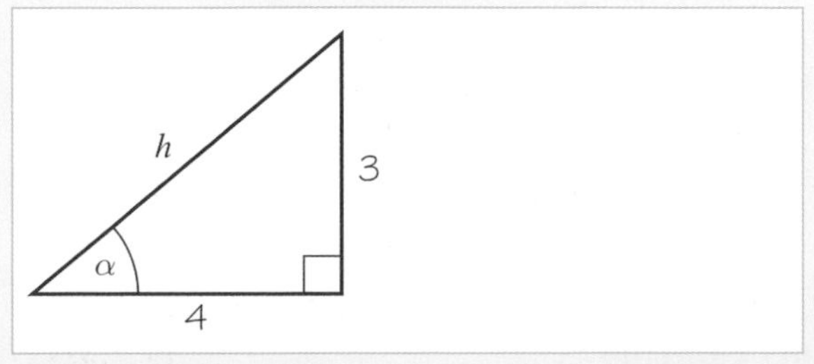

Problem-solving

When you are given a value for $\tan\alpha$ you can find the values of $\cos\alpha$ and $\sin\alpha$ without working out the value of α. Here $\tan\alpha = \frac{3}{4} = \frac{\text{opp}}{\text{adj}}$, so sketch a right-angled triangle with opposite 3 and adjacent 4.

a $\tan\alpha = \frac{3}{4}$ so $h = \sqrt{3^2 + 4^2} = 5$

$\sin\alpha = \frac{3}{5}$ $\cos\alpha = \frac{4}{5}$

R(→), $u_x = u\cos\alpha = 40 \times \frac{4}{5} = 32\,\text{m s}^{-1}$

R(↑), $u_y = u\sin\alpha = 40 \times \frac{3}{5} = 24\,\text{m s}^{-1}$

b $\mathbf{U} = (32\mathbf{i} + 24\mathbf{j})\,\text{m s}^{-1}$

Online Find cos α and sin α using your calculator.

You can write velocity as a vector using **i–j** notation. Remember to include units.

Example 5

A particle is projected with velocity $\mathbf{U} = (3\mathbf{i} + 5\mathbf{j})\,\text{m s}^{-1}$, where **i** and **j** are the unit vectors in the horizontal and vertical directions respectively. Find the initial speed of the particle and its angle of projection.

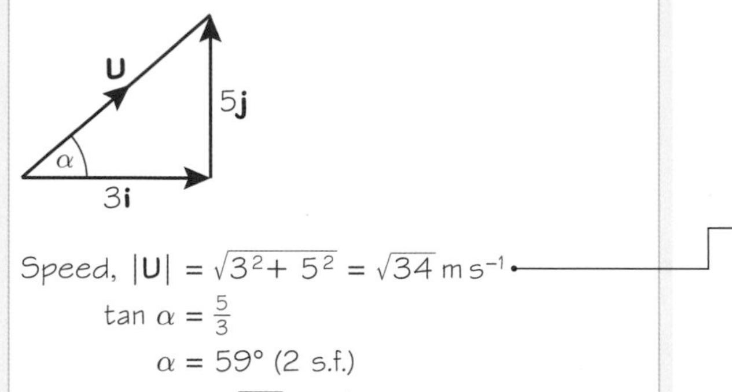

Speed, $|\mathbf{U}| = \sqrt{3^2 + 5^2} = \sqrt{34}\,\text{m s}^{-1}$

$\tan\alpha = \frac{5}{3}$

$\alpha = 59°$ (2 s.f.)

Initial speed is $\sqrt{34}\,\text{m s}^{-1}$ and the particle is projected at an angle of 59° above the horizontal.

Speed is the magnitude of the velocity vector. If the initial velocity is $p\mathbf{i} + q\mathbf{j}\,\text{m s}^{-1}$, the initial speed is $\sqrt{p^2 + q^2}$.

When an initial velocity is given in the form $p\mathbf{i} + q\mathbf{j}\,\text{m s}^{-1}$, the values of p and q are the horizontal and vertical components of the velocity respectively.

Exercise 6B

In this exercise **i** and **j** are unit vectors acting in a vertical plane, horizontally and vertically respectively.

1 A particle is projected from a point on a horizontal plane with an initial velocity of $25\,\text{m s}^{-1}$ at an angle of 40° above the horizontal.

a Find the horizontal and vertical components of the initial velocity.

b Express the initial velocity as a vector in the form $p\mathbf{i} + q\mathbf{j}\,\text{m s}^{-1}$.

2 A particle is projected from a cliff top with an initial velocity of $18\,\text{m s}^{-1}$ at an angle of 20° below the horizontal.

a Find the horizontal and vertical components of the initial velocity.

b Express the initial velocity as a vector in the form $p\mathbf{i} + q\mathbf{j}\,\text{m s}^{-1}$.

3 A particle is projected from a point on level ground with an initial velocity of $35\,\text{m}\,\text{s}^{-1}$ at an angle α above the horizontal, where $\tan\alpha = \frac{5}{12}$.

a Find the horizontal and vertical components of the initial velocity.

b Express the initial velocity as a vector in terms of **i** and **j**.

4 A particle is projected from the top of a building with an initial velocity of $28\,\text{m}\,\text{s}^{-1}$ at an angle θ below the horizontal, where $\tan\theta = \frac{7}{24}$.

a Find the horizontal and vertical components of the initial velocity.

b Express the initial velocity as a vector in terms of **i** and **j**.

5 A particle is projected with initial velocity $\mathbf{U} = (6\mathbf{i} + 9\mathbf{j})\,\text{m}\,\text{s}^{-1}$. Find the initial speed of the particle and its angle of projection.

6 A particle is projected with initial velocity $\mathbf{U} = (4\mathbf{i} - 5\mathbf{j})\,\text{m}\,\text{s}^{-1}$. Find the initial speed of the particle and its angle of projection.

(P) 7 A particle is projected with initial velocity $\mathbf{U} = 3k\mathbf{i} + 2k\mathbf{j}\,\text{m}\,\text{s}^{-1}$.

a Find the angle of projection.

Given the initial speed is $3\sqrt{13}\,\text{m}\,\text{s}^{-1}$,

b find the value of k.

6.3 Projection at any angle

You can solve problems involving particles projected at any angle by resolving the initial velocity into horizontal and vertical components.

The distance from the point from which the particle was projected to the point where it strikes the horizontal plane is called the **range**.

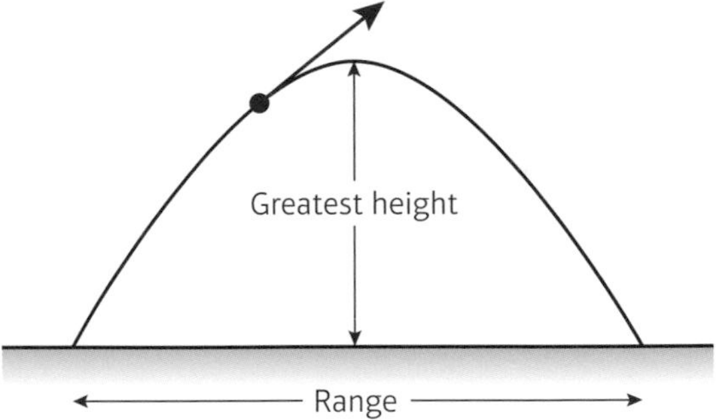

The time the particle takes to move from its point of projection to the point where it strikes the horizontal plane is called the **time of flight** of the projectile.

- **A projectile reaches its point of greatest height when the vertical component of its velocity is equal to 0.**

Example 6

A particle P is projected from a point O on a horizontal plane with speed $28\,\text{m}\,\text{s}^{-1}$ and with angle of elevation 30°. After projection, the particle moves freely under gravity until it strikes the plane at a point A. Find:

a the greatest height above the plane reached by P **b** the time of flight of P **c** the distance OA.

Resolving the velocity of projection horizontally and vertically:

R(→), $u_x = 28\cos 30° = 24.248...$

R(↑), $u_y = 28\sin 30° = 14$

28 sin 30°, 28, 30°, 28 cos 30°

a Taking the upwards direction as positive:

R(↑), $u = 14$, $v = 0$, $a = -9.8$, $s = ?$

$v^2 = u^2 + 2as$

$0^2 = 14^2 - 2 \times 9.8 \times s$

$s = \frac{14^2}{2 \times 9.8} = 10$

The greatest height above the plane reached by P is 10 m.

At the highest point the vertical component of the velocity is zero.

The vertical motion is motion with constant acceleration.

b R(↑), $s = 0$, $u = 14$, $a = -9.8$, $t = ?$

$s = ut + \frac{1}{2}at^2$

$0 = 14t - 4.9t^2$

$= t(14 - 4.9t)$

$t = 0$ or $t = \frac{14}{4.9} = 2.857...$

The time of flight is 2.9 s (2 s.f.)

When the particle strikes the plane, it is at the same height (zero) as when it started.

$t = 0$ corresponds to the point from which P was projected and can be ignored.

Watch out In this example the value for g was given to 2 significant figures so your answers should be given to 2 significant figures.

c R(→), distance = speed × time

$= 28\cos 30° \times 2.857...$

$= 69.282...$

$OA = 69$ m (2 s.f.)

There is no horizontal acceleration.

Use the unrounded value for the time of flight.

Example 7

A particle is projected from a point O with speed V m s^{-1} and at an angle of elevation of θ, where $\tan\theta = \frac{4}{3}$. The point O is 42.5 m above a horizontal plane. The particle strikes the plane at a point A, 5 s after it is projected.

a Show that $V = 20$. **b** Find the distance between O and A.

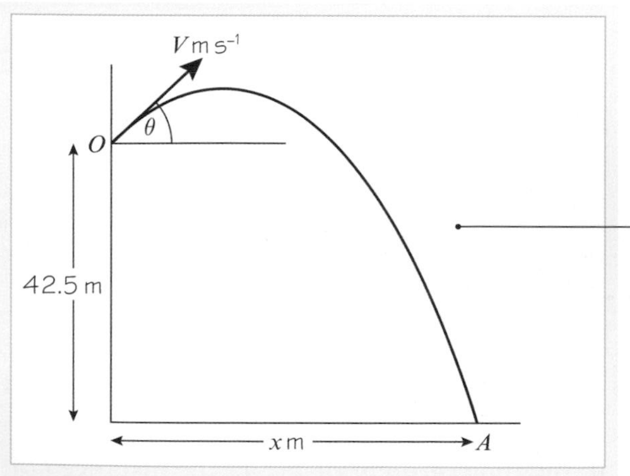

Start by drawing a diagram.

Resolving the velocity of projection horizontally and vertically:

R(→), $u_x = V\cos\theta = \frac{3}{5}V$

R(↑), $u_y = V\sin\theta = \frac{4}{5}V$

You will need $\sin\theta$ and $\cos\theta$ to resolve the initial velocity.
When you know $\tan\theta$ you can draw a triangle to find $\cos\theta$ and $\sin\theta$.

$\tan\theta = \frac{4}{3}$

$\sin\theta = \frac{4}{5}$

$\cos\theta = \frac{3}{5}$

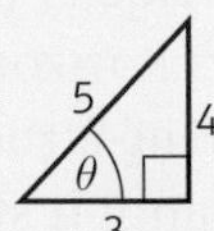

a Taking the upwards direction as positive:

R(↑), $s = -42.5$, $u = \frac{4}{5}V$, $g = -9.8$, $t = 5$

$s = ut + \frac{1}{2}at^2$

$-42.5 = \frac{4}{5}V \times 5 - 4.9 \times 25$

$4V = 4.9 \times 25 - 42.5 = 80$

$V = \frac{80}{4} = 20$, as required.

Use the formula $s = ut + \frac{1}{2}at^2$ to obtain an equation in V.

b Let the horizontal distance moved be x m:

R(→), distance = speed × time

$x = \frac{3}{5}V \times 5 = 3V = 60$

Use the value of V found in part **a** to find the horizontal distance moved by the particle.

Using Pythagoras' Theorem:

$OA^2 = 42.5^2 + 60^2 = 5406.25$

$OA = \sqrt{5406.25} = 73.527...$

The distance between O and A is 74 m, to 2 significant figures.

Example 8

A particle is projected from a point O with speed 35 m s^{-1} at an angle of elevation of 30°. The particle moves freely under gravity.

Find the length of time for which the particle is 15 m or more above O.

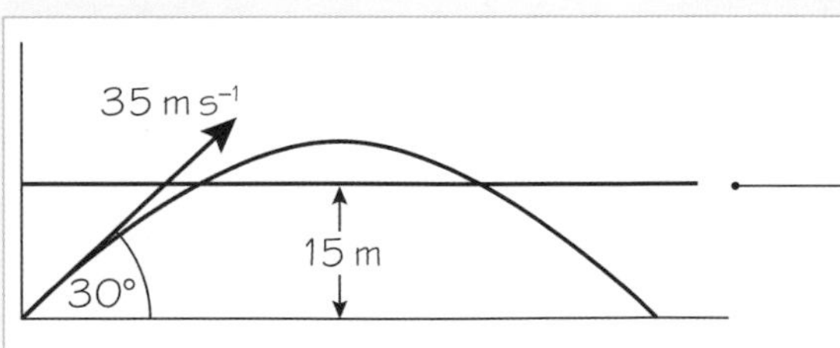

The particle is 15 m above O twice. First on the way up and then on the way down.

Resolving the initial velocity vertically:

In this example the horizontal component of the initial velocity is not used.

R(↑), $u_y = 35\sin 30° = 17.5$

$s = 15$, $u = 17.5$, $a = -9.8$, $t = ?$

$s = ut + \frac{1}{2}at^2$

$15 = 17.5t - 4.9t^2$

$4.9t^2 - 17.5t + 15 = 0$

Form a quadratic equation in t to find the two times when the particle is 15 m above O. Between these two times, the particle will be more than 15 m above O.

Multiplying by 10:

$49t^2 - 175t + 150 = 0$

$(7t - 10)(7t - 15) = 0$

$t = \frac{10}{7}, \frac{15}{7}$

$\frac{15}{7} - \frac{10}{7} = \frac{5}{7}$

Online Use your calculator to solve a quadratic equation.

The particle is 15 m or more above O for $\frac{5}{7}$ s.

You could also give this answer as a decimal to 2 significant figures, 0.71 s.

Example 9

A ball is struck by a racket at a point A which is 2 m above horizontal ground. Immediately after being struck, the ball has velocity $(5\mathbf{i} + 8\mathbf{j})\,\text{m s}^{-1}$, where $\mathbf{i}$ and $\mathbf{j}$ are unit vectors horizontally and vertically respectively. After being struck, the ball travels freely under gravity until it strikes the ground at the point B, as shown in the diagram above.

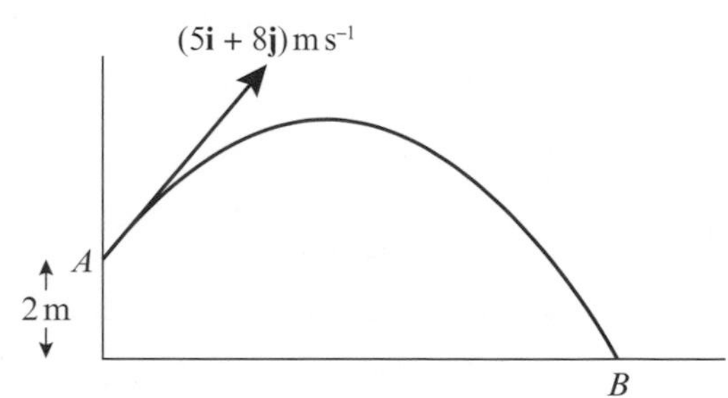

Find:

a the greatest height above the ground reached by the ball

b the speed of the ball as it reaches B

c the angle the velocity of the ball makes with the ground as the ball reaches B.

a Taking the upwards direction as positive:

R(↑), $u = 8$, $v = 0$, $a = -9.8$, $s = ?$

$v^2 = u^2 + 2as$

$0^2 = 8^2 - 2 \times 9.8 \times s$

$s = \frac{64}{19.6} = 3.265\ldots$

The greatest height above the ground reached by the ball is $2 + 3.265\ldots = 5.3$ m, to 2 significant figures.

> The velocity of projection has been given as a vector in terms of **i** and **j**. The horizontal component is 5 and the vertical component is 8.

> This is the greatest height above the point of projection. You need to add 2 m to find the height above the ground.

b The horizontal component of the velocity of the ball at B is $5\,\text{m s}^{-1}$.

The vertical component of the velocity of the ball at B is given by:

R(↑), $s = -2$, $u = 8$, $a = -9.8$, $v = ?$

$v^2 = u^2 + 2as$

$= 8^2 + 2 \times (-9.8) \times (-2) = 103.2$

The speed at B is given by:

$v^2 = 5^2 + 103.2 = 128.2$

$v = \sqrt{128.2}$

The speed of the ball as it reaches B is $11\,\text{m s}^{-1}$, to 2 significant figures.

> The horizontal motion is motion with constant speed, so the horizontal component of the velocity never changes.

> There is no need to find the square root of 103.2 at this point, as you need v^2 in the next stage of the calculation.

c The angle is given by:

$\tan\theta = \frac{\sqrt{103.2}}{5} \Rightarrow \theta = 64°$ (2 s.f.)

The angle the velocity of the ball makes with the ground as the ball reaches B is 64°, to the nearest degree.

> As the ball reaches B, its velocity has two components as shown below.

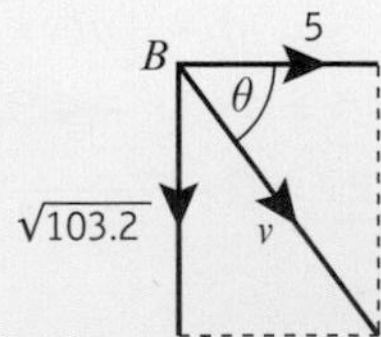

> The magnitude (speed) and direction of the velocity are found using trigonometry and Pythagoras' Theorem.

Exercise 6C

In this exercise **i** and **j** are unit vectors acting in a vertical plane, horizontally and vertically respectively.

Whenever a numerical value of g is required, take $g = 9.8\,\text{m}\,\text{s}^{-2}$ unless otherwise stated.

1 A particle is projected with speed $35\,\text{m}\,\text{s}^{-1}$ at an angle of elevation of 60°. Find the time the particle takes to reach its greatest height.

2 A ball is projected from a point 5 m above horizontal ground with speed $18\,\text{m}\,\text{s}^{-1}$ at an angle of elevation of 40°. Find the height of the ball above the ground 2 s after projection.

3 A stone is projected from a point above horizontal ground with speed $32\,\text{m}\,\text{s}^{-1}$, at an angle of 10° below the horizontal. The stone takes 2.5 s to reach the ground. Find:

a the height of the point of projection above the ground

b the distance from the point on the ground vertically below the point of projection to the point where the stone reaches the ground.

4 A projectile is launched from a point on horizontal ground with speed $150\,\text{m}\,\text{s}^{-1}$ at an angle of 10° above the horizontal. Find:

a the time the projectile takes to reach its highest point above the ground

b the range of the projectile.

5 A particle is projected from a point O on a horizontal plane with speed $20\,\text{m}\,\text{s}^{-1}$ at an angle of elevation of 45°. The particle moves freely under gravity until it strikes the ground at a point X. Find:

a the greatest height above the plane reached by the particle

b the distance OX.

(P) **6** A ball is projected from a point A on level ground with speed $24\,\text{m}\,\text{s}^{-1}$. The ball is projected at an angle θ to the horizontal where $\sin\theta = \frac{4}{5}$. The ball moves freely under gravity until it strikes the ground at a point B. Find:

a the time of flight of the ball

b the distance from A to B.

(P) **7** A particle is projected with speed $21\,\text{m}\,\text{s}^{-1}$ at an angle of elevation α. Given that the greatest height reached above the point of projection is 15 m, find the value of α, giving your answer to the nearest degree.

8 A particle P is projected from the origin with velocity $(12\mathbf{i} + 24\mathbf{j})\,\text{m}\,\text{s}^{-1}$, where **i** and **j** are horizontal and vertical unit vectors respectively. The particle moves freely under gravity. Find:

a the position vector of P after 3 s

b the speed of P after 3 s.

(P) **9** A stone is thrown with speed $30\,\text{m s}^{-1}$ from a window which is 20 m above horizontal ground. The stone hits the ground 3.5 s later. Find:

a the angle of projection of the stone

b the horizontal distance from the window to the point where the stone hits the ground.

(E/P) **10** A ball is thrown from a point O on horizontal ground with speed $U\,\text{m s}^{-1}$ at an angle of elevation of θ, where $\tan\theta = \frac{3}{4}$. The ball strikes a vertical wall which is 20 m from O at a point which is 3 m above the ground. Find:

a the value of U **(6 marks)**

b the time from the instant the ball is thrown to the instant that it strikes the wall. **(2 marks)**

(E/P) **11** A particle P is projected from a point A with position vector $20\mathbf{j}$ m with respect to a fixed origin O. The velocity of projection is $(5u\mathbf{i} + 4u\mathbf{j})\,\text{m s}^{-1}$. The particle moves freely under gravity, passing through a point B, which has position vector $(k\mathbf{i} + 12\mathbf{j})$ m, where k is a constant, before reaching the point C on the x-axis, as shown in the diagram. The particle takes 4 s to move from A to B. Find:

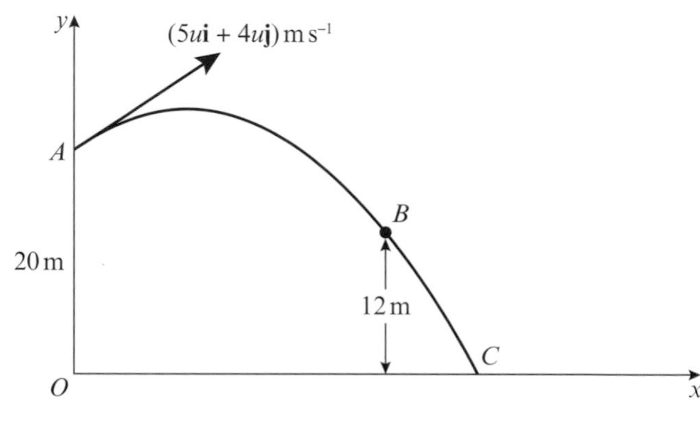

a the value of u **(4 marks)**

b the value of k **(2 marks)**

c the angle the velocity of P makes with the x-axis as it reaches C. **(6 marks)**

Watch out When finding a square root involving use of $g = 9.8\,\text{m s}^{-2}$ to work out an answer, an exact surd answer is **not** acceptable.

(E) **12** A stone is thrown from a point A with speed $30\,\text{m s}^{-1}$ at an angle of 15° below the horizontal. The point A is 14 m above horizontal ground. The stone strikes the ground at the point B, as shown in the diagram. Find:

a the time the stone takes to travel from A to B **(6 marks)**

b the distance AB. **(2 marks)**

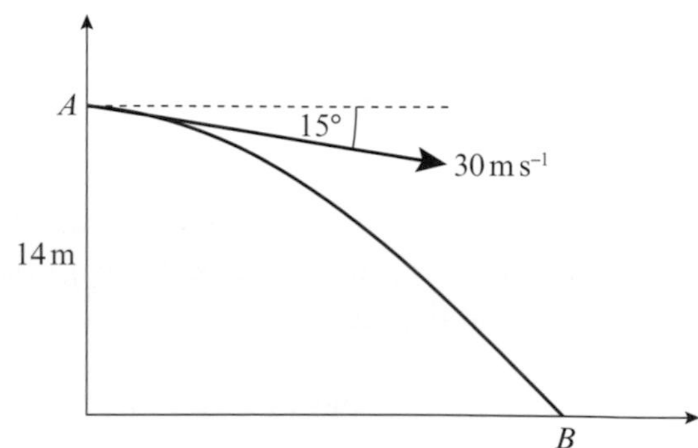

(E/P) **13** A particle is projected from a point on level ground with speed $U\,\text{m s}^{-1}$ and angle of elevation α. The maximum height reached by the particle is 42 m above the ground and the particle hits the ground 196 m from its point of projection.

Find the value of α and the value of U. **(9 marks)**

(E/P) **14** In this question use $g = 10\,\text{m s}^{-2}$.
An object is projected with speed $U\,\text{m s}^{-1}$ from a point A at the top of a vertical building. The point A is 25 m above the ground. The object is projected at an angle α above the horizontal, where $\tan\alpha = \frac{5}{12}$. The object hits the ground at the point B, which is at a horizontal distance of 42 m from the foot of the building, as shown in the diagram. The object is modelled as a particle moving freely under gravity.

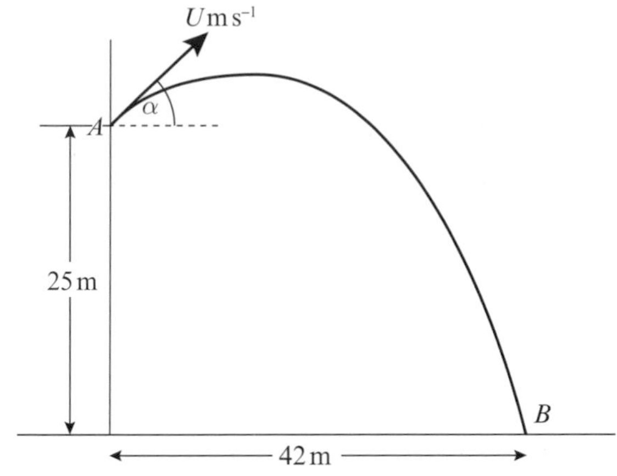

Find:

a the value of U **(6 marks)**

b the time taken by the object to travel from A to B **(2 marks)**

c the speed of the object when it is 12.4 m above the ground, giving your answer to 2 significant figures. **(5 marks)**

(E/P) **15** An object is projected from a fixed origin O with velocity $(4\mathbf{i} + 5\mathbf{j})\,\text{m s}^{-1}$. The particle moves freely under gravity and passes through the point P with position vector $k(\mathbf{i} - \mathbf{j})$ m, where k is a positive constant.

a Find the value of k. **(6 marks)**

b Find:

i the speed of the object at the instant when it passes through P

ii the direction of motion of the object at the instant when it passes through P. **(7 marks)**

(E/P) **16** A basketball player is standing on the floor 10 m from the basket. The height of the basket is 3.05 m, and he shoots the ball from a height of 2 m, at an angle of 40° above the horizontal.

The basketball can be modelled as a particle moving in a vertical plane. Given that the ball passes through the basket,

a find the speed with which the basketball is thrown. **(6 marks)**

b State two factors that can be ignored by modelling the basketball as a particle. **(2 marks)**

Challenge

A vertical tower is 85 m high. A stone is projected at a speed of 20 m s^{-1} from the top of a tower at an angle of α below the horizontal. At the same time, a second stone is projected horizontally at a speed of 12 m s^{-1} from a window in the tower 45 m above the ground.

Given that the two stones move freely under gravity in the same vertical plane, and that they collide in mid-air, show that the time that elapses between the moment they are projected and the moment they collide is 2.5 s.

6.4 Projectile motion formulae

You need to be able to derive general formulae related to the motion of a particle which is projected from a point on a horizontal plane and moves freely under gravity.

Example 10

A particle is projected from a point on a horizontal plane with an initial velocity U at an angle α above the horizontal and moves freely under gravity until it hits the plane at point B. Given that the acceleration due to gravity is g, find expressions for:

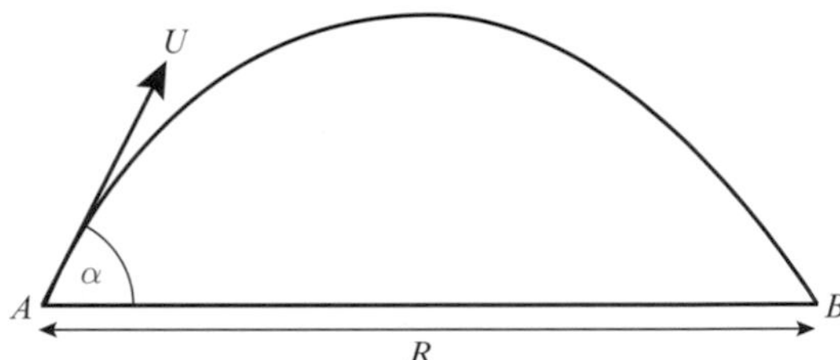

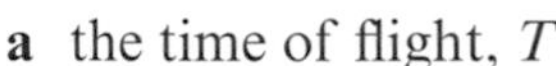

a the time of flight, T

b the range, R, on the horizontal plane.

Taking the upwards direction as positive and resolving the velocity of projection:

R(↑), $u_y = U\sin\alpha$

R(→), $u_x = U\cos\alpha$

a Considering vertical motion:

R(↑), $u = U\sin\alpha,\ s = 0,\ a = -g,\ t = T$

$$s = ut + \tfrac{1}{2}at^2$$

$$0 = (U\sin\alpha)T - \tfrac{1}{2} \times g \times T^2$$

$$0 = T\left(U\sin\alpha - \frac{gT}{2}\right)$$

either $T = 0$ (at A) or $U\sin\alpha - \dfrac{gT}{2} = 0$

so $T = \dfrac{2U\sin\alpha}{g}$

When the particle reaches the horizontal plane, the vertical displacement is 0.

Taking out the factor T, one solution is $T = 0$ which is at the start of the motion.

Online Explore the parametric equations for the path of the particle and their Cartesian form, both algebraically and graphically using technology.

Problem-solving

Follow the same steps as you would if you were given values of U and α and asked to find the time of flight and the range. The answer will be an algebraic expression in terms of U and α instead of a numerical value.

b Considering horizontal motion:

R(→), $v_x = U\cos\alpha$, $s = R$, $t = T$

$s = v_x t$

$R = U\cos\alpha \times T$ using $T = \frac{2U\sin\alpha}{g}$

$R = U\cos\alpha \times \frac{2U\sin\alpha}{g} = \frac{2U^2\sin\alpha\cos\alpha}{g}$

Using $2\sin\alpha\cos\alpha \equiv \sin 2\alpha$:

$R = \frac{U^2\sin 2\alpha}{g}$

Substitute for T in the equation $R = U\cos\alpha \times T$

$U\cos\alpha \times \frac{2U\sin\alpha}{g} = \frac{U\cos\alpha}{1} \times \frac{2U\sin\alpha}{g}$

Use the double-angle formula for sin 2α.

← **Pure Year 2, Section 7.3**

Notation g is usually left as a letter in the formulae for projectile motion.

Example 11

A particle is projected from a point with speed U at an angle of elevation α and moves freely under gravity. When the particle has moved a horizontal distance x, its height above the point of projection is y.

a Show that $y = x\tan\alpha - \frac{gx^2}{2u^2}(1 + \tan^2\alpha)$.

A particle is projected from a point O on a horizontal plane, with speed 28 m s^{-1} at an angle of elevation α. The particle passes through a point B, which is at a horizontal distance of 32 m from O and at a height of 8 m above the plane.

b Find the two possible values of α, giving your answers to the nearest degree.

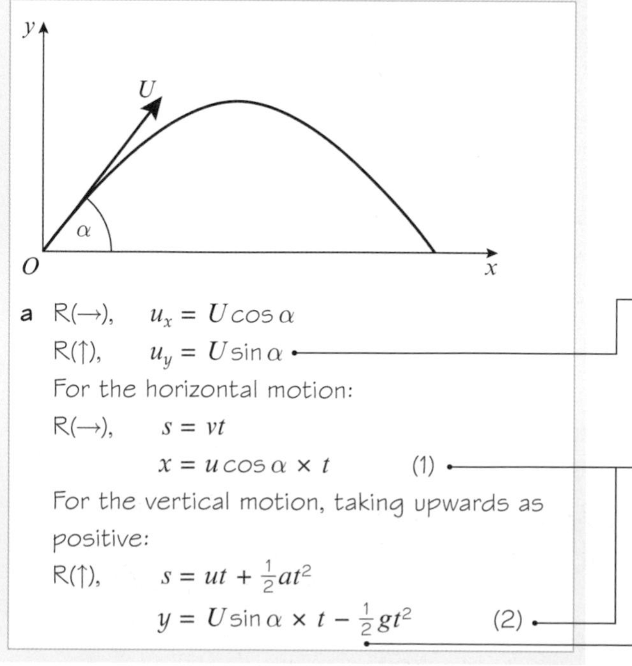

a R(→), $u_x = U\cos\alpha$

R(↑), $u_y = U\sin\alpha$

For the horizontal motion:

R(→), $s = vt$

$x = u\cos\alpha \times t$ (1)

For the vertical motion, taking upwards as positive:

R(↑), $s = ut + \frac{1}{2}at^2$

$y = U\sin\alpha \times t - \frac{1}{2}gt^2$ (2)

Resolve the velocity of projection horizontally and vertically.

You have obtained two equations, labelled (1) and (2). Both equations contain t and the result you have been asked to show has no t in it. You must eliminate t using substitution.

If the upwards direction is taken as positive, the vertical acceleration is $-g$.

Rearranging (1) to make t the subject of the formula:

$$t = \frac{x}{U\cos\alpha} \quad (3)$$

Substituting (3) into (2):

$$y = U\sin\alpha \times \frac{x}{U\cos\alpha} - \frac{1}{2}g\left(\frac{x}{U\cos\alpha}\right)^2$$

Using $\tan\alpha \equiv \frac{\sin\alpha}{\cos\alpha}$ and $\frac{1}{\cos\alpha} \equiv \sec\alpha$,

$$y = x\tan\alpha - \frac{gx^2}{2U^2}\sec^2\alpha$$

Using $\sec^2\alpha \equiv 1 + \tan^2\alpha$,

$$y = x\tan\alpha - \frac{gx^2}{2u^2}(1 + \tan^2\alpha), \text{ as required.}$$

b Using the result in **a** with $U = 28$, $x = 32$, $y = 8$ and $g = 9.8$

$$8 = 32\tan\alpha - 6.4(1 + \tan^2\alpha)$$

Rearranging as a quadratic in $\tan\alpha$:

$$6.4\tan^2\alpha - 32\tan\alpha + 14.4 = 0$$
$$4\tan^2\alpha - 20\tan\alpha + 9 = 0$$
$$(2\tan\alpha - 1)(2\tan\alpha - 9) = 0$$
$$\tan\alpha = \tfrac{1}{2}, \tfrac{9}{2}$$

$\alpha = 27°$ and $77°$, to the nearest degree

(1) and (2) are parametric equations describing the path of the particle. You can eliminate the parameter, t, to find the Cartesian form of the path. ← Pure Year 2, Section 8.5

To obtain a quadratic expression in $\tan\alpha$, you need to use the identity $\sec^2\alpha \equiv 1 + \tan^2\alpha$. ← Pure Year 2, Section 6.4

You could use your calculator to solve this equation.

There are two possible angles of elevation for which the particle will pass through B. This sketch illustrates the two paths.

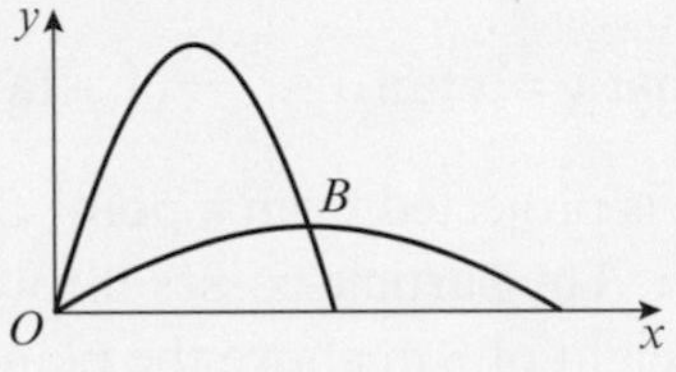

- **For a particle which is projected from a point on a horizontal plane with an initial velocity U at an angle α above the horizontal, and that moves freely under gravity:**
  - **Time of flight = $\frac{2U\sin\alpha}{g}$**
  - **Time to reach greatest height = $\frac{U\sin\alpha}{g}$**
  - **Range on horizontal plane = $\frac{U^2\sin 2\alpha}{g}$**
  - **Equation of trajectory: $y = x\tan\alpha - gx^2\frac{(1 + \tan^2\alpha)}{2U^2}$**

  where y is the vertical height of the particle, x is the horizontal distance from the point of projection, and g is the acceleration due to gravity.

Watch out You need to know how to derive the equations. But be careful of using them in projectile problems. They are hard to memorise, and it is usually safer to answer projectile problems using the techniques covered in Section 16.3.

Hint The equation for the trajectory of the particle is a **quadratic** equation for y in x. This proves that the path of a projectile moving freely under gravity is a quadratic curve, or **parabola**.

Exercise 6D

Whenever a numerical value of g is required, take $g = 9.8\,\text{m s}^{-2}$ unless otherwise stated.

(P) **1** A particle is launched from a point on a horizontal plane with initial velocity $U\,\text{m s}^{-1}$ at an angle of elevation α. The particle moves freely under gravity until it strikes the plane.
The greatest height of the particle is h m.

Show that $h = \dfrac{U^2 \sin^2 \alpha}{2g}$

(P) **2** A particle is projected from a point with speed $21\,\text{m s}^{-1}$ at an angle of elevation α and moves freely under gravity. When the particle has moved a horizontal distance x m, its height above the point of projection is y m.

a Show that $y = x \tan \alpha - \dfrac{x^2}{90 \cos^2 \alpha}$

b Given that $y = 8.1$ when $x = 36$, find the value of $\tan \alpha$.

(P) **3** A projectile is launched from a point on a horizontal plane with initial speed $U\,\text{m s}^{-1}$ at an angle of elevation α. The particle moves freely under gravity until it strikes the plane. The range of the projectile is R m.

a Show that the time of flight of the particle is $\dfrac{2U \sin \alpha}{g}$ seconds.

b Show that $R = \dfrac{U^2 \sin 2\alpha}{g}$.

c Deduce that, for a fixed u, the greatest possible range is when $\alpha = 45°$.

d Given that $R = \dfrac{2U^2}{5g}$, find the two possible values of the angle of elevation at which the projectile could have been launched.

(P) **4** A firework is launched vertically with a speed of $v\,\text{m s}^{-1}$. When it reaches its maximum height, the firework explodes into two parts, which are projected horizontally in opposite directions, each with speed $2v\,\text{m s}^{-1}$.
Show that the two parts of the firework land a distance $\dfrac{4v^2}{g}$ m apart.

(E/P) **5** In this question use $g = 10\,\text{m s}^{-2}$.
A particle is projected from a point O with speed U at an angle of elevation α above the horizontal and moves freely under gravity. When the particle has moved a horizontal distance x, its height above O is y.

a Show that $y = x \tan \alpha - \dfrac{gx^2}{2U^2 \cos^2 \alpha}$ **(4 marks)**

A boy throws a stone from a point P at the end of a pier. The point P is 15 m above sea level. The stone is projected with a speed of $8\,\text{m s}^{-1}$ at an angle of elevation of 40°. By modelling the ball as a particle moving freely under gravity,

b find the horizontal distance of the stone from P when the ball is 2 m above sea level. **(5 marks)**

(E/P) **6** A particle is projected from a point with speed U at an angle of elevation α above the horizontal and moves freely under gravity. When it has moved a horizontal distance x, its height above the point of projection is y.

a Show that $y = x\tan\alpha - \dfrac{gx^2}{2U^2}(1 + \tan^2\alpha)$ **(5 marks)**

An athlete throws a javelin from a point P at a height of 2 m above horizontal ground. The javelin is projected at an angle of elevation of 45° with a speed of $30\,\text{m s}^{-1}$. By modelling the javelin as a particle moving freely under gravity,

b find, to 3 significant figures, the horizontal distance of the javelin from P when it hits the ground **(5 marks)**

c find, to 2 significant figures, the time elapsed from the point the javelin is thrown to the point it hits the ground. **(2 marks)**

(E/P) **7** A girl playing volleyball on horizontal ground hits the ball towards the net 9 m away from a point 1.5 m above the ground. The ball moves in a vertical plane which is perpendicular to the net. The ball just passes over the top of the net, which is 2.4 m above the ground, as shown in the diagram.

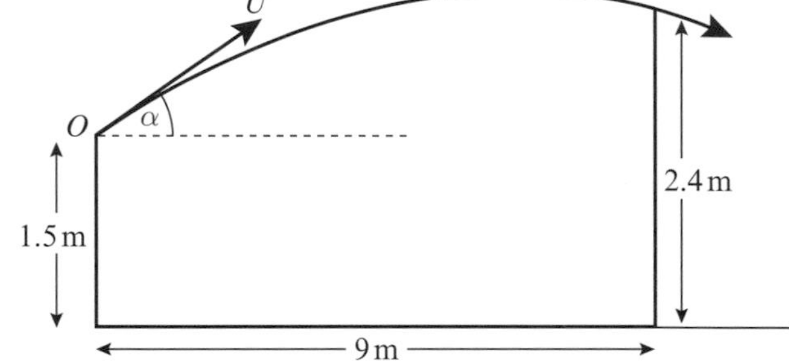

The ball is modelled as a particle projected with initial speed $U\,\text{m s}^{-1}$ from point O, 1.5 m above the ground at an angle α to the horizontal.

a By writing down expressions for the horizontal and vertical distances from O to the ball, t seconds after it was hit, show that when the ball passes over the net

$$0.9 = 9\tan\alpha - \frac{81g}{2\,U^2\cos^2\alpha}$$ **(6 marks)**

Given that $\alpha = 30°$,

b find the speed of the ball as it passes over the net. **(6 marks)**

(E/P) **8** In this question **i** and **j** are unit vectors in a horizontal and upward vertical direction respectively. An object is projected from a fixed point A on horizontal ground with velocity $(k\mathbf{i} + 2k\mathbf{j})\,\text{m s}^{-1}$, where k is a positive constant. The object moves freely under gravity until it strikes the ground at B, where it immediately comes to rest. Relative to O, the position vector of a point on the path of the object is $(x\mathbf{i} + y\mathbf{j})$ m.

a Show that $y = 2x - \dfrac{gx^2}{2k^2}$ **(5 marks)**

Given that $AB = R$ m and the maximum vertical height of the object above the ground is H m,

b using the result in part **a**, or otherwise, find, in terms of k and g,

i R **ii** H **(6 marks)**

Challenge

A stone is projected from a point on a straight sloping hill. Given that the hill slopes downwards at an angle of 45°, and that the stone is projected at an angle of 45° above the horizontal with speed U m s^{-1}.

Show that the stone lands a distance $\frac{2\sqrt{2}U^2}{g}$ m down the hill.

Mixed exercise 6

Whenever a numerical value of g is required, take $g = 9.8$ m s^{-2} unless otherwise stated.

1 A particle P is projected from a point O on a horizontal plane with speed 42 m s^{-1} and with angle of elevation 45°. After projection, the particle moves freely under gravity until it strikes the plane. Find:

a the greatest height above the plane reached by P

b the time of flight of P.

2 A stone is thrown horizontally with speed 21 m s^{-1} from a point P on the edge of a cliff h metres above sea level. The stone lands in the sea at a point Q, where the horizontal distance of Q from the cliff is 56 m.

Calculate the value of h.

(E) **3** A ball is thrown from a window above a horizontal lawn. The velocity of projection is 15 m s^{-1} and the angle of elevation is α, where $\tan \alpha = \frac{4}{3}$. The ball takes 4 s to reach the lawn. Find:

a the horizontal distance between the point of projection and the point where the ball hits the lawn **(3 marks)**

b the vertical height above the lawn from which the ball was thrown. **(3 marks)**

(E) **4** A projectile is fired with velocity 40 m s^{-1} at an angle of elevation of 30° from a point A on horizontal ground. The projectile moves freely under gravity until it reaches the ground at the point B. Find:

a the distance AB **(5 marks)**

b the speed of the projectile at the first instant when it is 15 m above the ground. **(5 marks)**

(E/P) **5** A particle P is projected from a point on a horizontal plane with speed U at an angle of elevation θ.

a Show that the range of the projectile is $\frac{U^2 \sin 2\theta}{g}$. **(6 marks)**

b Hence find, as θ varies, the maximum range of the projectile. **(2 marks)**

c Given that the range of the projectile is $\frac{2U^2}{3g}$, find the two possible value of θ.

Give your answers to the nearest 0.1°. **(3 marks)**

(E) **6**

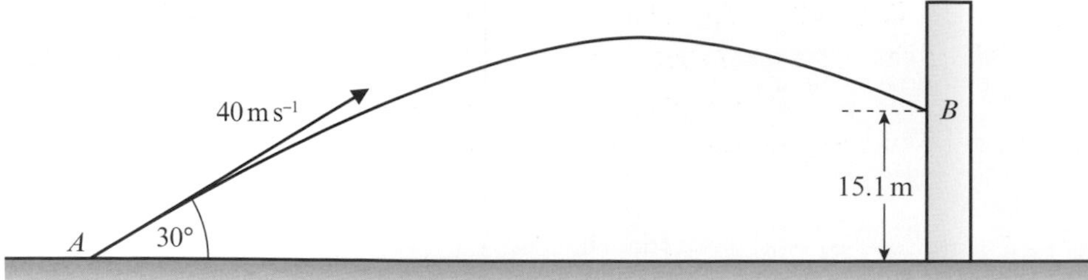

A golf ball is driven from a point A with a speed of $40\,\text{m}\,\text{s}^{-1}$ at an angle of elevation of 30°. On its downward flight, the ball hits an advertising hoarding at a height 15.1 m above the level of A, as shown in the diagram above. Find:

a the time taken by the ball to reach its greatest height above A **(3 marks)**

b the time taken by the ball to travel from A to B **(6 marks)**

c the speed with which the ball hits the hoarding. **(5 marks)**

(E/P) **7** In this question use $g = 10\,\text{m}\,\text{s}^{-2}$.
A boy plays a game at a fairground. He needs to throw a ball through a hole in a vertical target to win a prize. The motion of the ball is modelled as that of a particle moving freely under gravity. The ball moves in a vertical plane which is perpendicular to the plane of the target. The boy throws the ball horizontally at the same height as the hole with a speed of $10\,\text{m}\,\text{s}^{-1}$. It hits the target at a point 20 cm below the hole.

a Find the horizontal distance from the point where the ball was thrown to the target. **(4 marks)**

The boy throws the ball again with the same speed and at the same distance from the target.

b Work out the possible angles above the horizontal the boy could throw the ball so that it passes through the hole. **(6 marks)**

(E/P) **8** In this question use $g = 10\,\text{m}\,\text{s}^{-2}$.
A stone is thrown from a point P at a target, which is on horizontal ground. The point P is 10 m above the point O on the ground. The stone is thrown from P with speed $20\,\text{m}\,\text{s}^{-1}$ at an angle of α below the horizontal, where $\tan\alpha = \frac{3}{4}$.
The stone is modelled as a particle and the target as a point T. The distance OT is 9 m. The stone misses the target and hits the ground at the point Q, where OTQ is a straight line, as shown in the diagram. Find:

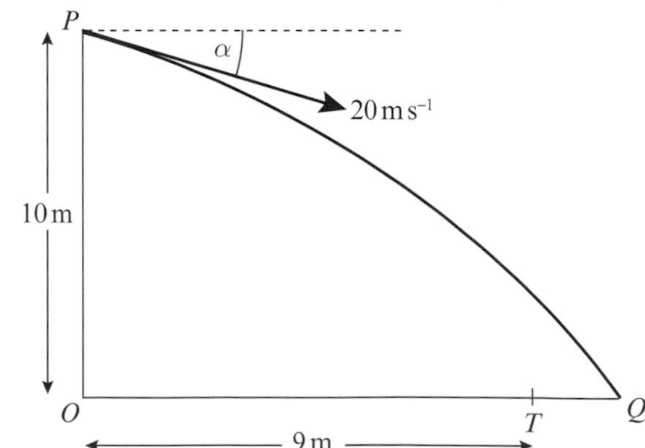

a the time taken by the ball to travel from P to Q **(5 marks)**

b the distance TQ. **(4 marks)**

The point A is on the path of the ball vertically above T.

c Find the speed of the ball at A. **(5 marks)**

9 A vertical mast is 32 m high. Two balls P and Q are projected simultaneously. Ball P is projected horizontally from the top of the mast with speed $18\,\text{m}\,\text{s}^{-1}$. Ball Q is projected from the bottom of the mast with speed $30\,\text{m}\,\text{s}^{-1}$ at an angle α above the horizontal. The balls move freely under gravity in the same vertical plane and collide in mid-air. By considering the horizontal motion of each ball,

a prove that $\cos\alpha = \dfrac{3}{5}$ **(4 marks)**

b Find the time which elapses between the instant when the balls are projected and the instant when they collide. **(4 marks)**

Challenge

A cruise ship is 250 m long, and is accelerating forwards in a straight line at a constant rate of $1.5\,\text{m}\,\text{s}^{-2}$. A golfer stands at the stern (back) of the cruise ship and hits a golf ball towards the bow (front). Given that the golfer hits the golf ball at an angle of elevation of 60°, and that the ball lands directly on the bow of the cruise ship, find the speed, v, with which the golfer hits the ball.

Summary of key points

1 The **horizontal** motion of a projectile is modelled as having **constant velocity** ($a = 0$). You can use the formula $s = vt$.

2 The **vertical** motion of a projectile is modelled as having **constant acceleration** due to gravity ($a = g$).

3 When a particle is projected with initial velocity U, at an angle α above the horizontal:
- The **horizontal component** of the initial velocity is $U\cos\alpha$
- The **vertical component** of the initial velocity is $U\sin\alpha$

4 A projectile reaches its point of greatest height when the vertical component of its velocity is equal to 0.

5 For a particle which is projected from a point on a horizontal plane with an initial velocity U at an angle α above the horizontal, and that moves freely under gravity:
- Time of flight $= \dfrac{2U\sin\alpha}{g}$
- Time to reach greatest height $= \dfrac{U\sin\alpha}{g}$
- Range on horizontal plane $= \dfrac{U^2\sin 2\alpha}{g}$
- Equation of trajectory: $y = x\tan\alpha - gx^2\dfrac{(1+\tan^2\alpha)}{2U^2}$

where y is the vertical height of the particle, x is the horizontal distance from the point of projection, and g is the acceleration due to gravity.

7 Applications of forces

Objectives

After completing this chapter you should be able to:

- Find an unknown force when a system is in equilibrium → pages 129–132
- Solve statics problems involving weight, tension and pulleys → pages 132–137
- Understand and solve problems involving limiting equilibrium → pages 137–146
- Solve problems involving motion on rough or smooth inclined planes → pages 147–150
- Solve problems involving connected particles that require the resolution of forces → pages 150–154

Prior knowledge check

1 A particle of mass 2 kg sits on a rough plane that is inclined at 45° to the horizontal. A force of 10 N acts parallel to and up the plane. Given that the particle is on the point of moving, work out the coefficient of friction μ. ← Section 5.3

2 A uniform rod AB of length 2 m and mass 5 kg rests in equilibrium at an angle of 60° to a horizontal surface. The rod is pivoted at A and a force of magnitude X N acts perpendicular to the rod at B. Find the value of X.

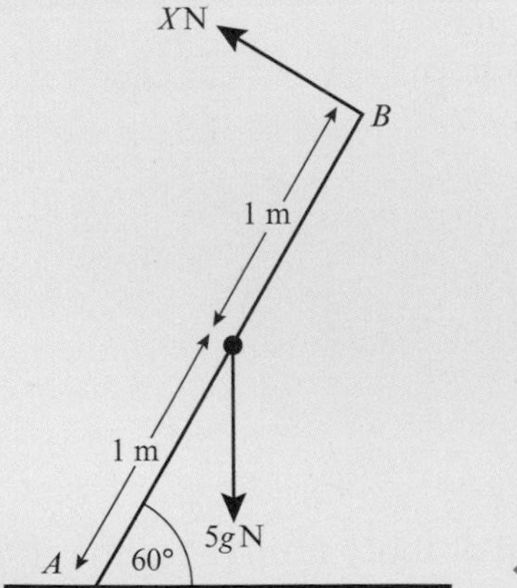

← Section 4.3

A tightrope walker uses a mathematical model to calculate the tension in his wire. This allows him to make sure that the wire is strong enough to hold his weight safely. → Mixed exercise, Q3

7.1 Static particles

■ **A particle or rigid body is in static equilibrium if it is at rest and the resultant force acting on the particle is zero.**

To solve problems in statics you should:

- Draw a diagram showing clearly the forces acting on the particle.
- Resolve the forces into horizontal and vertical components or, if the particle is on an inclined plane, into components parallel and perpendicular to the plane.
- Set the sum of the components in each direction equal to zero.
- Solve the resulting equations to find the unknown force(s).

Hint The particle is not accelerating, so $a = 0$. $F = ma = 0$

Example 1

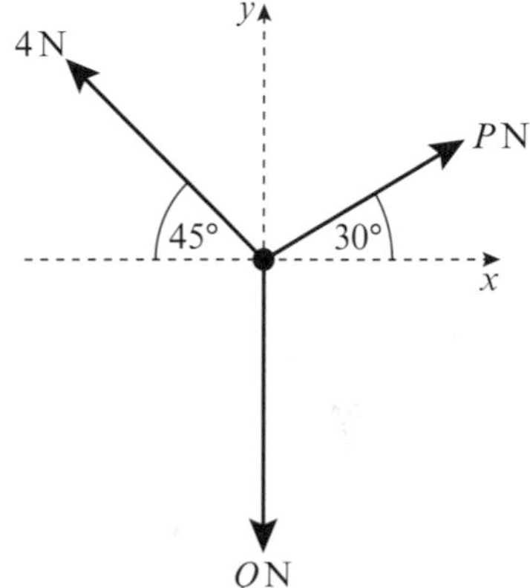

The diagram shows a particle in equilibrium under the forces shown. By resolving horizontally and vertically find the magnitudes of the forces P and Q.

Method 1:

R(→), $P\cos 30° - 4\cos 45° = 0$

R(↑), $P\sin 30° + 4\sin 45° - Q = 0$

Resolve horizontally and vertically. Equate the sum of the forces to zero as there is no acceleration (the particle is in equilibrium).

$P = \dfrac{4\cos 45°}{\cos 30°} = 3.27$ (3 s.f.)

$Q = P\sin 30° + 4\sin 45° = 4.46$ (3 s.f.)

Solve the first equation to find P (as there is only one unknown quantity), and then use your value for P in the second equation to find Q.

If exact answers are required these would be $P = \dfrac{4\sqrt{6}}{3}$ and $Q = \dfrac{2(\sqrt{6} + 3\sqrt{2})}{3}$

Method 2:

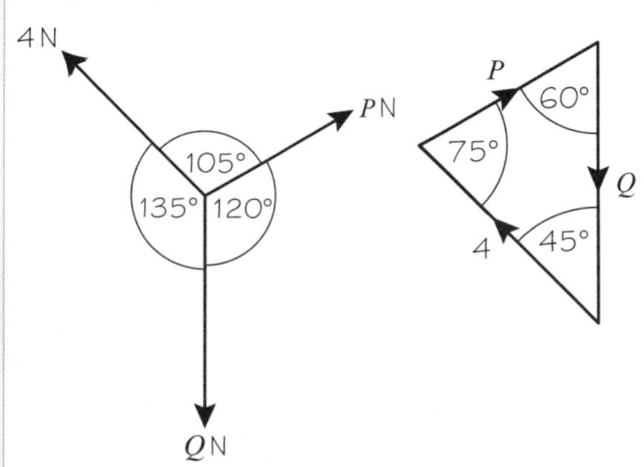

Problem-solving

You can use a vector diagram to solve equilibrium problems involving three forces. Because the particle is in equilibrium, the three forces will form a **closed triangle**.

If the angle between forces on the force diagram is θ, the angle between those forces on the **triangle of forces** is $180° - \theta$.

The length of each side of the triangle is the magnitude of the force.

So $\dfrac{4}{\sin 60°} = \dfrac{P}{\sin 45°} = \dfrac{Q}{\sin 75°}$

$P = \dfrac{4 \sin 45°}{\sin 60°} = \dfrac{4\sqrt{6}}{3}$ N

$Q = \dfrac{4 \sin 75°}{\sin 60°} = \dfrac{2(\sqrt{6} + 3\sqrt{2})}{3}$ N

Use the sine rule.

Watch out This method only works for a particle in equilibrium. If the resultant force is not zero, the vector diagram will not be a closed triangle.

Example 2

The diagram shows a particle in equilibrium on an inclined plane under the forces shown. Find the magnitude of the force P and the size of the angle α.

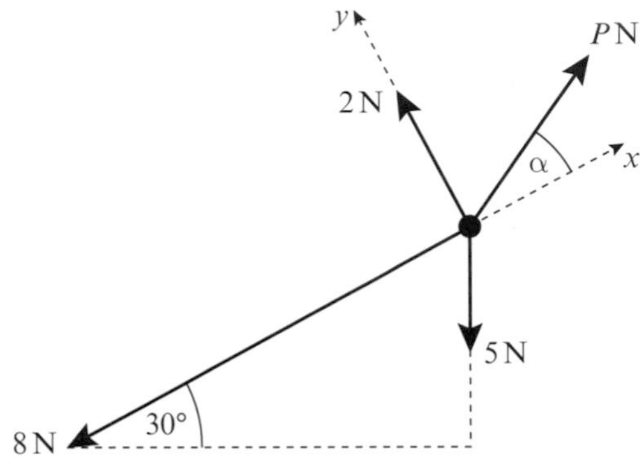

R(↗), $P \cos \alpha - 8 - 5 \sin 30° = 0$

$\therefore \quad P \cos \alpha = 8 + 5 \sin 30° \qquad (1)$

R(↖), $P \sin \alpha + 2 - 5 \cos 30° = 0$

$\therefore \quad P \sin \alpha = 5 \cos 30° - 2 \qquad (2)$

Divide equation (2) by equation (1) to give:

$\tan \alpha = \dfrac{5 \cos 30° - 2}{8 + 5 \sin 30°} = \dfrac{2.330}{10.5} = 0.222$

$\therefore \quad \alpha = 12.5°$ (3 s.f.)

Substitute into equation (1):

$P \sin 12.5°... = 5 \cos 30° - 2$

$\therefore \quad P = 10.8$ (3 s.f.)

Resolve parallel to the plane. Take the direction up the plane as positive.

Rearrange the equation to make $P \cos \alpha$ the subject.

Resolve perpendicular to the plane. Rearrange the second equation to make $P \sin \alpha$ the subject.

After division use $\dfrac{\sin \alpha}{\cos \alpha} \equiv \tan \alpha$.

Use $\tan^{-1}$ and give your answer to three significant figures.

You could check your answers by substituting into equation (2).

Exercise 7A

1 Each of the following diagrams shows a particle in static equilibrium. For each particle:

i resolve the components in the x-direction

ii resolve the components in the y-direction

iii find the magnitude of any unknown forces (marked P and Q) and the size of any unknown angles (marked θ).

a

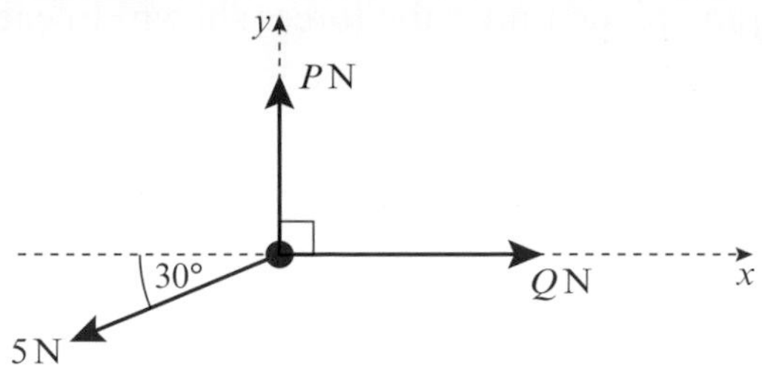

b

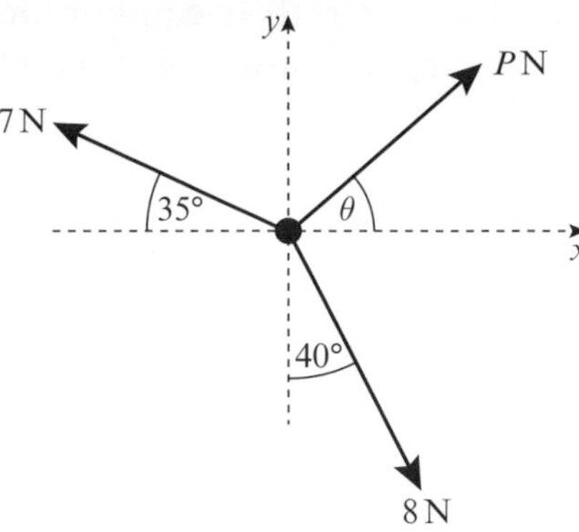

c

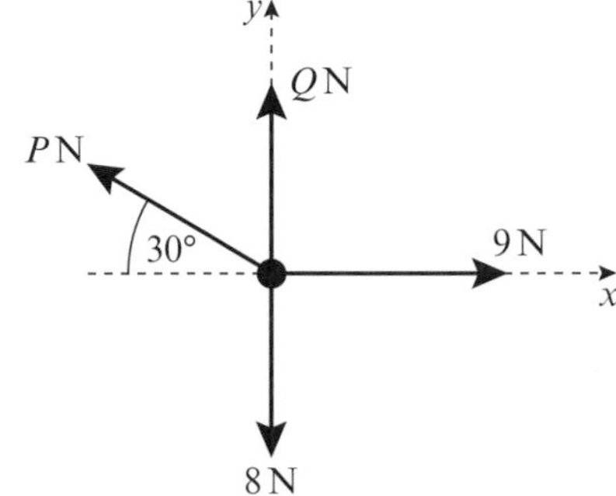

d

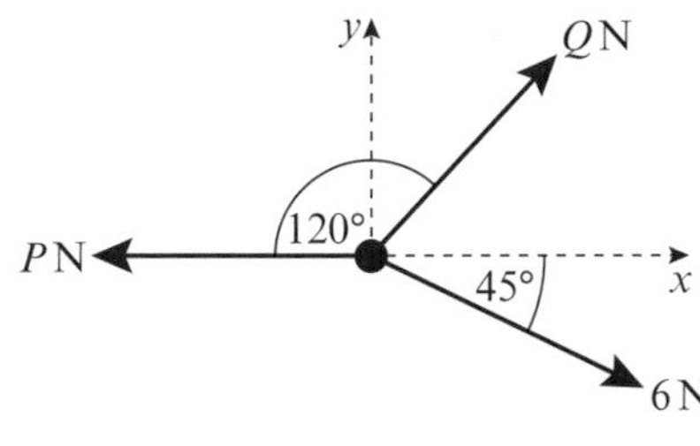

e

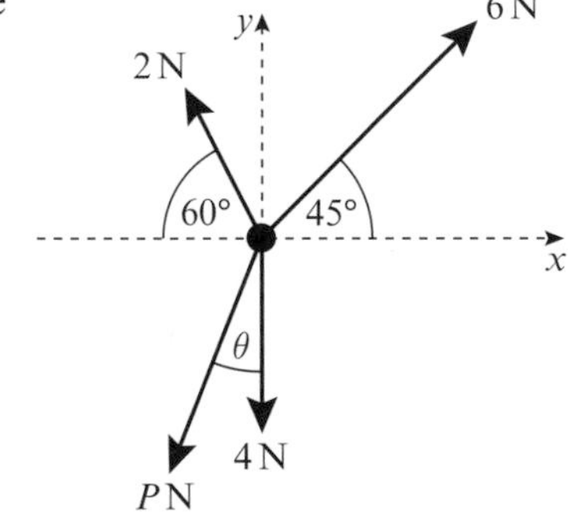

f

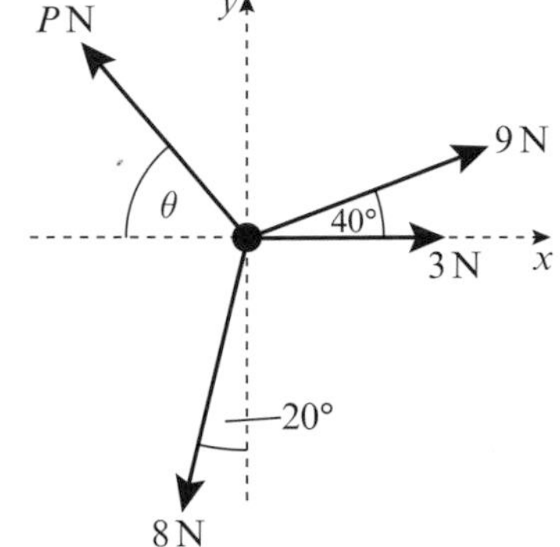

2 For each of the following particles in static equilibrium:

i draw a triangle of forces diagram.

ii Use trigonometry to find the magnitude of any unknown forces (marked P and Q) and the size of any unknown angles (marked θ).

a

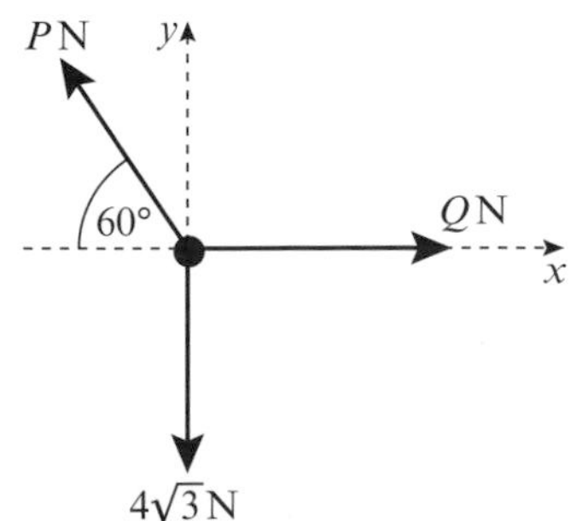

b

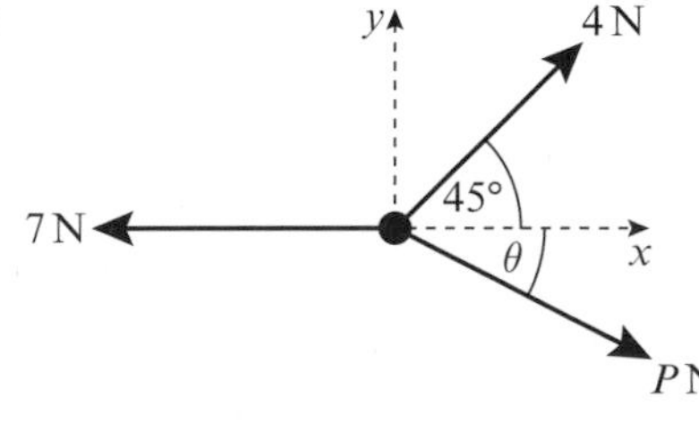

Hint The triangle of forces diagram for part **a** is:

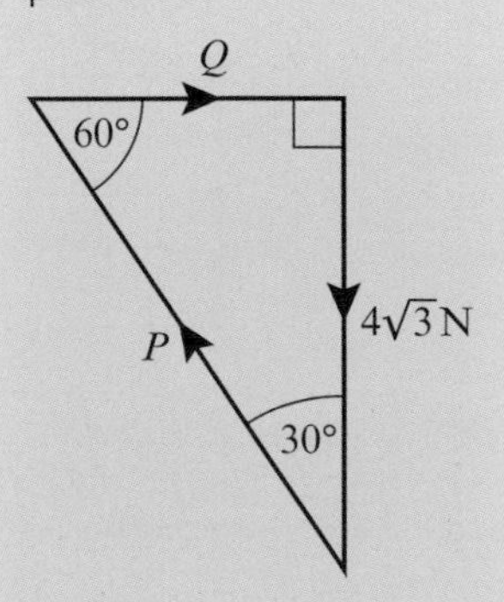

3 Each of these particles rests in equilibrium on a sloping plane under the forces shown. In each case, find the magnitude of forces P and Q.

a

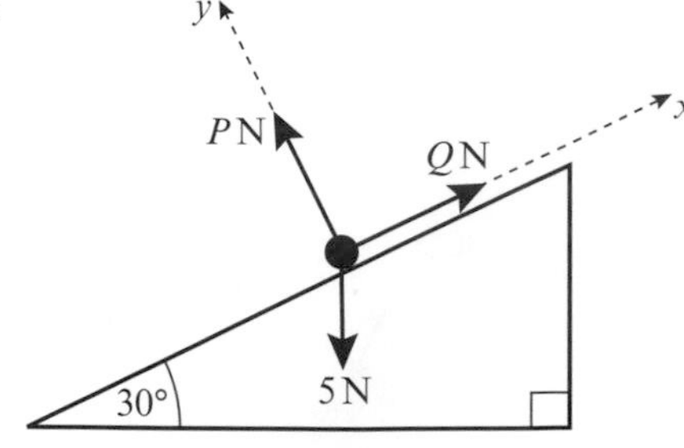

b

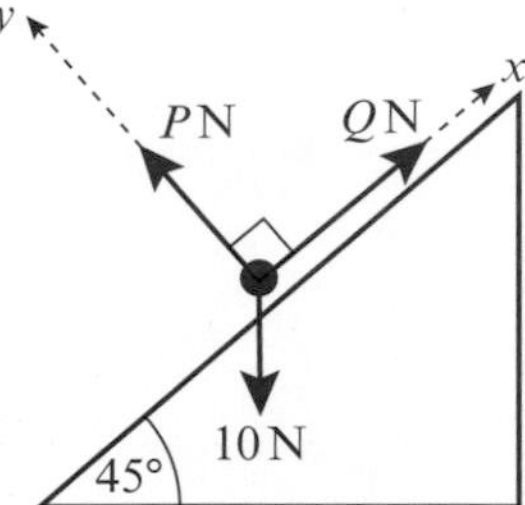

c

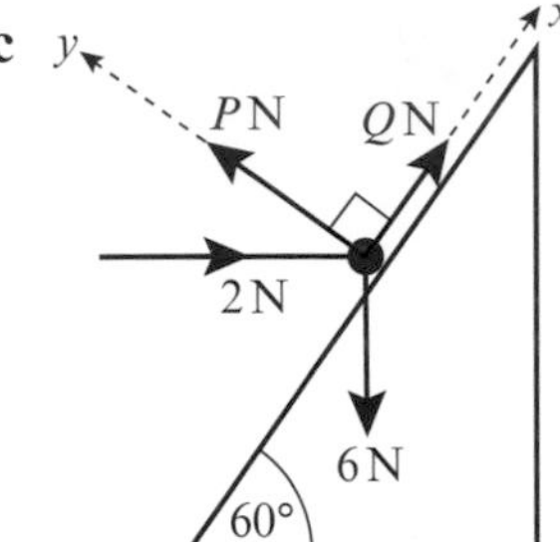

d

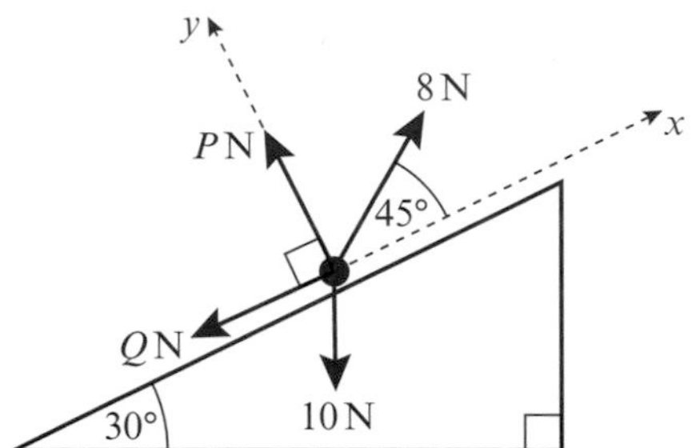

e

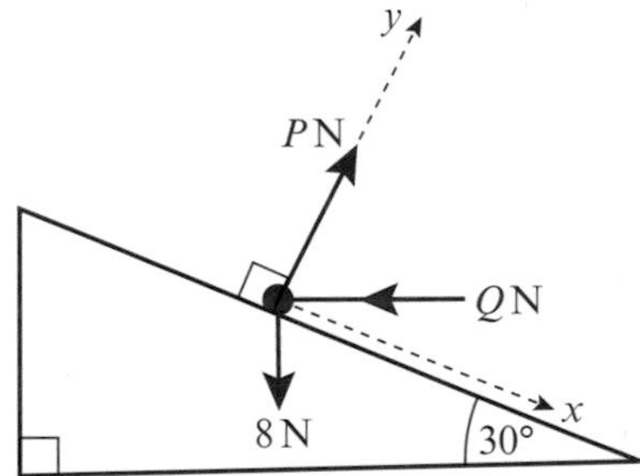

Challenge

The diagram shows three coplanar forces of A, B and C acting on a particle in equilibrium.

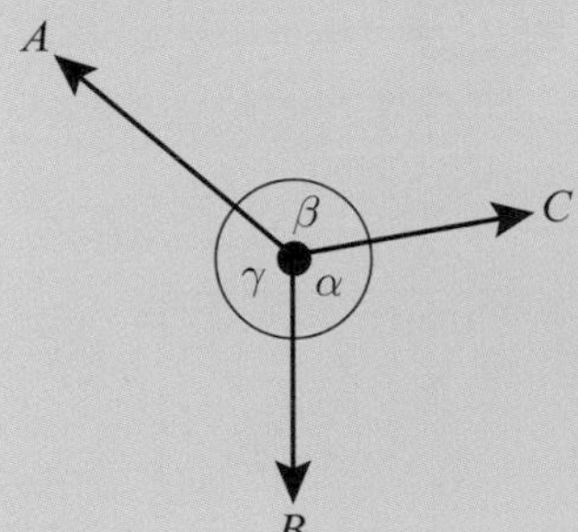

Show that $\dfrac{A}{\sin\alpha} = \dfrac{B}{\sin\beta} = \dfrac{C}{\sin\gamma}$

Notation This result is known as **Lami's Theorem**.

7.2 Modelling with statics

You can use force diagrams to model objects in static equilibrium, and to solve problems involving weight, tension and pulleys.

Example 3

A smooth bead Y is threaded on a light inextensible string. The ends of the string are attached to two fixed points, X and Z, on the same horizontal level. The bead is held in equilibrium by a horizontal force of magnitude 8 N acting parallel to ZX. The bead Y is vertically below X and $\angle XZY = 30°$ as shown in the diagram.

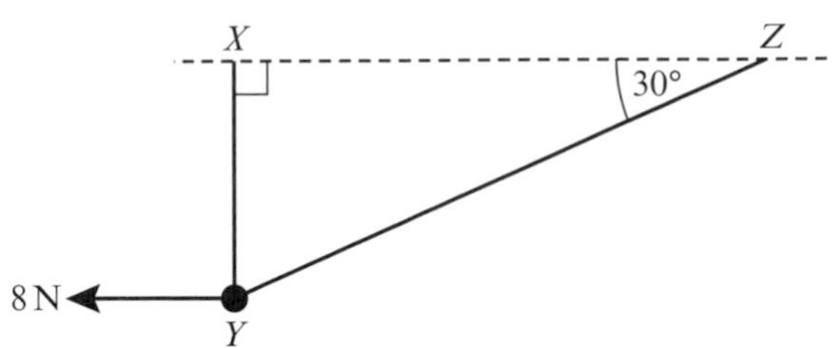

Find the tension in the string and the weight of the bead.

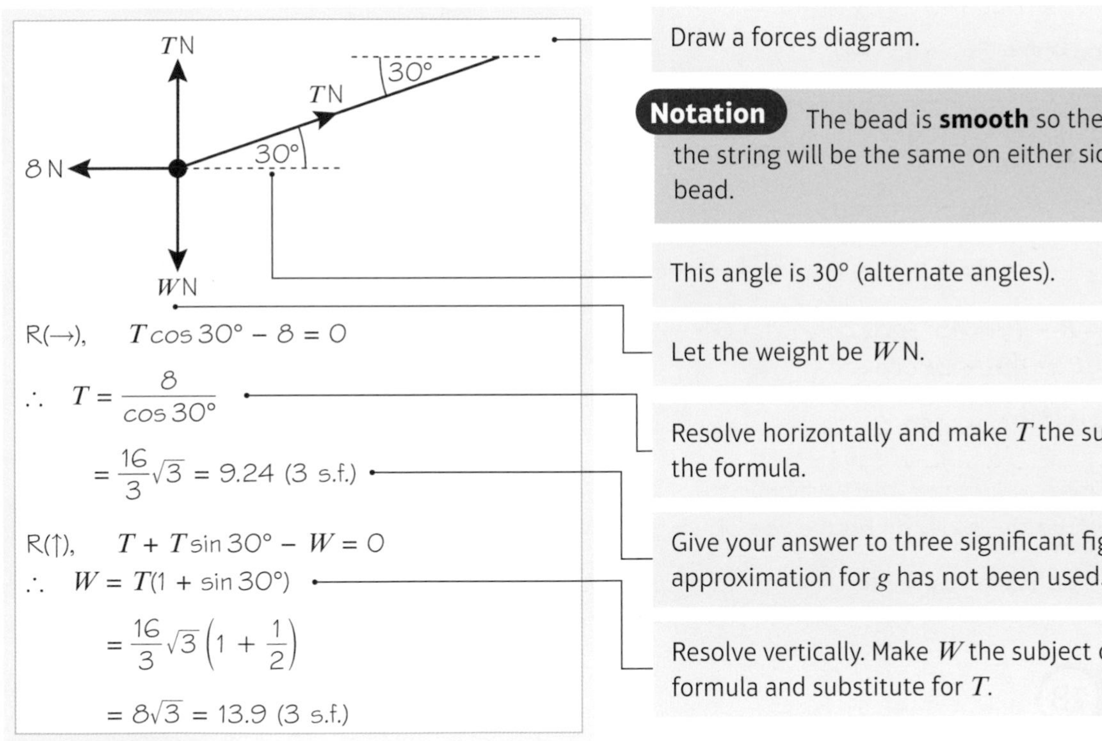

R(→), $T\cos 30° - 8 = 0$

$\therefore \quad T = \dfrac{8}{\cos 30°}$

$= \dfrac{16}{3}\sqrt{3} = 9.24$ (3 s.f.)

R(↑), $T + T\sin 30° - W = 0$

$\therefore \quad W = T(1 + \sin 30°)$

$= \dfrac{16}{3}\sqrt{3}\left(1 + \dfrac{1}{2}\right)$

$= 8\sqrt{3} = 13.9$ (3 s.f.)

Draw a forces diagram.

Notation The bead is **smooth** so the tension in the string will be the same on either side of the bead.

This angle is 30° (alternate angles).

Let the weight be W N.

Resolve horizontally and make T the subject of the formula.

Give your answer to three significant figures as an approximation for g has not been used.

Resolve vertically. Make W the subject of the formula and substitute for T.

Example 4

A mass of 3 kg rests on the surface of a smooth plane which is inclined at an angle of 45° to the horizontal. The mass is attached to a cable which passes up the plane along the line of greatest slope and then passes over a smooth pulley at the top of the plane. The cable carries a mass of 1 kg freely suspended at the other end. The masses are modelled as particles, and the cable as a light inextensible string. There is a force of P N acting horizontally on the 3 kg mass and the system is in equilibrium.

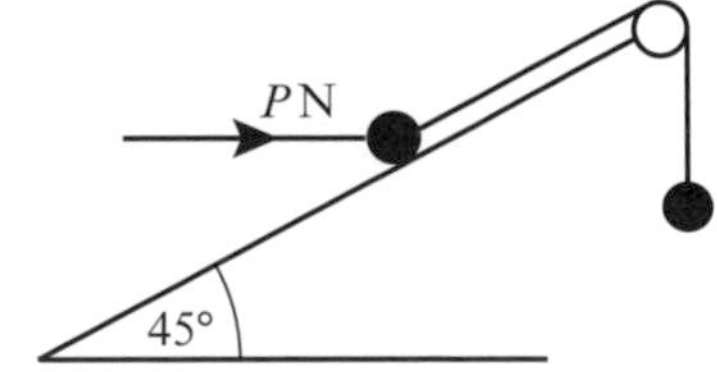

Calculate: **a** the magnitude of P **b** the normal reaction between the mass and the plane.
c State how you have used the assumption that the pulley is smooth in your calculations.

R N, T N, P N, T N, 45°, $3g$ N, $1g$ N

Draw a diagram showing the forces acting on each particle. The tension, T N, will be the same throughout the string. The normal reaction, R N acts perpendicular to the plane. Show the weights $3g$ N and $1g$ N.

a Consider the 1 kg mass:

R(↑), $T - 1g = 0$ — Resolve vertically to obtain T.

$\therefore\ T = g = 9.8$

Consider the 3 kg mass: — Resolve up the plane.

R(↗), $T + P\cos 45° - 3g\sin 45° = 0$ — R has no component in this direction as R is perpendicular to the plane.

$\therefore\ P\cos 45° = 3g\sin 45° - T$

But $T = g$ — Substitute the value for T you found earlier.

$\therefore\ P\cos 45° = 3g\sin 45° - g$ — Divide this equation by cos 45° and use the fact that $\frac{\sin 45°}{\cos 45°} = \tan 45° = 1$.

$$P = 3g - \frac{g}{\cos 45°}$$

$$= 3g - g\sqrt{2} = 16 \text{ (2 s.f.)}$$ — Use the result that $\cos 45° = \sin 45° = \frac{1}{\sqrt{2}}$

b R(↖), $R - P\sin 45° - 3g\cos 45° = 0$ — Resolve perpendicular to the plane.

$\therefore\ R = P\sin 45° + 3g\cos 45°$

$$= 6g\frac{\sqrt{2}}{2} - g = 32 \text{ (2 s.f.)}$$ — Substitute the value of P which you have found to evaluate R.

c The pulley is smooth so the tension in the string will be the same on both sides of the pulley.

Exercise 7B

1 A picture of mass 5 kg is suspended by two light inextensible strings, each inclined at 45° to the horizontal as shown. By modelling the picture as a particle find the tension in the strings when the system is in equilibrium.

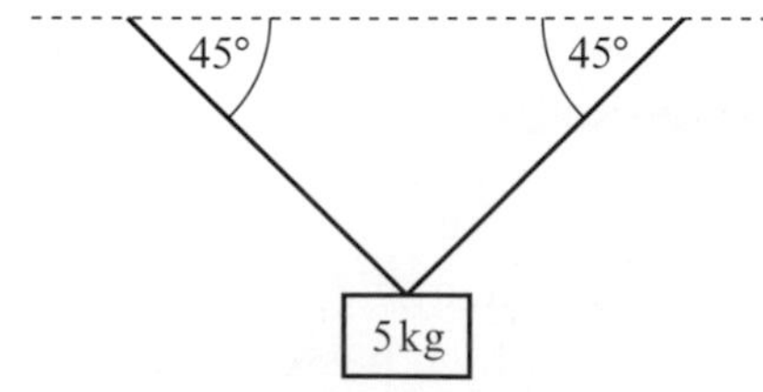

Problem-solving

This is a three-force problem involving an object in static equilibrium, so you could use a triangle of forces.

2 A particle of mass m kg is suspended by a single light inextensible string. The string is inclined at an angle of 30° to the vertical and the other end of the string is attached to a fixed point O. Equilibrium is maintained by a horizontal force of magnitude 10 N which acts on the particle, as shown in the diagram. Find:

a the tension in the string **b** the value of m.

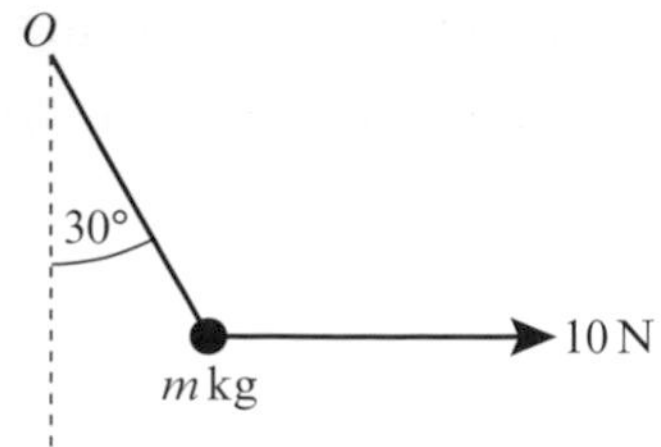

3 A particle of weight 12 N is suspended by a light inextensible string from a fixed point O. A horizontal force of 8 N is applied to the particle and the particle remains in equilibrium with the string at an angle θ to the vertical. Find:

a the angle θ **b** the tension in the string.

4 A particle of mass 6 kg hangs in equilibrium, suspended by two light inextensible strings, inclined at 60° and 45° to the horizontal, as shown. Find the tension in each of the strings.

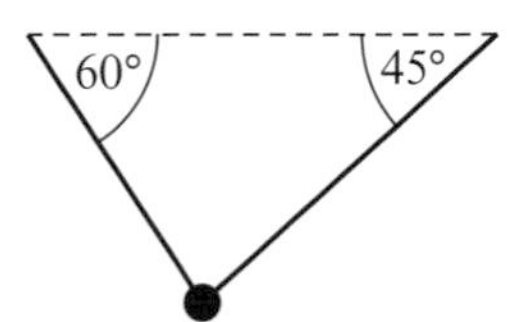

Hint The particle is attached **separately** to each string, so the tension in the two strings can be different.

(E) 5 A smooth bead B is threaded on a light inextensible string. The ends of the string are attached to two fixed points, A and C, on the same horizontal level. The bead is held in equilibrium by a horizontal force of magnitude 2 N acting parallel to CA. The sections of string make angles of 60° and 30° with the horizontal. Find:

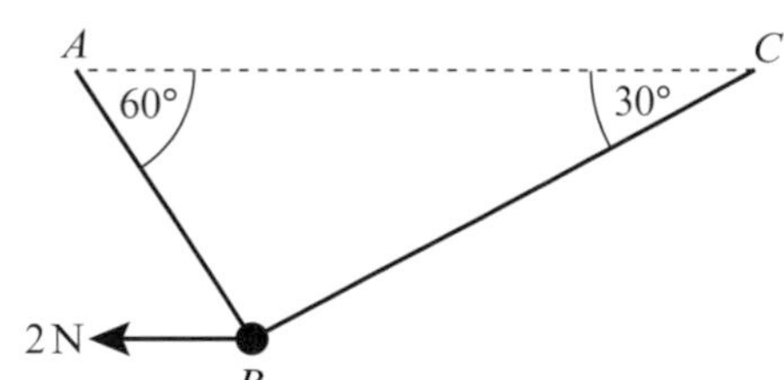

a the tension in the string **(3 marks)**

b the mass of the bead. **(4 marks)**

c State how you have used the modelling assumption that the bead is smooth in your calculations. **(1 mark)**

(E) 6 A smooth bead B is threaded on a light inextensible string. The ends of the string are attached to two fixed points A and C where A is vertically above C. The bead is held in equilibrium by a horizontal force of magnitude 2 N. The sections AB and BC of the string make angles of 30° and 60° with the vertical respectively. Find:

a the tension in the string **(3 marks)**

b the mass of the bead, giving your answer to the nearest gram. **(4 marks)**

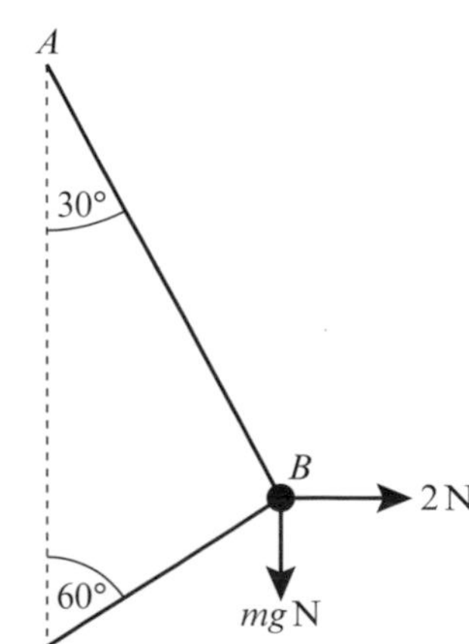

(E) **7** A particle of weight 2 N rests on a smooth horizontal surface and remains in equilibrium under the action of the two external forces shown in the diagram. One is a horizontal force of magnitude 1 N and the other is a force of magnitude P N which acts at an angle θ to the horizontal, where $\tan\theta = \frac{12}{5}$. Find:

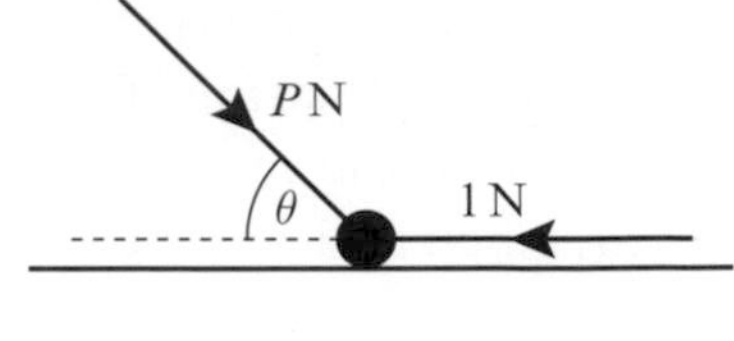

a the value of P **(3 marks)**

b the normal reaction between the particle and the surface. **(2 marks)**

(P) **8** A particle A of mass m kg rests on a smooth horizontal table. The particle is attached by a light inextensible string to another particle B of mass $2m$ kg, which hangs over the edge of the table. The string passes over a smooth pulley, which is fixed at the edge of the table so that the string is horizontal between A and the pulley and then is vertical between the pulley and B. A horizontal force F N applied to A maintains equilibrium. The normal reaction between A and the table is R N.

a Find the values of F and R in terms of m.

The pulley is now raised to a position above the edge of the table so that the string is inclined at 30° to the horizontal between A and the pulley. The string still hangs vertically between the pulley and B. A horizontal force F' N applied to A maintains equilibrium in this new situation. The normal reaction between A and the table is now R' N.

b Find, in terms of m, the values of F' and R'.

9 A particle of mass 2 kg rests on a smooth inclined plane, which makes an angle of 45° with the horizontal. The particle is maintained in equilibrium by a force P N acting up the line of greatest slope of the inclined plane, as shown in the diagram. Find the value of P.

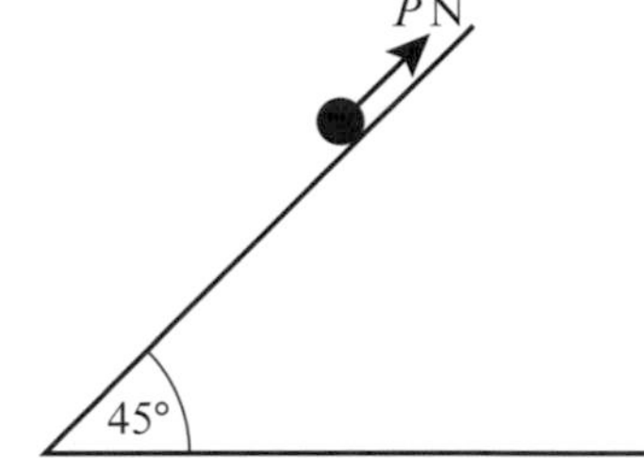

10 A particle of mass 4 kg is held in equilibrium on a smooth plane which is inclined at 45° to the horizontal by a horizontal force of magnitude P N, as shown in the diagram. Find the value of P.

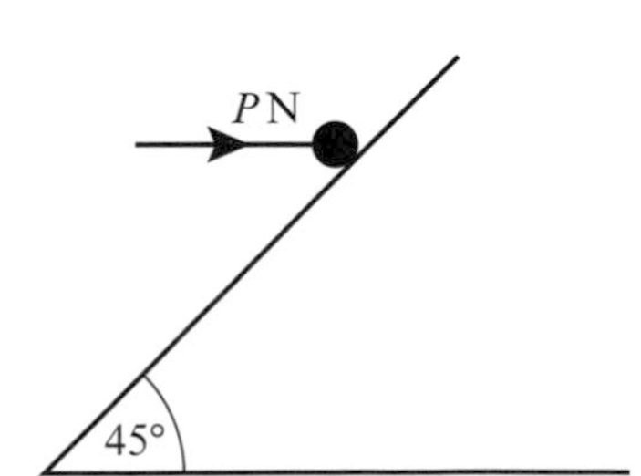

(E/P) **11** A particle A of mass 2 kg rests in equilibrium on a smooth inclined plane. The plane makes an angle θ with the horizontal, where $\tan\theta = \frac{3}{4}$.

The particle is attached to one end of a light inextensible string which passes over a smooth pulley, as shown in the diagram. The other end of the string is attached to a particle B of mass 5 kg. Particle A is also acted upon by a force of magnitude F N down the plane, along a line of greatest slope.

Find:

a the magnitude of the normal reaction between A and the plane **(5 marks)**

b the value of F. **(3 marks)**

c State how you have used the fact that the pulley is smooth in your calculations. **(1 mark)**

E/P **12** A particle of weight 20 N rests in equilibrium on a smooth inclined plane. It is maintained in equilibrium by the application of two external forces as shown in the diagram. One of the forces is a horizontal force of 5 N, the other is a force P N acting at an angle of 30° to the plane, as shown in the diagram. Find the magnitude of the normal reaction between the particle and the plane. **(8 marks)**

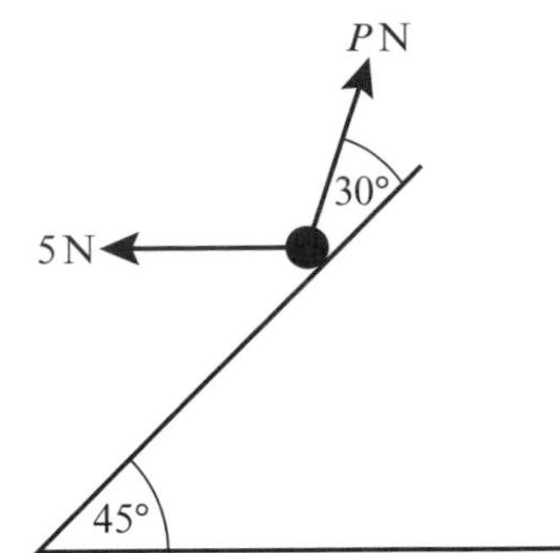

7.3 Friction and static particles

When a body is in static equilibrium under the action of a number of forces, including friction, you need to consider whether the body is on the point of moving or not.

In many cases the force of friction will be less than μR, as a smaller force is sufficient to prevent motion and to maintain static equilibrium. In these situations the equilibrium is not limiting.

- **The maximum value of the frictional force $F_{MAX} = \mu R$ is reached when the body you are considering is on the point of moving. The body is then said to be in limiting equilibrium.**
- **In general, the force of friction F is such that $F \leqslant \mu R$, and the direction of the frictional force is opposite to the direction in which the body would move if the frictional force were absent.**

Example 5

A mass of 8 kg rests on a rough horizontal plane. The mass may be modelled as a particle, and the coefficient of friction between the mass and the plane is 0.5. Find the magnitude of the maximum force P N which acts on this mass without causing it to move if:

a the force P is horizontal

b the force P acts at an angle 60° above the horizontal.

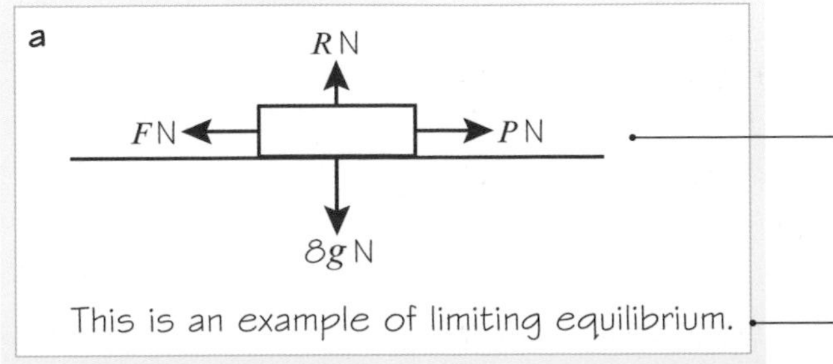

Draw a diagram showing the weight $8g$ N, the normal reaction R N, the force P N and the friction F N. The friction is in the opposite direction to force P N.

The question asks you for the maximum force before movement takes place.

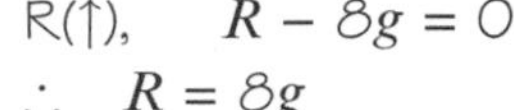

R(↑), $R - 8g = 0$

$\therefore \quad R = 8g$

As friction is limiting, $F = \mu R$

$\therefore \quad F = 0.5 \times 8g$

$= 39.2$

R(→), $P - F = 0$

$\therefore \quad P = F = 39$ (2 s.f.)

b

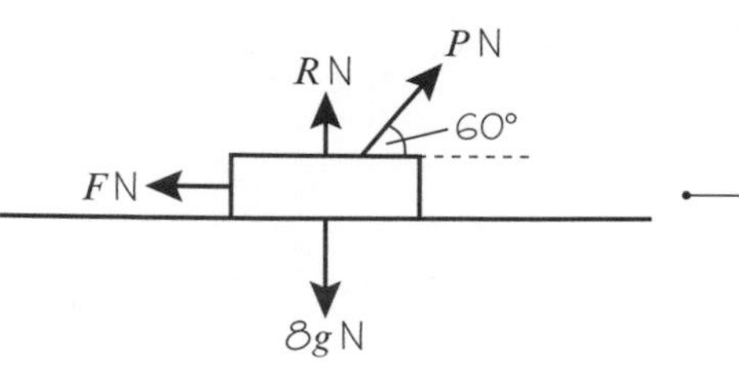

Again this is limiting equilibrium.

R(↑), $R + P\sin 60° - 8g = 0$

$\therefore \quad R = 8g - P\sin 60°$

As friction is limiting, $F = \mu R$

$\therefore \quad F = 0.5\,(8g - P\sin 60°)$

R(→), $P\cos 60° - F = 0$

$\therefore \quad P\cos 60° = 0.5\,(8g - P\sin 60°)$

$\therefore \quad P\cos 60° + 0.5\,P\sin 60° = 0.5 \times 8g$

$\therefore \quad P(\cos 60° + 0.5\sin 60°) = 4g$

$$\therefore \quad P = \frac{4g}{\cos 60° + 0.5\sin 60°}$$

$P = 42$ (2 s.f.)

For an object in limiting equilibrium, $F = F_{MAX}$

Give your answer to two significant figures.

Draw another diagram showing P at 60° above the horizontal.

Express R in terms of P.

Use $F = \mu R$ with $\mu = 0.5$

As $F = P\cos 60°$ eliminate F from the previous equation.

Collect the terms in P and factorise to make P the subject.

Example 6

A box of mass 10 kg rests in limiting equilibrium on a rough plane inclined at 20° above the horizontal.

a Find the coefficient of friction between the box and the plane.

A horizontal force of magnitude P N is applied to the box. Given that the box remains in equilibrium,

b find the maximum possible value of P.

a

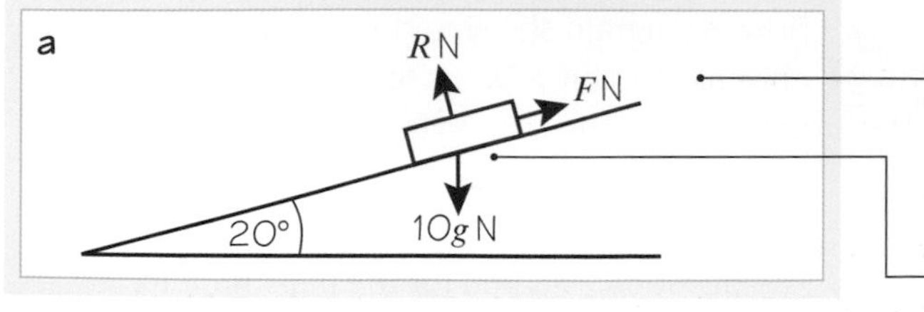

Model the box as a particle and draw a diagram showing the weight, the normal reaction and the force of friction.

The friction acts up the plane, as it acts in an opposite direction to the motion that would take place if there was no friction.

R(↖), $R - 10g\cos 20° = 0$

$\therefore\ R = 92.089...$

R(↗), $F - 10g\sin 20° = 0$

$\therefore\ F = 33.517...$

As the friction is limiting

$F = \mu R$

$\therefore\ 33.517... = \mu \times 92.089...$

$\therefore\ \mu = \dfrac{33.517...}{92.089...} = 0.36$ (to 2 s.f.)

Resolve perpendicular and parallel to the plane.

Online Use the **STO** function to store exact values on your calculator.

Find R and F, then use $F = \mu R$ to find μ.

Give your answer to two significant figures and note that $\mu = \tan 20°$.

b

R N

P N

$0.36R$

$10g$ N

20°

Watch out For the maximum possible value of P, the box will be on the point of moving up the slope, so the friction will act down the slope.

When P is at its maximum value,

$F = \mu R = 0.36R$

R(↗), $P\cos 20° - 0.36R - 10g\sin 20° = 0$ (1)

R(↖), $R - 10g\cos 20° - P\sin 20° = 0$ (2)

From (1): $R = \dfrac{P\cos 20° - 10g\sin 20°}{0.36}$

From (2) $R = P\sin 20° + 10g\cos 20°$

So $\dfrac{P\cos 20° - 10g\sin 20°}{0.36}$

$= P\sin 20° + 10g\cos 20°$

Eliminate R to find P.

$P = \dfrac{3.6g\cos 20° + 10g\sin 20°}{\cos 20° - 0.36\sin 20°}$

$= 82$ N (2 s.f.)

You have used $g = 9.8\ \text{m s}^{-2}$ in your calculations, so round your final answer to 2 significant figures.

Exercise 7C

1 A book of mass 2 kg rests on a rough horizontal table. When a force of magnitude 8 N acts on the book, at an angle of 20° to the horizontal in an upward direction, the book is on the point of slipping.

Calculate, to three significant figures, the value of the coefficient of friction between the book and the table.

Hint 'On the point of slipping' means that the book is in limiting equilibrium.

2 A block of mass 4 kg rests on a rough horizontal table. When a force of 6 N acts on the block, at an angle of 30° to the horizontal in a downward direction, the block is on the point of slipping. Find the value of the coefficient of friction between the block and the table.

3 A block of weight 10 N is at rest on a rough horizontal surface. A force of magnitude 3 N is applied to the block at an angle of 60° above the horizontal in an upward direction. The coefficient of friction between the block and the surface is 0.3.

a Calculate the force of friction. **b** Determine whether the friction is limiting.

(P) 4 A packing crate of mass 10 kg rests on rough horizontal ground. It is filled with books which are evenly distributed through the crate. The coefficient of friction between the crate and the ground is 0.3.

a Find the mass of the books if the crate is in limiting equilibrium under the effect of a horizontal force of magnitude 147 N.

b State what modelling assumptions you have made.

5 A block of mass 2 kg rests on a rough horizontal plane. A force P acts on the block at an angle of 45° to the horizontal. The equilibrium is limiting, with $\mu = 0.3$.

Find the magnitude of P if:

a P acts in a downward direction **b** P acts in an upward direction.

6 A particle of mass 0.3 kg is on a rough plane which is inclined at an angle 30° to the horizontal. The particle is held at rest on the plane by a force of magnitude 3 N acting up the plane, in a direction parallel to a line of greatest slope of the plane. The particle is on the point of slipping up the plane. Find the coefficient of friction between the particle and the plane.

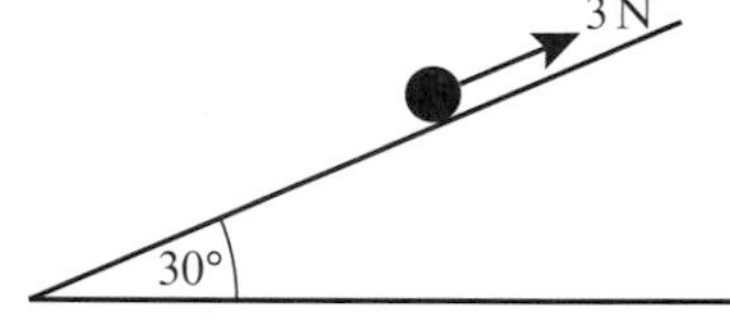

7 A particle of mass 1.5 kg rests in equilibrium on a rough plane under the action of a force of magnitude X N acting up a line of greatest slope of the plane. The plane is inclined at 25° to the horizontal. The particle is in limiting equilibrium and on the point of moving up the plane. The coefficient of friction between the particle and the plane is 0.25. Calculate:

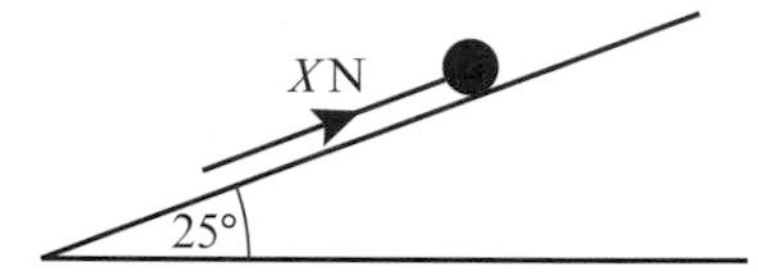

a the normal reaction of the plane on the particle **b** the value of X.

8 A horizontal force of magnitude 20 N acts on a block of mass 1.5 kg, which is in equilibrium resting on a rough plane inclined at 30° to the horizontal. The line of action of the force is in the same vertical plane as the line of greatest slope of the inclined plane.

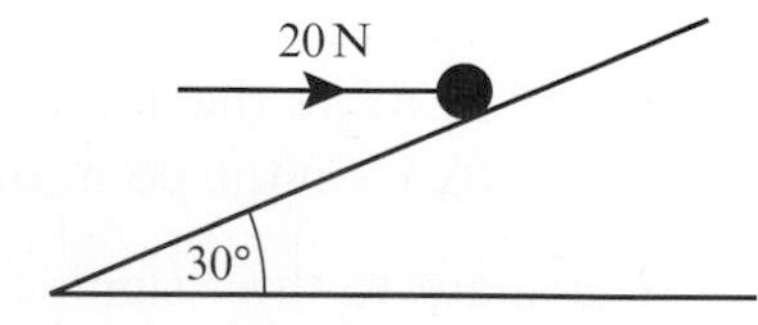

a Find the normal reaction between the block and the plane. **(4 marks)**

b Find the magnitude and direction of the frictional force acting on the block. **(3 marks)**

c Hence find the minimum value of the coefficient of friction between the block and the plane. **(2 marks)**

(E) **9** A box of mass 3 kg lies on a rough plane inclined at 40° to the horizontal. The box is held in equilibrium by means of a horizontal force of magnitude X N. The line of action of the force is in the same vertical plane as the line of greatest slope of the inclined plane. The coefficient of friction between the box and the plane is 0.3 and the box is in limiting equilibrium and is about to move up the plane.

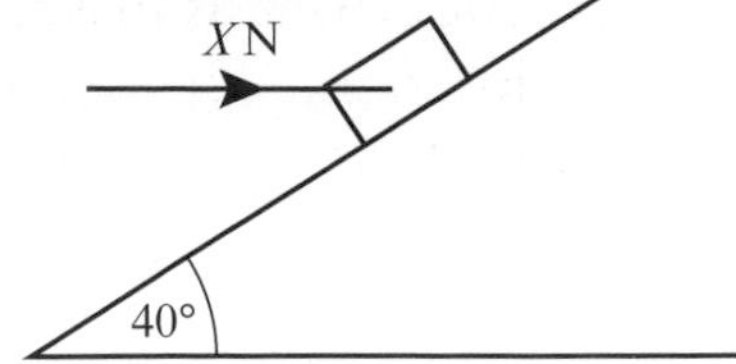

a Find X. **(6 marks)**

b Find the normal reaction between the box and the plane. **(2 marks)**

10 A small child, sitting on a sledge, rests in equilibrium on an inclined slope. The sledge is held by a rope which lies along the slope and is under tension. The sledge is on the point of slipping down the plane. Modelling the child and sledge as a particle and the rope as a light inextensible string, calculate the tension in the rope, given that the mass of the child and sledge is 22 kg, the coefficient of friction is 0.125 and that the slope is a plane inclined at 35° to the horizontal.

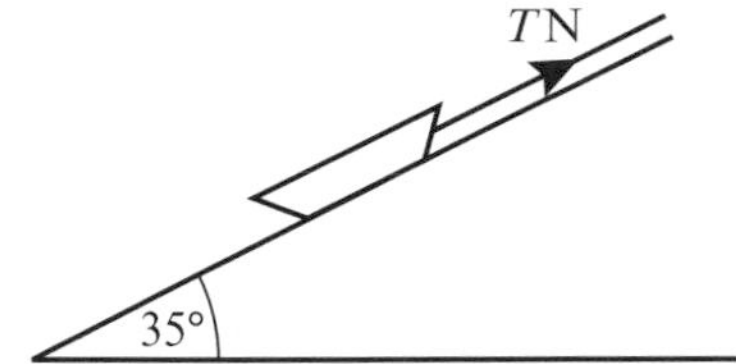

(E/P) **11** A box of mass 0.5 kg is placed on a plane which is inclined at an angle of 40° to the horizontal. The coefficient of friction between the box and the plane is $\frac{1}{5}$. The box is kept in equilibrium by a light string which lies in a vertical plane containing a line of greatest slope of the plane. The string makes an angle of 20° with the plane, as shown in the diagram. The box is in limiting equilibrium and may be modelled as a particle. The tension in the string is T N.

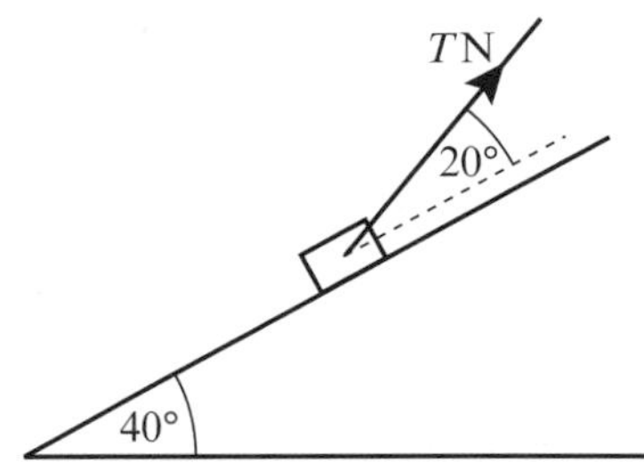

Find the range of possible values of T. **(8 marks)**

Problem-solving

The box might be about to move up or down the slope.

(E/P) **12** A box of mass 1 kg is placed on a plane, which is inclined at an angle of 40° to the horizontal. The box is kept in equilibrium on the point of moving up the plane by a light string, which lies in a vertical plane containing a line of greatest slope of the plane. The string makes an angle of 20° with the plane, as shown in the diagram. The box is in limiting equilibrium and may be modelled as a particle. The tension in the string is 10 N and the coefficient of friction between the box and the plane is μ. Find μ.

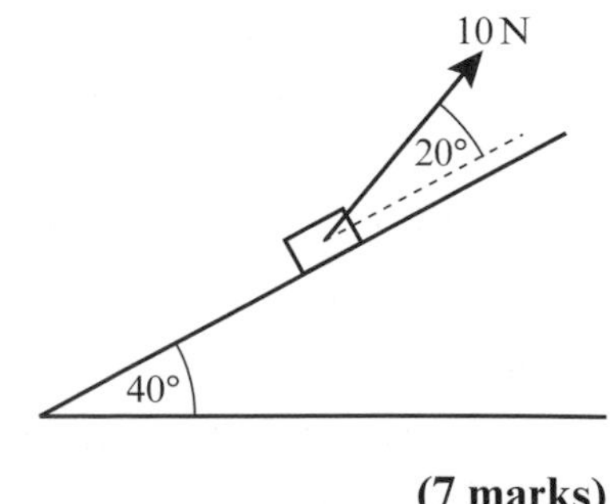

(7 marks)

E/P **13** A particle of mass 2 kg rests in limiting equilibrium on a rough plane angled at θ above the horizontal where $\tan\theta = \frac{3}{4}$. A horizontal force of magnitude P N acting into the plane is applied to the box. Given that the box remains in equilibrium, find the maximum possible value of P. **(8 marks)**

> **Problem-solving**
> First find the coefficient of friction between the box and the plane.

7.4 Static rigid bodies

If you need to consider the rotational forces acting on an object you can model it as a **rigid body**.

- **For a rigid body in static equilibrium:**
  - **the body is stationary**
  - **the resultant force in any direction is zero**
  - **the resultant moment is zero**

> **Links** The moment of a force of magnitude F N about a point P is Fd, where d is the **perpendicular distance** from the line of action of the force to P. ← Section 4.1

You will sometimes need to consider the moments acting on the body and the resultant force acting on the body separately.

Example 7

A uniform rod AB of mass 40 kg and length 10 m rests with the end A on rough horizontal ground. The rod rests against a smooth peg C where $AC = 8$ m. The rod is in limiting equilibrium at an angle of 15° to the horizontal. Find:

a the magnitude of the reaction at C

b the coefficient of friction between the rod and the ground.

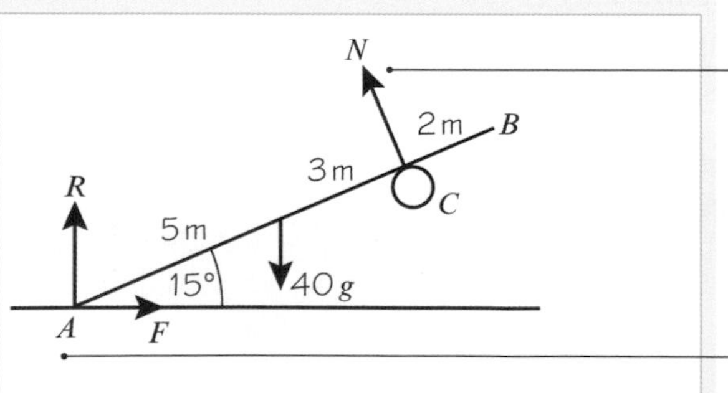

> Start with a diagram showing all the forces.

> N, the reaction at C, is perpendicular to the rod. The peg is smooth, so there is no friction here.

> At A there is a normal reaction and a frictional force. The peg is smooth so the only force stopping the rod from sliding is the frictional force at A.

a Taking moments about A:

$40g \times 5\cos 15° = N \times 8$

$$N = \frac{200g\cos 15°}{8}$$

$= 25g\cos 15°$

$= 236.65\ldots$ N

The reaction at C has magnitude 240 N (2 s.f.).

> **Problem-solving**
> Solve part **a** by taking moments. If you take moments about A you can ignore the frictional force. For part **b** you can resolve forces horizontally and vertically for the whole body.

b R(→), $F = N\cos 75° = 61.25$ N

R(↑), $R + N\cos 15° = 40g$

$R = 40g - N\cos 15° = 163.41\ldots$ N

The rod is in limiting equilibrium, so

$F = \mu R$, $\mu = \dfrac{F}{R} = 0.37$ (2 s.f.)

Resolve horizontally and vertically.

Example 8

A ladder AB, of mass m and length $3a$, has one end A resting on rough horizontal ground. The other end B rests against a smooth vertical wall. A load of mass $2m$ is fixed on the ladder at the point C, where $AC = a$. The ladder is modelled as a uniform rod in a vertical plane perpendicular to the wall and the load is modelled as a particle. The ladder rests in limiting equilibrium at an angle of 60° with the ground.

a Find the coefficient of friction between the ladder and the ground.

b State how you have used the assumption that the ladder is uniform in your calculations.

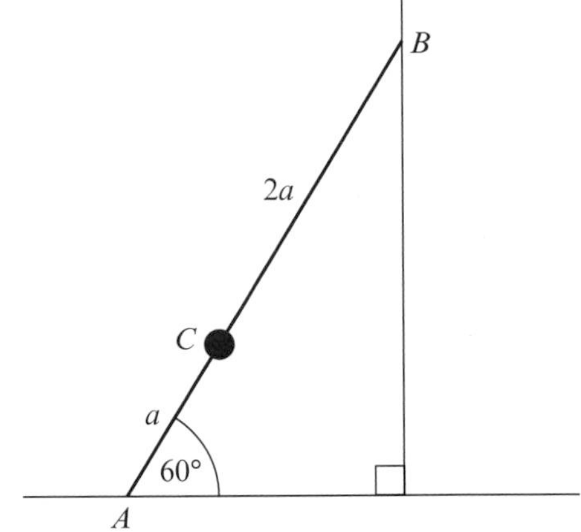

a

P
B
1.5a
0.5a
R
C
mg
a
2 mg
60°
A
F

R(→), $F = P$

R(↑), $R = 2mg + mg = 3mg$

Taking moments about B:

$2mg \times 2a\cos 60° + mg \times 1.5a\cos 60°$
$+ F \times 3a\sin 60° = R \times 3a\cos 60°$

$5.5mg\cos 60° + 3F\sin 60°$
$= 3R\cos 60°$

The reaction at B is perpendicular to the wall. The wall is smooth, so there is no friction at B.

Watch out The reactions at the wall and the floor are different and so must be labelled with different letters.

Online Explore the forces in this question in a more detailed diagram using GeoGebra.

Resolve vertically and horizontally for the whole system.

You want to find $\mu = \dfrac{F}{R}$ so take moments at B. Remember to use the **perpendicular distance** from each force to B.

You can divide both sides by a.

$$2.75mg + \frac{3\sqrt{3}}{2}F = 1.5R$$

Since $R = 3mg$, and $F = \mu R$ (as the ladder is in limiting equilibrium)

$$2.75mg + \frac{3\sqrt{3}}{2}\mu \times 3mg = 1.5 \times 3mg$$

Divide through by mg and solve to find μ.
The answer is independent of g so round to 3 s.f.

$$\mu = \frac{4.5 - 2.75}{\left(\frac{9\sqrt{3}}{2}\right)} = \frac{7}{18\sqrt{3}} = 0.225 \text{ (3 s.f.)}$$

b The assumption that the ladder is uniform allows you to assume that its weight acts at its midpoint.

Problem-solving

There are other options for points to take moments about. For example, if you were to take moments about the point where the lines of action of R and P meet you would eliminate R and P from your working and simplify your calculation.

Exercise 7D

Whenever a numerical value of g is required, take $g = 9.8\,\text{m}\,\text{s}^{-2}$.

1 A uniform rod AB of weight 80 N rests with its lower end A on a rough horizontal floor. A string attached to end B keeps the rod in equilibrium. The string is held at 90° to the rod. The tension in the string is T. The coefficient of friction between the rod and the ground is μ. R is the normal reaction at A and F is the frictional force at A. Find the magnitudes of T, R and F, and the least possible value of μ.

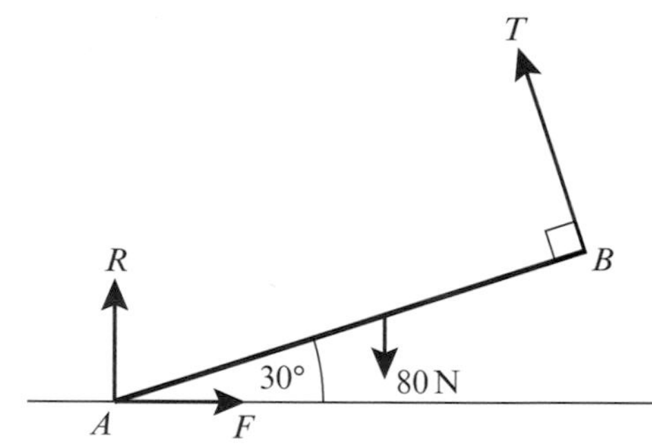

2 A uniform ladder of mass 10 kg and length 5 m rests against a smooth vertical wall with its lower end on rough horizontal ground. The ladder rests in equilibrium at an angle of 65° to the horizontal. Find:

a the magnitude of the normal reaction S at the wall

b the magnitude of the normal reaction R at the ground and the frictional force at the ground

c the least possible value of the coefficient of friction between the ladder and the ground.

d State how you have used the assumption that the ladder is uniform in your calculations.

(P) **3** A uniform ladder AB of mass 20 kg rests with its top A against a smooth vertical wall and its base B on rough horizontal ground. The coefficient of friction between the ladder and the ground is $\frac{3}{4}$. A mass of 10 kg is attached to the ladder. Given that the ladder is about to slip, find the inclination of the ladder to the horizontal,

a if the 10 kg mass is attached at A

b if the 10 kg mass is attached at B.

c State how you have used the assumption that the wall is smooth in your calculations.

E/P **4** A uniform ladder of mass 20 kg and length 8 m rests against a smooth vertical wall with its lower end on rough horizontal ground. The coefficient of friction between the ground and the ladder is 0.3. The ladder is inclined at an angle θ to the horizontal, where $\tan\theta = 2$. A boy of mass 30 kg climbs up the ladder. By modelling the ladder as a uniform rod, the boy as a particle and the wall as smooth and vertical,

a find how far up the ladder the boy can climb before the ladder slips. **(8 marks)**

b Criticise this model with respect to:
i the ladder **ii** the wall. **(2 marks)**

5 A smooth horizontal rail is fixed at a height of 3 m above a rough horizontal surface. A uniform pole AB of weight 4 N and length 6 m is resting with end A on the rough ground and touching the rail at point C. The vertical plane containing the pole is perpendicular to the rail. The distance AC is 4.5 m and the pole is in limiting equilibrium. Calculate:

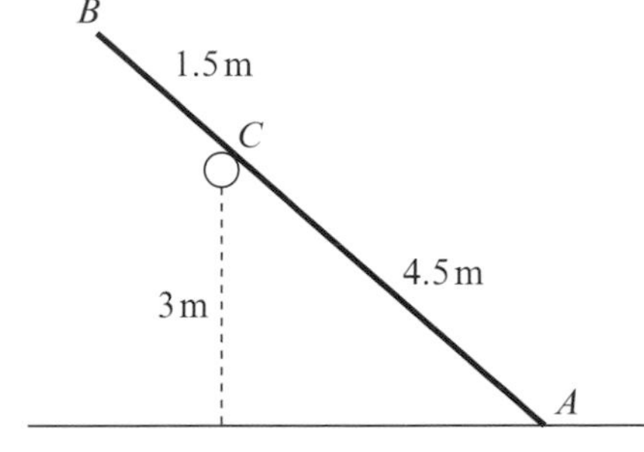

a the magnitude of the force exerted by the rail on the pole

b the coefficient of friction between the pole and the ground.

c State how you have used the assumption that the rail is smooth in your calculations

P **6** A uniform ladder rests in limiting equilibrium with its top against a smooth vertical wall and its base on a rough horizontal floor. The coefficient of friction between the ladder and the floor is μ. Given that the ladder makes an angle θ with the floor, show that $2\mu\tan\theta = 1$.

7 A uniform ladder AB has length 7 m and mass 20 kg. The ladder is resting against a smooth cylindrical drum at P, where AP is 5 m, with end A in contact with rough horizontal ground. The ladder is inclined at 35° to the horizontal.

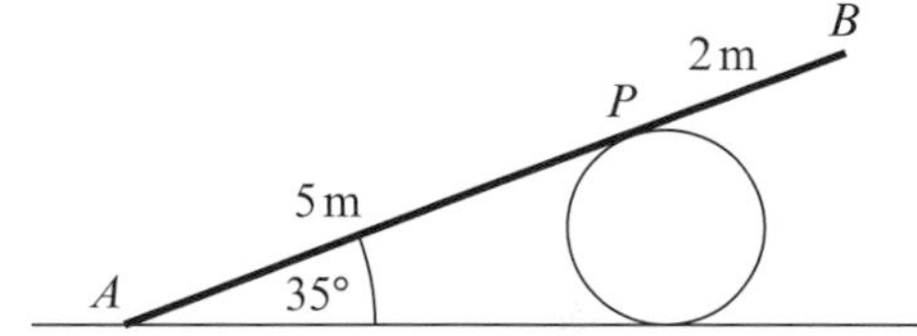

Find the normal and frictional components of the contact force at A, and hence find the least possible value of the coefficient of friction between the ladder and the ground.

E/P **8** A uniform ladder rests in limiting equilibrium with one end on rough horizontal ground and the other end against a rough vertical wall. The coefficient of friction between the ladder and the ground is μ_1 and the coefficient of friction between the ladder and the wall is μ_2. Given that the ladder makes an angle θ with the horizontal, show that $\tan\theta = \frac{1 - \mu_1\mu_2}{2\mu_1}$. **(8 marks)**

E/P **9** A uniform ladder of weight W rests in equilibrium with one end on rough horizontal ground and the other resting against a smooth vertical wall. The vertical plane containing the ladder is at right angles to the wall and the ladder is inclined at 60° to the horizontal. The coefficient of friction between the ladder and the ground is μ.

a Find, in terms of W, the magnitude of the force exerted by the wall on the ladder. **(6 marks)**

b Show that $\mu \geqslant \frac{\sqrt{3}}{6}$. **(3 marks)**

A load of weight w is attached to the ladder at its upper end (resting against the wall).

c Given that $\mu = \frac{\sqrt{3}}{5}$ and that the equilibrium is limiting, find w in terms of W. **(8 marks)**

E/P **10** A uniform rod XY has weight 20 N and length 90 cm. The rod rests on two parallel pegs, with X above Y, in a vertical plane which is perpendicular to the axes of the pegs, as shown in the diagram. The rod makes an angle of 30° to the horizontal and touches the two pegs at P and Q, where $XP = 20$ cm and $XQ = 60$ cm.

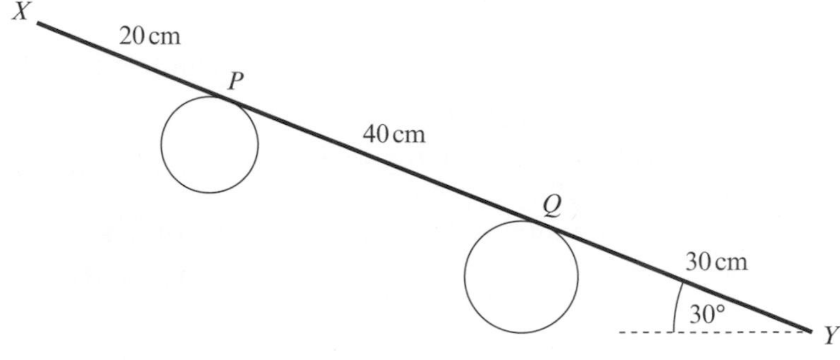

a Calculate the normal components of the forces on the rod at P and at Q. **(8 marks)**

The coefficient of friction between the rod and each peg is μ.

b Given that the rod is about to slip, find μ. **(2 marks)**

E **11** A ladder XY, of length l and weight W, has its end X on rough horizontal ground. The coefficient of friction between the ladder and the ground is $\frac{1}{5}$. The end Y of the ladder is resting against a smooth vertical wall. A window cleaner of weight $9W$ stands at the top of the ladder. To stop the ladder from slipping, the window cleaner's assistant applies a horizontal force of magnitude P to the ladder at X, towards the wall. The force acts in a direction which is perpendicular to the wall. The ladder rests in equilibrium in a vertical plane perpendicular to the wall and makes an angle θ with the horizontal ground, where $\tan \theta = \sqrt{3}$. The window cleaner is modelled as a particle and the ladder is modelled as a uniform rod.

a Find, in terms of W, the reaction of the wall on the ladder at Y. **(5 marks)**

b Find, in terms of W, the range of possible values of P for which the ladder remains in equilibrium. **(5 marks)**

c State how you have used the modelling assumption that the ladder is uniform in your calculations. **(1 mark)**

In practice, the ladder is wider and heavier at the bottom. The model is adjusted so the ladder is modelled as a non-uniform rod with its centre of mass closer to the base.

d State, with a reason, the effect this will have on

i the magnitude of the reaction of the wall on the ladder at Y

ii the range of possible values of P for which the ladder remains in equilibrium. **(4 marks)**

7.5 Dynamics and inclined planes

When a particle is moving along a rough plane, the force of friction is equal to μR, and acts so as to oppose the direction of motion.

Example 9

A particle is held at rest on a rough plane which is inclined to the horizontal at an angle α, where $\tan\alpha = 0.75$. The coefficient of friction between the particle and the plane is 0.5. The particle is released and slides down the plane. Find:

a the acceleration of the particle

b the distance it slides in the first 2 seconds.

a

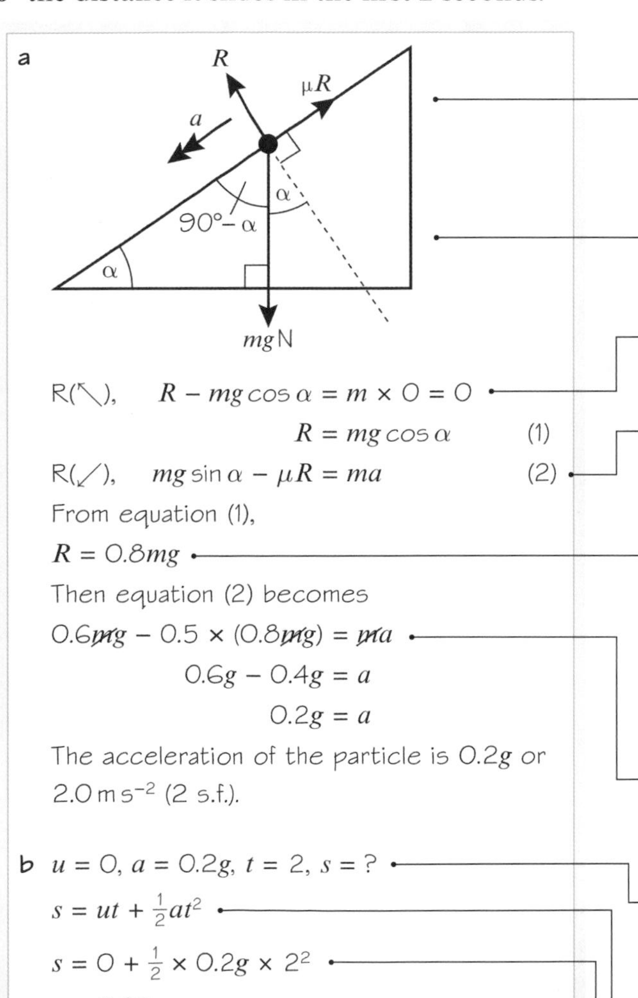

Draw a diagram showing all the forces and the acceleration. Note that you are not given the mass of the particle so call it m.

Since the particle slides down the plane friction will be limiting, so $F = \mu R = 0.5R$.

R(↖), $R - mg\cos\alpha = m \times 0 = 0$

$R = mg\cos\alpha$ (1)

R(↙), $mg\sin\alpha - \mu R = ma$ (2)

Resolve perpendicular to the acceleration.

Resolve in the direction of the acceleration and use $F = ma$.

From equation (1),

$R = 0.8mg$

If $\tan\alpha = 0.75 = \frac{3}{4}$,

$\cos\alpha = \frac{4}{5} = 0.8$

and $\sin\alpha = \frac{3}{5} = 0.6$

(or use your calculator).

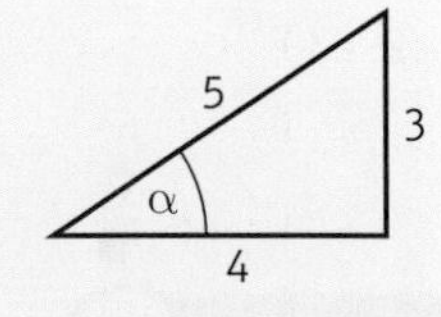

Then equation (2) becomes

$0.6\cancel{m}g - 0.5 \times (0.8\cancel{m}g) = \cancel{m}a$

$0.6g - 0.4g = a$

$0.2g = a$

Substitute for $\sin\alpha$, μ and R and divide through by m.

The acceleration of the particle is $0.2g$ or $2.0\,\text{m s}^{-2}$ (2 s.f.).

b $u = 0,\ a = 0.2g,\ t = 2,\ s = ?$

Use the *suvat* equations for motion with constant acceleration. ← **Year 1, Chapter 9**

$s = ut + \frac{1}{2}at^2$

Choose the appropriate formula.

$s = 0 + \frac{1}{2} \times 0.2g \times 2^2$

$= 3.92$

$= 3.9$ (2 s.f.)

Substitute in the values.

The particle slides 3.9 m (2 s.f.) down the plane.

Example 10

A box of mass 2 kg is pushed up a rough plane by a horizontal force of magnitude 25 N. The plane is inclined to the horizontal at an angle of 10°. Given that the coefficient of friction between the box and the plane is 0.3, find the acceleration of the box.

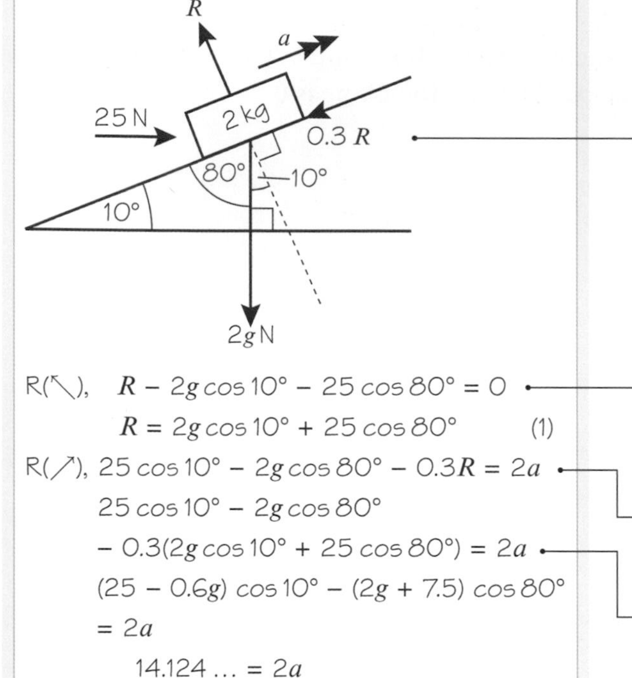

Draw a diagram showing all the forces and the acceleration. The box is moving so $F = F_{MAX}$.

R(↖), $R - 2g\cos 10° - 25\cos 80° = 0$

$R = 2g\cos 10° + 25\cos 80°$ (1)

Resolve perpendicular to the slope. The box is moving up the slope so there is no resultant force in this direction.

R(↗), $25\cos 10° - 2g\cos 80° - 0.3R = 2a$

Resolve up the slope and write an equation of motion for the box.

$25\cos 10° - 2g\cos 80°$

$- 0.3(2g\cos 10° + 25\cos 80°) = 2a$

$(25 - 0.6g)\cos 10° - (2g + 7.5)\cos 80°$

$= 2a$

Substitute for R from equation (1) and simplify.

$14.124\ldots = 2a$

$a = 7.1$ (2 s.f.)

The box accelerates up the plane at $7.1\,\text{m s}^{-2}$ (2 s.f.).

Use unrounded values in your calculations, but round your final answer to 2 s.f.

Exercise 7E

1 A particle of mass 0.5 kg is placed on a smooth inclined plane. Given that the plane makes an angle of 20° with the horizontal, find the acceleration of the particle.

2 The diagram shows a box of mass 2 kg being pushed up a smooth plane by a horizontal force of magnitude 20 N. The plane is inclined to the horizontal at an angle α, where $\tan\alpha = \frac{3}{4}$.

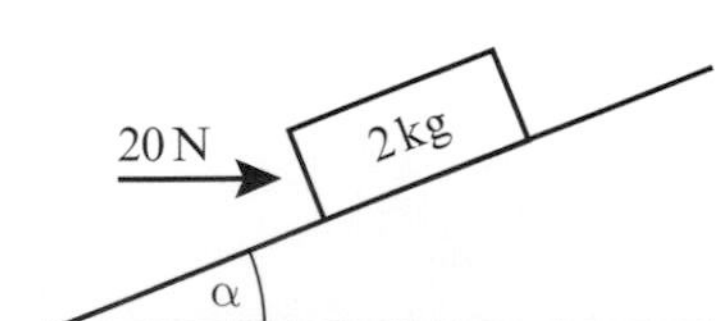

Find:

a the normal reaction between the box and the plane

b the acceleration of the box up the plane.

(P) **3** A boy of mass 40 kg slides from rest down a straight slide of length 5 m. The slide is inclined to the horizontal at an angle of 20°. The coefficient of friction between the boy and the slide is 0.1. By modelling the boy as a particle, find:

a the acceleration of the boy

b the speed of the boy at the bottom of the slide.

(P) **4** A block of mass 20 kg is released from rest at the top of a rough slope. The slope is inclined to the horizontal at an angle of 30°. After 6 s the speed of the block is 21 m s^{-1}. Find the coefficient of friction between the block and the slope.

5 A book of mass 2 kg slides down a rough plane inclined at 20° to the horizontal. The acceleration of the book is 1.5 m s^{-2}. Find the coefficient of friction between the book and the plane.

6 A block of mass 4 kg is pulled up a rough slope, inclined at 25° to the horizontal, by means of a rope. The rope lies along the line of the slope. The tension in the rope is 30 N. Given that the acceleration of the block is 2 m s^{-2} find the coefficient of friction between the block and the plane.

(E/P) **7** A parcel of mass 10 kg is released from rest on a rough plane which is inclined at 25° to the horizontal.

a Find the normal reaction between the parcel and the plane. **(2 marks)**

Two seconds after being released the parcel has moved 4 m down the plane.

b Find the coefficient of friction between the parcel and the plane. **(2 marks)**

(E/P) **8** A particle P is projected up a rough plane which is inclined at an angle a to the horizontal, where $\tan\alpha = \frac{3}{4}$. The coefficient of friction between the particle and the plane is $\frac{1}{3}$. The particle is projected from the point A with speed 20 m s^{-1} and comes to instantaneous rest at the point B.

a Show that while P is moving up the plane its deceleration is $\frac{13g}{15}$. **(5 marks)**

b Find, to two significant figures, the distance AB. **(2 marks)**

c Find, to two significant figures, the time taken for P to move from A to B. **(2 marks)**

d Find the speed of P when it returns to A. **(7 marks)**

(E/P) **9** A particle of mass 2 kg is released from rest on a rough slope that is angled at α to the horizontal where $\tan\alpha = \frac{2}{5}$. After 3 seconds the speed of the particle is 6 m s^{-1}. Work out the coefficient of friction μ. **(8 marks)**

(P) **10** A particle of mass m kg is released from rest on a rough slope that is angled at α to the horizontal. The particle begins to accelerate down the slope. Show that the acceleration of the particle is independent of its mass.

E/P **11** A particle of mass 5 kg is projected up a rough slope at $16\,\text{m s}^{-1}$ and comes to rest at a point P after 5 s. Given that the slope is inclined at 10° to the horizontal,

a work out the coefficient of friction μ. **(7 marks)**

b State, with supporting calculations, whether the particle will remain at rest at P or will begin to slide back down the slope. **(2 marks)**

7.6 Connected particles

You need to be able to solve problems about connected particles on inclined and rough surfaces.

Links Unless connected particles are moving in the same direction they must be considered separately. ← **Year 1, Chapter 10**

Example 11

Two particles P and Q of masses 5 kg and 10 kg respectively are connected by a light inextensible string. The string passes over a small smooth pulley which is fixed at the top of a rough inclined plane. P rests on the inclined plane and Q hangs on the edge of the plane with the string vertical and taut. The plane is inclined to the horizontal at an angle α where $\tan\alpha = 0.75$, as shown in the diagram. The coefficient of friction between P and the plane is 0.2. The system is released from rest.

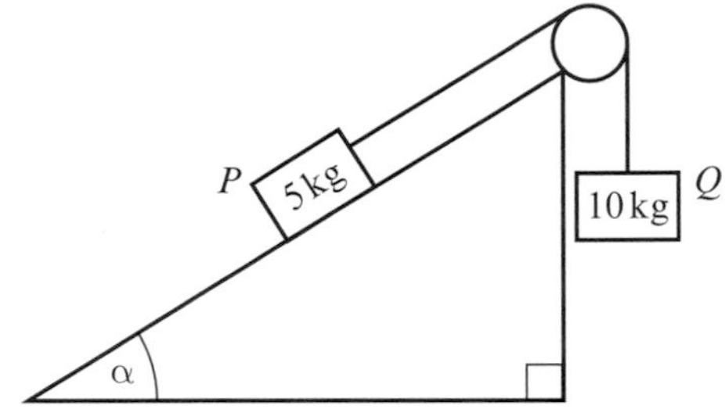

a Find the acceleration of the system. **b** Find the tension in the string.

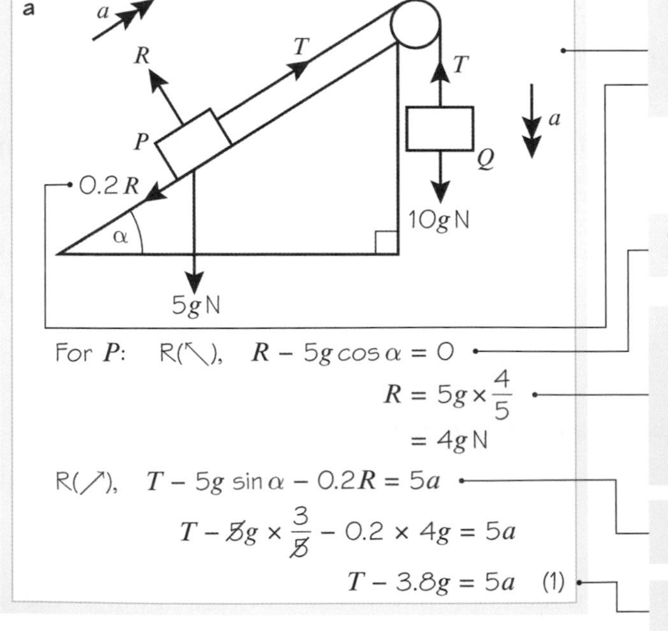

Draw a diagram showing all the forces acting on each particle and their accelerations. Friction is limiting.

For P: $R(\nwarrow)$, $R - 5g\cos\alpha = 0$

$R = 5g \times \frac{4}{5}$

$= 4g\,\text{N}$

There is no acceleration perpendicular to the plane.

If $\tan\alpha = \frac{3}{4}$, $\cos\alpha = \frac{4}{5}$ and $\sin\alpha = \frac{3}{5}$

$R(\nearrow)$, $T - 5g\sin\alpha - 0.2R = 5a$

$T - 5g \times \frac{3}{5} - 0.2 \times 4g = 5a$

$T - 3.8g = 5a \quad (1)$

Resolve in the direction of the acceleration.

Substitute for R and simplify.

For Q: R(↓), $10g - T = 10a$ (2)

$10g - T + T - 3.8g = 10a + 5a$

$6.2g = 15a$

$a = \frac{31g}{75}$ or 4.1 m s^{-2} (2 s.f.)

b $T - 3.8g = 5 \times \frac{31g}{75}$

$T = 3.8g + \frac{31g}{15} = 57\text{ N}$ (2 s.f.)

Resolve in the direction of the acceleration.

Add equations (1) and (2) to eliminate T.

Either of these answers would be acceptable.

Substitute for a in equation (1), using an unrounded value of a.

Example 12

One end of a light inextensible string is attached to a block A of mass 2 kg. The block A is held at rest on a smooth fixed plane which is inclined to the horizontal at an angle of 30°. The string lies along the line of greatest slope of the plane and passes over a smooth light pulley which is fixed at the top of the plane. The other end of the string is attached to a block B of mass 5 kg. The system is released from rest. By modelling the blocks as particles and ignoring air resistance,

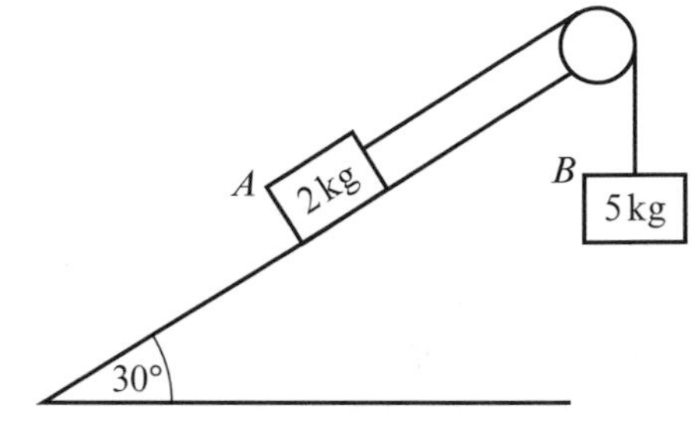

a **i** show that the acceleration of block B is $\frac{4}{7}g$

ii find the tension in the string.

b State how you have used the fact that the string is inextensible in your calculations.

c Calculate the magnitude of the force exerted on the pulley by the string.

a i

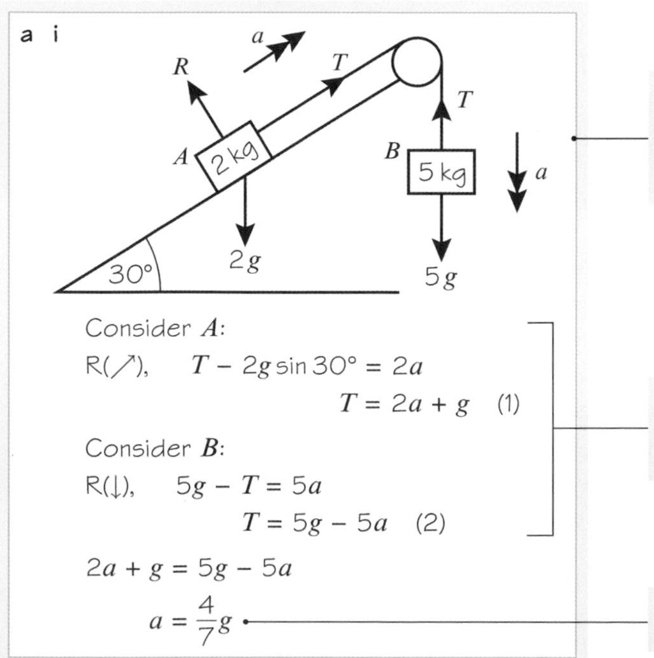

Draw a diagram showing all the forces. The pulley is smooth so the tension will be the same on each side.

Consider A:

R(↗), $T - 2g\sin 30° = 2a$

$T = 2a + g$ (1)

Consider B:

R(↓), $5g - T = 5a$

$T = 5g - 5a$ (2)

Write equations of motion for block A and block B separately.

$2a + g = 5g - 5a$

$a = \frac{4}{7}g$

Eliminate T from equations (1) and (2) to find a.

ii $T = 5g - 5a$

$T = 5g - 5 \times \frac{4}{7}g = \frac{15}{7}g$

Substitute your value of a back into one of the equations to find T.

b The string is inextensible so the acceleration of A and B is the same.

Don't just write that the string does not stretch. You need to state how this fact affects your calculations.

c

$T = \frac{15}{7}g$, $T = \frac{15}{7}g$, θ, θ, R, 30°

$\theta = 30°$

R(↙), $R = 2T\cos\theta$

$R = 2 \times \frac{15}{7}g \times \frac{\sqrt{3}}{2} = \frac{15\sqrt{3}}{7}g$ N

You can leave answers in exact form.

Problem-solving

The force exerted on the pulley by the string is the resultant of the two tensions. The magnitudes are the same so the line of action of the resultant will bisect the lines of actions of the two forces.

Exercise 7F

(P) **1** Two particles P and Q of equal mass are connected by a light inextensible string. The string passes over a small smooth pulley which is fixed at the top of a smooth inclined plane. The plane is inclined to the horizontal at an angle α where $\tan\alpha = 0.75$. Particle P is held at rest on the inclined plane at a distance of 2 m from the pulley and Q hangs freely on the edge of the plane at a distance of 3 m above the ground with the string vertical and taut. Particle P is released. Find the speed with which it hits the pulley.

2 A van of mass 900 kg is towing a trailer of mass 500 kg up a straight road which is inclined to the horizontal at an angle α where $\tan\alpha = 0.75$. The van and the trailer are connected by a light inextensible tow-bar. The engine of the van exerts a driving force of magnitude 12 kN and the van and the trailer experience constant resistances to motion of magnitudes 1600 N and 600 N respectively.

a Find the acceleration of the van.

b Find the tension in the tow-bar.

c Comment on the modelling assumption that the resistances to motion of the van and trailer are constant.

(E/P) **3** Two particles P and Q of mass 2 kg and 3 kg respectively are connected by a light inextensible string. The string passes over a small smooth pulley which is fixed at the top of a rough inclined plane. The plane is inclined to the horizontal at an angle of 30°. Particle P is held at rest on the inclined plane and Q hangs freely with the string vertical and taut. Particle P is released and it accelerates up the plane at 2.5 m s^{-2}. Find:

a the tension in the string **(2 marks)**

b the coefficient of friction between P and the plane **(4 marks)**

c the force exerted by the string on the pulley. **(3 marks)**

E/P **4** Two particles A and B, of mass m kg and 3 kg respectively, are connected by a light inextensible string. The particle A is held resting on a smooth fixed plane inclined at 30° to the horizontal. The string passes over a smooth pulley P fixed at the top of the plane. The portion AP of the string lies along a line of greatest slope of the plane and B hangs freely from the pulley, as shown in the figure. The system is released from rest with B at a height of 0.25 m above horizontal ground. Immediately after release, B descends with an acceleration of $\frac{2}{5}g$. Given that A does not reach P, calculate:

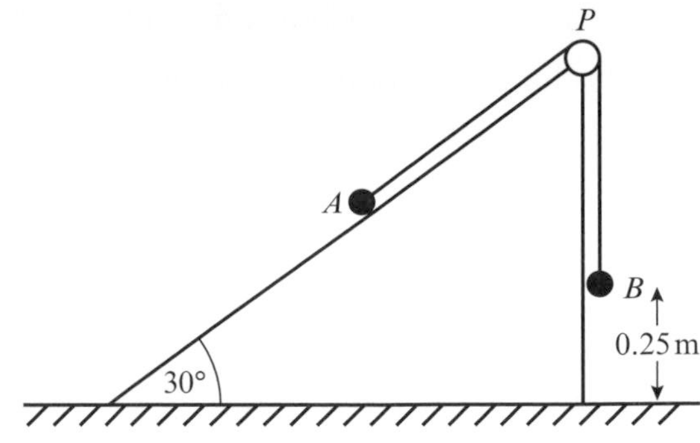

a the tension in the string while B is descending **(2 marks)**

b the value of m. **(4 marks)**

The particle B strikes the ground and does not rebound. Find:

c the time between the instant when B strikes the ground and the instant when A reaches its highest point on the plane. **(6 marks)**

E/P **5** Two particles A and B on back-to-back rough slopes are connected by a light inextensible string that passes over a smooth pulley as shown in the diagram. A has mass 2 kg and B has mass m kg.

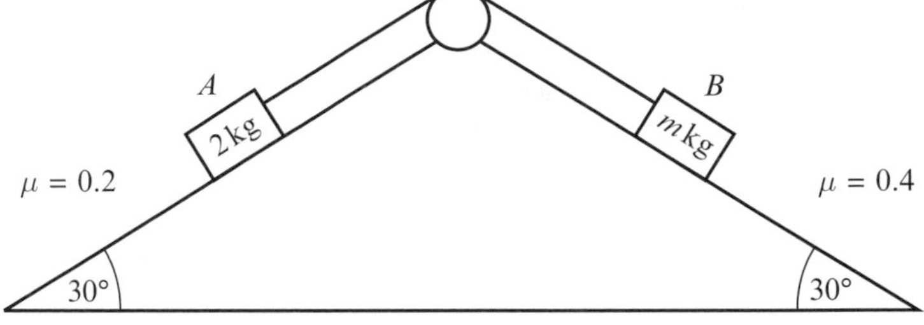

The coefficient of friction between A and the slope is 0.2 and the coefficient of friction between B and the slope is 0.4.

a Show that the maximum value that m can take before the particles begin to move is

$$\frac{10 + 2\sqrt{3}}{5 - 2\sqrt{3}}$$ **(6 marks)**

b Given that $m = 10$, find the acceleration of the particles. **(6 marks)**

E/P **6** A block of metal P of mass 1.5 kg rests on a rough horizontal work bench and is attached to one end of a light inextensible string. The string passes over a small smooth pulley fixed at the edge of the bench. The other end of the string is attached to a box Q of mass 1.6 kg which hangs freely below the pulley, as shown in the diagram. The coefficient of friction between P and the table is μ. The system is released from rest with the string taut. Two seconds after release, Q has velocity $6\,\text{m s}^{-1}$. Modelling P and Q as particles,

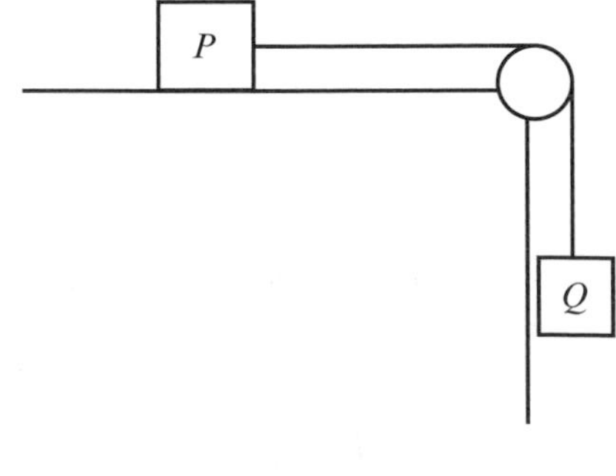

a calculate the acceleration of Q **(3 marks)**

b find the tension in the string **(4 marks)**

c show that μ is 0.434 (3 s.f.). **(5 marks)**

d State how in your calculations you have used the information that the string is inextensible. **(1 mark)**

Challenge

Two particles of mass m_1 and m_2 lie in static equilibrium on a triangular wedge as shown in the diagram. The particles are connected by a light inextensible string that passes over a smooth pulley.

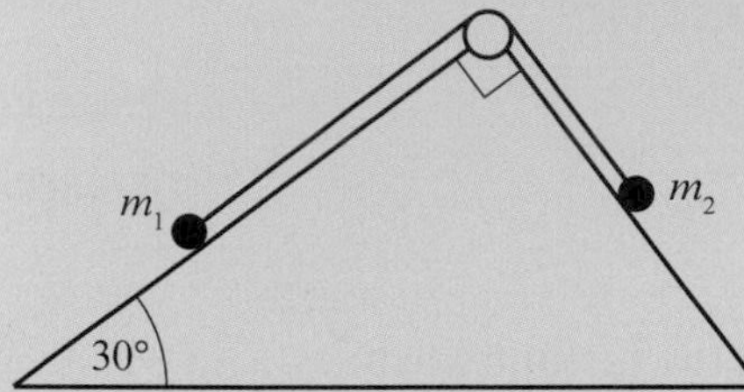

a Given that the wedge is smooth, show that $\frac{m_1}{m_2} = \sqrt{3}$.

b Given instead that the wedge is rough, and that the coefficient of friction between each particle and the wedge is μ, show that

$$\frac{\sqrt{3} - \mu}{1 + \mu\sqrt{3}} \leqslant \frac{m_1}{m_2} \leqslant \frac{\sqrt{3} + \mu}{1 - \mu\sqrt{3}}$$

Mixed exercise 7

1 A particle is acted upon by three forces as shown in the diagram.

Given that the particle is in equilibrium, work out:

a the size of angle θ

b the magnitude of P.

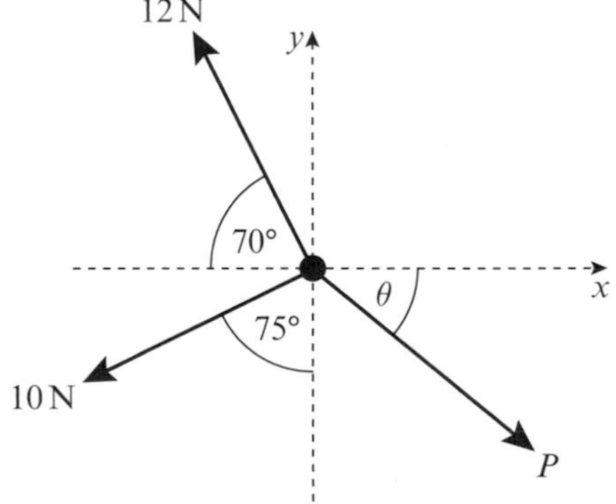

2 A particle is acted upon by three forces as shown in the diagram. Given that it is in equilibrium find:

a the size of angle θ

b the magnitude of W.

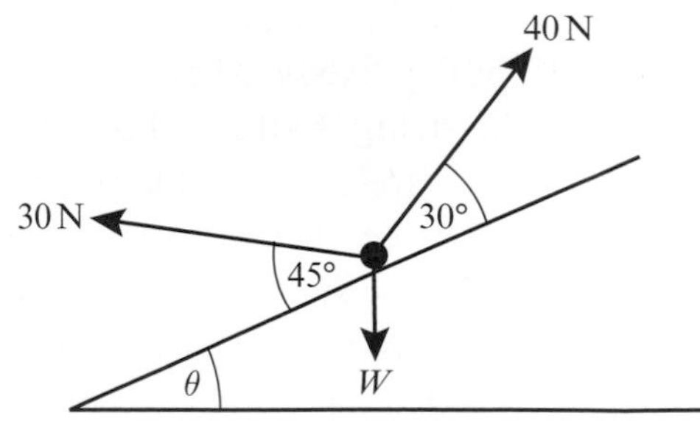

3 An acrobat of mass 55 kg stands on a tightrope. By modelling the acrobat as a particle and the tightrope as two inextensible strings, calculate the tension in the tightrope on each side of the rope.

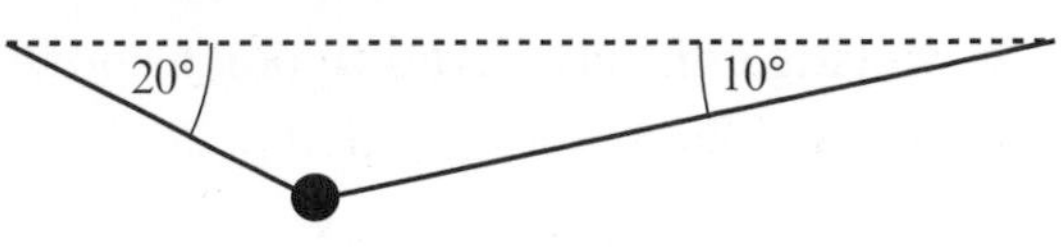

(E/P) 4 A box of mass 5 kg sits on a smooth slope that is angled at 30° to the horizontal. It is attached to a light scale-pan by a light inextensible string which passes over a smooth pulley, as shown in the diagram. The scale-pan carries two masses A and B. The mass of A is 2 kg and the mass of B is 5 kg. Work out the force exerted by A on B. **(8 marks)**

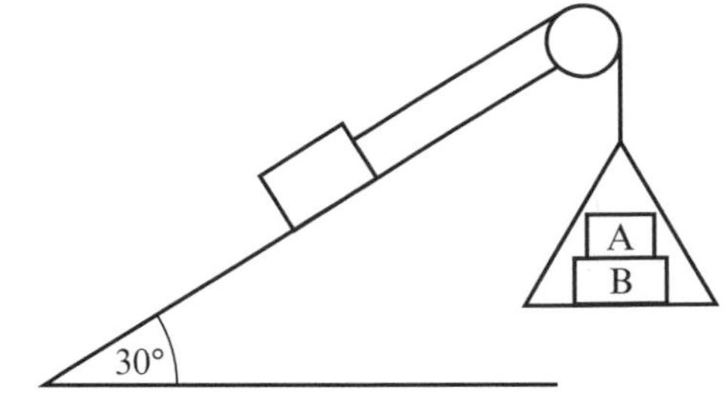

(E/P) 5 A particle Q of mass 5 kg rests in equilibrium on a smooth inclined plane. The plane makes an angle θ with the horizontal, where $\tan\theta = \frac{3}{4}$.

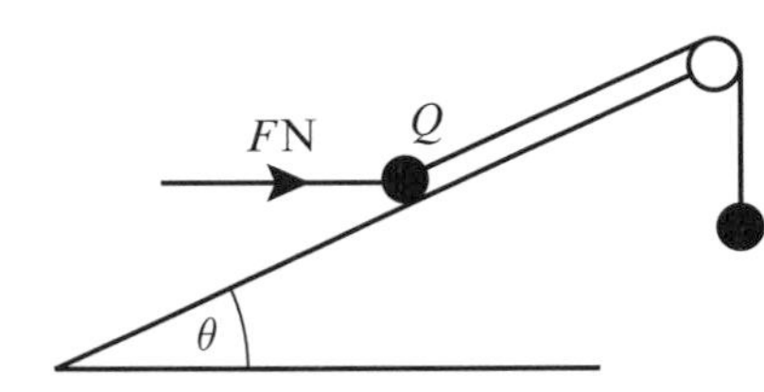

Q is attached to one end of a light inextensible string which passes over a smooth pulley as shown. The other end of the string is attached to a particle of mass 2 kg.

The particle Q is also acted upon by a force of magnitude F N acting horizontally, as shown in the diagram.

Find the magnitude of:

a the force F **(5 marks)**

b the normal reaction between particle Q and the plane. **(3 marks)**

The plane is now assumed to be rough.

c State, with a reason, which of the following statements is true:

1. F will be larger 2. F will be smaller 3. F could be either larger or smaller. **(2 marks)**

(E) 6 A smooth bead B of mass 2 kg is threaded on a light inextensible string. The ends of the string are attached to two fixed points A and C where A is vertically above C. The bead is held in equilibrium by a horizontal force of magnitude P N. The sections AB and BC make angles of 20° and 70° with the vertical as shown.

a Show that the tension in the string is 33 N (2 s.f.). **(3 marks)**

b Calculate the value of P. **(3 marks)**

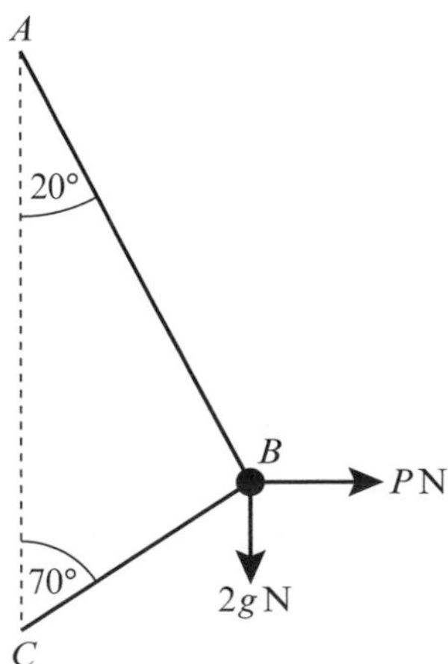

(E) **7** A sledge of mass 50 kg sits on a snowy hill that is angled at 40° to the horizontal. The sledge is held in place by a rope that is angled at 30° above the line of greatest slope of the hill.

a By modelling the sledge as a particle, the hill as a smooth slope and the rope as a light inextensible string, work out the tension in the rope. **(4 marks)**

b Give one criticism of this model. **(1 mark)**

(E/P) **8** A uniform ladder AB has one end A on smooth horizontal ground. The other end B rests against a smooth vertical wall. The ladder is modelled as a uniform rod of mass m and length $5a$. The ladder is kept in equilibrium by a horizontal force F acting at a point C of the ladder where $AC = a$. The force F and the ladder lie in a vertical plane perpendicular to the wall. The ladder is inclined to the horizontal at an angle θ, where $\tan\theta = \frac{9}{5}$, as shown in the diagram.

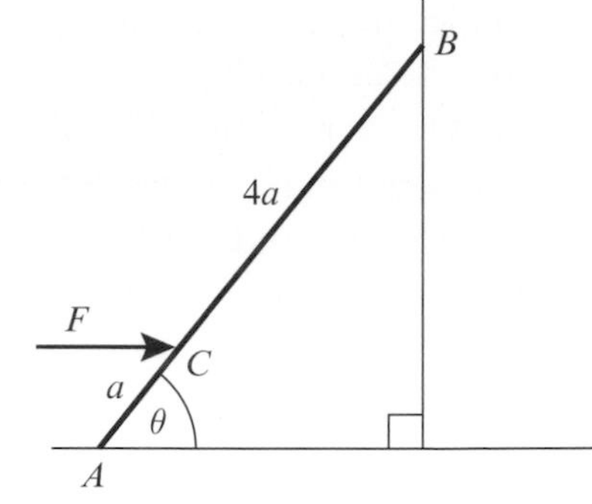

Show that $F = \frac{25mg}{72}$. **(8 marks)**

(E/P) **9** A uniform ladder AB, of mass m and length $2a$, has one end A on rough horizontal ground. The other end B rests against a smooth vertical wall. The ladder is in a vertical plane perpendicular to the wall. The ladder makes an angle α with the vertical, where $\tan\alpha = \frac{3}{4}$. A child of mass $2m$ stands on the ladder at C where $AC = \frac{2}{3}a$, as shown in the diagram. The ladder and the child are in equilibrium.

By modelling the ladder as a rod and the child as a particle, calculate the least possible value of the coefficient of friction between the ladder and the ground.

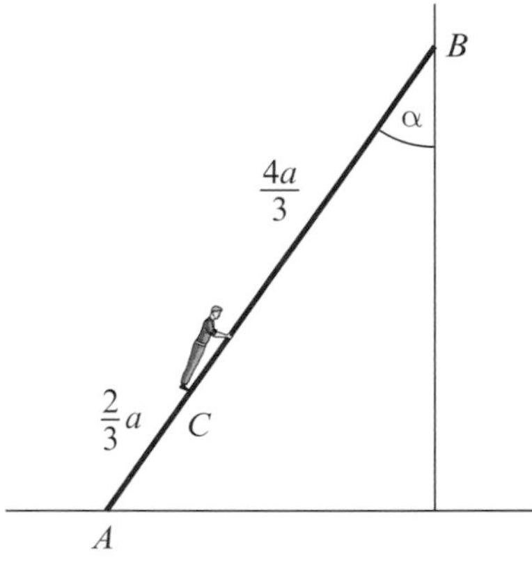

(8 marks)

(E/P) **10** A uniform ladder, of weight W and length $2a$, rests in equilibrium with one end A on a smooth horizontal floor and the other end B against a rough vertical wall. The ladder is in a vertical plane perpendicular to the wall. The coefficient of friction between the wall and the ladder is μ. The ladder makes an angle θ with the floor, where $\tan\theta = \frac{4}{3}$. A horizontal light inextensible string CD is attached to the ladder at the point C, where $AC = \frac{1}{4}a$. The string is attached to the wall at the point D, with BD vertical, as shown in the diagram.

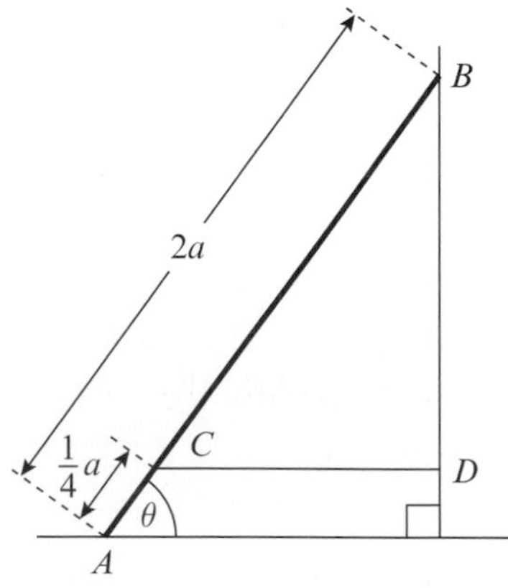

The tension in the string is $\frac{1}{3}W$. By modelling the ladder as a rod,

a find the magnitude of the force of the floor on the ladder **(5 marks)**

b show that $\mu \geqslant \frac{1}{3}$. **(3 marks)**

c State how you have used the modelling assumption that the ladder is a rod. **(1 mark)**

(E/P) **11** A uniform ladder, of weight W and length 5 m, has one end on rough horizontal ground and the other touching a smooth vertical wall. The coefficient of friction between the ladder and the ground is 0.3.

The top of the ladder touches the wall at a point 4 m vertically above the level of the ground.

a Show that the ladder can not rest in equilibrium in this position. **(6 marks)**

In order to enable the ladder to rest in equilibrium in the position described above, a brick is attached to the bottom of the ladder.

Assuming that this brick is at the lowest point of the ladder, but not touching the ground,

b show that the horizontal frictional force exerted by the ladder on the ground is independent of the mass of the brick **(4 marks)**

c find, in terms of W and g, the smallest mass of the brick for which the ladder will rest in equilibrium. **(3 marks)**

(E/P) **12** A non-uniform ladder PQ of mass 20 kg and length 4 metres, rests with P on smooth horizontal ground and Q against a rough vertical wall. The coefficient of friction between the ladder and the wall is 0.2. The centre of mass of the ladder is 1 m from P. The ladder is inclined at an angle α to the horizontal, where $\tan \alpha = \frac{5}{2}$. A horizontal force F applied to the base of the ladder can just prevent it from slipping. By modelling the ladder as a rod determine the value of F. **(10 marks)**

(E/P) **13** A particle of mass 3 kg is released from rest on a rough slope that is angled at α to the horizontal where $\tan \alpha = \frac{3}{4}$. After 1.5 seconds the particle has travelled 6 m. Work out the coefficient of friction μ. **(6 marks)**

(E) **14** A particle of mass 5 kg is pushed up a rough slope, inclined at 30° to the horizontal, by a force of 80 N applied at an angle of 10° slope. Given that the coefficient of friction of the slope is 0.4, find the acceleration of the particle. **(6 marks)**

(E) **15** Two particles, A of mass m_1 kg and B of mass m_2 kg are connected by a light inextensible string. The string passes over a smooth pulley, P. A sits on a rough horizontal table, where the coefficient of friction between A and the table is μ, and B lies directly below P. Given that $m_2 > \mu m_1$, show that the acceleration of the system is $\frac{g(m_2 - \mu m_1)}{m_1 + m_2}$. **(5 marks)**

(E/P) **16** Two particles of masses m_1 and m_2 are connected by a light inextensible string that passes over a smooth pulley. The particles are released from rest on smooth slopes angled at 30° and 45° to the horizontal as shown in the diagram. Given that m_2 is accelerating down the 45° slope at $\frac{1}{2}$ m s^{-2}, show that

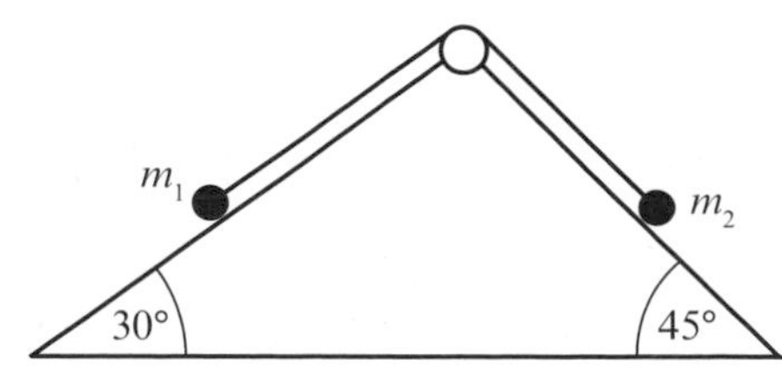

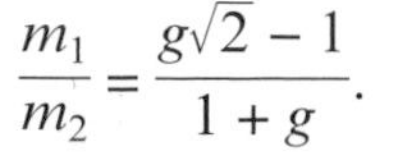
$\frac{m_1}{m_2} = \frac{g\sqrt{2} - 1}{1 + g}$. **(6 marks)**

Challenge

The diagram shows a uniform rod AB of length 3 m and of mass 10 kg. The rod is smoothly hinged at A which lies on a vertical wall. A particle of mass 5 kg is suspended 1 m from B. The rod is kept in a horizontal position by a light inextensible string BC, where C lies on the wall directly above A. The plane ABC is perpendicular to the wall and $\angle ABC$ is 60°.

a Calculate the tension in the string.

b Work out the magnitude and direction of the reaction at the hinge.

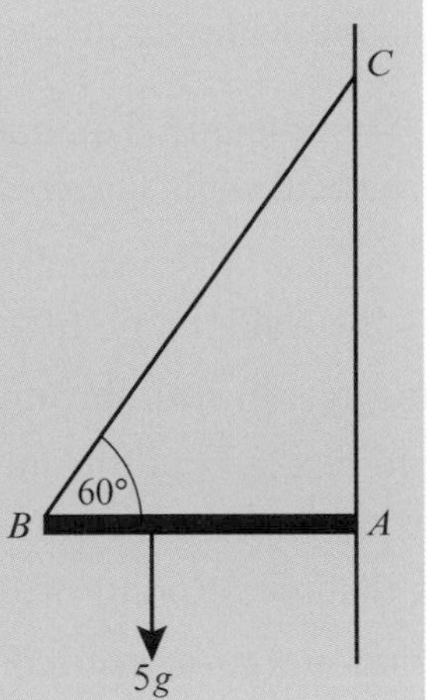

Watch out The reaction at the hinge does not have to be normal (perpendicular) to the wall.

Summary of key points

1 A particle or rigid body is in static equilibrium if it is at rest and the resultant force acting on the particle is zero.

2 The maximum value of the frictional force $F_{MAX} = \mu R$ is reached when the body you are considering is on the point of moving. The body is then said to be in limiting equilibrium.

3 In general, the force of friction F is such that $F \leqslant \mu R$, and the direction of the frictional force is opposite to the direction in which the body would move if the frictional force were absent.

4 For a rigid body in static equilibrium:
- the body is stationary
- the resultant force in any direction is zero
- the resultant moment is zero.

Further kinematics

8

Objectives

After completing this chapter you should be able to:

- Work with vectors for displacement, velocity and acceleration when using the vector equations of motion → pages 160–167
- Use calculus with harder functions of time involving variable acceleration → pages 167–170
- Differentiate and integrate vectors with respect to time → pages 171–177

Vectors are used to represent motion in two and three dimensions. The surface of the ocean can be modelled as a two-dimensional plane, and the velocity of a ship can be described using a vector.

→ Exercise 8A, Q12

Prior knowledge check

1 For the vectors $\mathbf{s} = \begin{pmatrix} 5 \\ -12 \end{pmatrix}$ and $\mathbf{t} = \begin{pmatrix} 4 \\ -7 \end{pmatrix}$, find:

a $3\mathbf{s} + \mathbf{t}$ **b** $2\mathbf{s} - 5\mathbf{t}$

c the unit vector in the direction of **s**.

Give your answers in the form $a\mathbf{i} + b\mathbf{j}$.

←Pure Year 1, Chapter 11

2 A particle moves in a straight line with acceleration 5 m s^{-2}. The initial velocity of the particle is 3 m s^{-1}. When $t = 4$ seconds, find:

a the velocity of the particle

b the displacement from the starting point.

←Year 1, Chapter 9

3 **a** Differentiate: **i** $3e^{2x}$ **ii** $2\sin 3x$

b Integrate: **i** $4e^{3x+1}$ **ii** $5\cos 2\pi x$

←Pure Year 2, Chapters 9, 11

8.1 Vectors in kinematics

You can use two-dimensional vectors to describe motion in a plane.

- **If a particle starts from the point with position vector $\mathbf{r}_0$ and moves with constant velocity $\mathbf{v}$, then its displacement from its initial position at time t is $\mathbf{v}t$ and its position vector $\mathbf{r}$ is given by $\mathbf{r} = \mathbf{r}_0 + \mathbf{v}t$.**

Notation In this equation $\mathbf{r}$, $\mathbf{r}_0$ and $\mathbf{v}$ are vectors and t is a scalar. Displacement, velocity and acceleration can be given using **i**-**j** notation, or as column vectors.

Unless otherwise informed, you should assume that **i** and **j** are unit vectors due east and north respectively.

Example 1

A particle starts from the point with position vector $(3\mathbf{i} + 7\mathbf{j})$ m and moves with constant velocity $(2\mathbf{i} - \mathbf{j})$ m s^{-1}.

a Find the position vector of the particle 4 seconds later.

b Find the time at which the particle is due east of the origin.

a $\mathbf{r} = \mathbf{r}_0 + \mathbf{v}t$ — Write down the formula.

$= (3\mathbf{i} + 7\mathbf{j}) + 4(2\mathbf{i} - \mathbf{j})$

$= (3\mathbf{i} + 7\mathbf{j}) + (8\mathbf{i} - 4\mathbf{j})$

$= (11\mathbf{i} + 3\mathbf{j})$ m — With a vector quantity you still need to give units.

b $\mathbf{r} = (3\mathbf{i} + 7\mathbf{j}) + (2\mathbf{i} - \mathbf{j})t$ — Use $\mathbf{r} = \mathbf{r}_0 + \mathbf{v}t$

$= (3 + 2t)\mathbf{i} + (7 - t)\mathbf{j}$

$7 - t = 0$ so $t = 7$ seconds.

Problem-solving When the particle is due east of the origin, the displacement will only have an **i**-component. Set the coefficient of **j** equal to 0 and solve the equation to find t.

Online Explore the solution to this example using technology.

You can solve questions involving constant acceleration in two dimensions using the vector equations of motion.

- **For an object moving in a plane with constant acceleration:**
  - $\mathbf{v} = \mathbf{u} + \mathbf{a}t$
  - $\mathbf{r} = \mathbf{u}t + \frac{1}{2}\mathbf{a}t^2$

  where
  - **u is the initial velocity**
  - **a is the acceleration**
  - **v is the velocity at time t**
  - **r is the displacement at time t**

Links These are the vector equivalents of the *suvat* formulae for motion in **one** dimension:

$v = u + at$

$s = ut + \frac{1}{2}at^2$ ← Year 1, Sections 9.3, 9.4

Example 2

A particle P has velocity $(-3\mathbf{i} + \mathbf{j})\,\text{m s}^{-1}$ at time $t = 0$. The particle moves with constant accleration $\mathbf{a} = (2\mathbf{i} + 3\mathbf{j})\,\text{m s}^{-2}$. Find the speed of the particle and the bearing on which it is travelling at time $t = 3$ seconds.

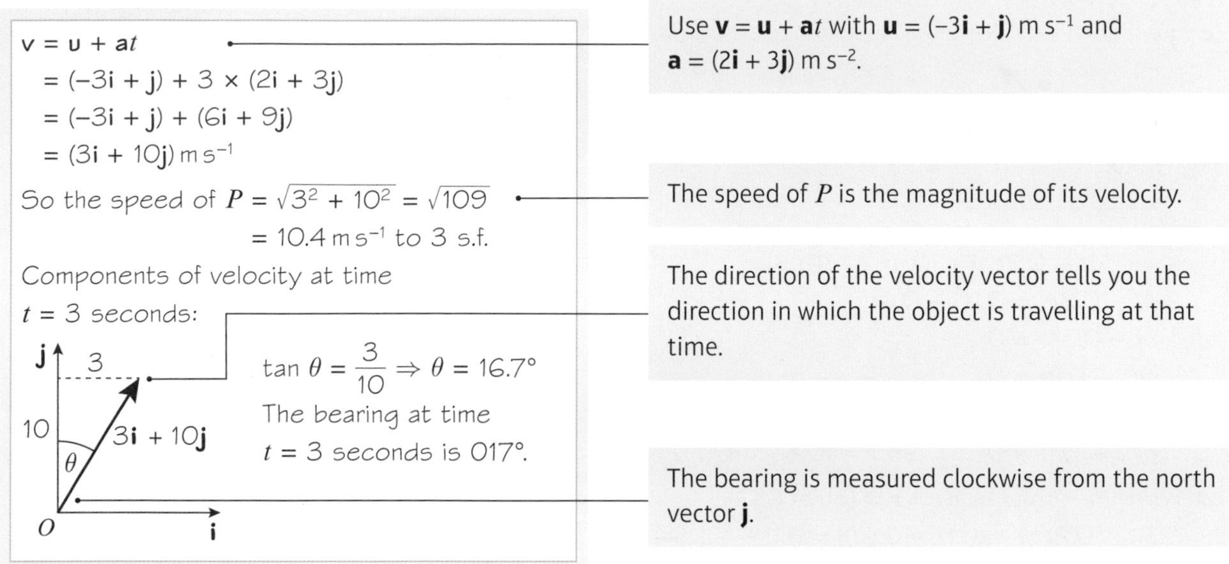

Example 3

An ice skater is skating on a large flat ice rink. At time $t = 0$ the skater is at a fixed point O and is travelling with velocity $(2.4\mathbf{i} - 0.6\mathbf{j})\,\text{m s}^{-1}$.

At time $t = 20$ s the skater is travelling with velocity $(-5.6\mathbf{i} + 3.4\mathbf{j})\,\text{m s}^{-1}$.

Relative to O, the skater has position vector **s** at time t seconds.

Modelling the ice skater as a particle with constant acceleration, find:

a the acceleration of the ice skater

b an expression for **s** in terms of t

c the time at which the skater is directly north-east of O.

A second skater travels so that she has position vector $\mathbf{r} = (1.1t - 6)\mathbf{j}$ m relative to O at time t.

d Show that the two skaters will meet.

a Using $\mathbf{v} = \mathbf{u} + \mathbf{a}t$,

$$\begin{pmatrix}-5.6\\3.4\end{pmatrix} = \begin{pmatrix}2.4\\-0.6\end{pmatrix} + 20\mathbf{a}$$

It is often easier to work using column vectors rather than **i**–**j** notation.

$$20\mathbf{a} = \begin{pmatrix}-5.6 - 2.4\\3.4 + 0.6\end{pmatrix} = \begin{pmatrix}-8\\4\end{pmatrix}$$

$$\mathbf{a} = \tfrac{1}{20}\begin{pmatrix}-8\\4\end{pmatrix} = \begin{pmatrix}-0.4\\0.2\end{pmatrix}\text{m s}^{-2}$$

or $\mathbf{a} = (-0.4\mathbf{i} + 0.2\mathbf{j})\,\text{m s}^{-2}$

Unless otherwise instructed you can give your answer either in column vector or **i**–**j** form. Remember to include units with your answer.

b Using $\mathbf{s} = \mathbf{u}t + \frac{1}{2}\mathbf{a}t^2$

$$\mathbf{s} = \begin{pmatrix} 2.4 \\ -0.6 \end{pmatrix} t + \frac{1}{2}\begin{pmatrix} -0.4 \\ 0.2 \end{pmatrix} t^2$$

$$= \begin{pmatrix} 2.4t - 0.2t^2 \\ -0.6t + 0.1t^2 \end{pmatrix} \text{m}$$

or $\mathbf{s} = ((2.4t - 0.2t^2)\mathbf{i} + (0.1t^2 - 0.6t)\mathbf{j})$ m

c

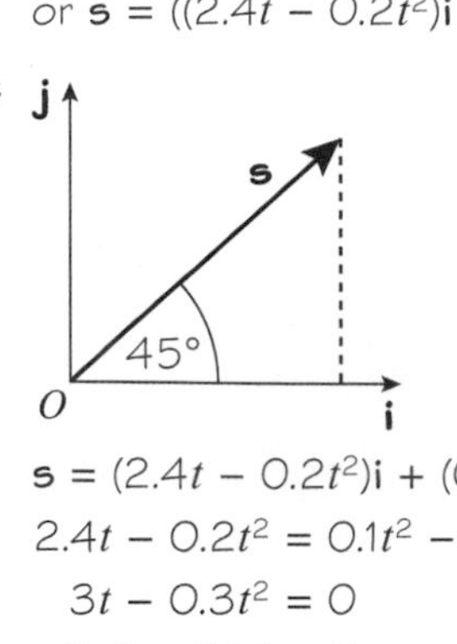

$\mathbf{s} = (2.4t - 0.2t^2)\mathbf{i} + (0.1t^2 - 0.6t)\mathbf{j}$

$2.4t - 0.2t^2 = 0.1t^2 - 0.6t$

$3t - 0.3t^2 = 0$

$3t(1 - 0.1t) = 0$

either $3t = 0$ so ~~$t = 0$~~

or $1 - 0.1t = 0$ so $t = 10$ seconds

d For the second skater: $\mathbf{r} = (1.1t - 6)\mathbf{j}$

$(2.4t - 0.2t^2)\mathbf{i} + (0.1t^2 - 0.6t)\mathbf{j} = (1.1t - 6)\mathbf{j}$

i: $2.4t - 0.2t^2 = 0$

$24t - 2t^2 = 0$

$2t(12 - t) = 0$ either $t = 0$ or $t = 12$

j: When $t = 0$, $0.1t^2 - 0.6t = 0$

and $1.1t - 6 = -6$

so skaters do not meet at $t = 0$.

When $t = 12$ seconds, $0.1t^2 - 0.6t = 7.2$

and $1.1t - 6 = 7.2$

So the two skaters will meet when $t = 12$ seconds.

Online Explore the solution to this example using technology.

Problem-solving

When the skater is directly north-east of O, the **i** and **j** components of the displacement must be equal. Set these components equal and solve the corresponding equation to find the value of t.

At $t = 0$, the skater is at O and both components are equal to 0, so reject this solution.

Equate the displacement vectors for each skater. If the skaters meet, there will be a value of t for which $\mathbf{s} = \mathbf{r}$.

Equate coefficients of **i**.

Check whether the **j** components are equal at either $t = 0$ or $t = 12$.

You could also equate the **j** coefficients and solve:

$0.1t^2 - 0.6t = 1.1t - 6$

$t^2 - 17t + 60 = 0$

$(t - 12)(t - 5) = 0$

So $t = 12$ or $t = 5$.

$t = 12$ satisfies both equations so the skaters meet at this time.

Exercise 8A

For all questions in this exercise, take **i** and **j** to be the unit vectors due east and north respectively.

1 A particle P starts at the point with position vector $\mathbf{r}_0$. P moves with constant velocity $\mathbf{v}$ m s^{-1}. After t seconds, P is at the point with position vector $\mathbf{r}$.

a Find $\mathbf{r}$ if $\mathbf{r}_0 = 2\mathbf{i}$, $\mathbf{v} = \mathbf{i} + 3\mathbf{j}$, and $t = 4$.

b Find $\mathbf{r}$ if $\mathbf{r}_0 = 3\mathbf{i} - \mathbf{j}$, $\mathbf{v} = -2\mathbf{i} + \mathbf{j}$, and $t = 5$.

c Find $\mathbf{r}_0$ if $\mathbf{r} = 4\mathbf{i} + 3\mathbf{j}$, $\mathbf{v} = 2\mathbf{i} - \mathbf{j}$, and $t = 3$.

d Find $\mathbf{r}_0$ if $\mathbf{r} = -2\mathbf{i} + 5\mathbf{j}$, $\mathbf{v} = -2\mathbf{i} + 3\mathbf{j}$, and $t = 6$.

e Find $\mathbf{v}$ if $\mathbf{r}_0 = 2\mathbf{i} + 2\mathbf{j}$, $\mathbf{r} = 8\mathbf{i} - 7\mathbf{j}$, and $t = 3$.

f Find t if $\mathbf{r}_0 = 4\mathbf{i} + \mathbf{j}$, $\mathbf{r} = 12\mathbf{i} - 11\mathbf{j}$, and $\mathbf{v} = 2\mathbf{i} - 3\mathbf{j}$.

2 A radio-controlled boat starts from position vector $(10\mathbf{i} - 5\mathbf{j})$ m relative to a fixed origin and travels with constant velocity, passing a point with position vector $(-2\mathbf{i} + 9\mathbf{j})$ m after 4 seconds. Find the speed and bearing of the boat.

3 A clockwork mouse starts from a point with position vector $(-2\mathbf{i} + 3\mathbf{j})$ m relative to a fixed origin and moves in a straight line with a constant speed of $4\,\text{m s}^{-1}$. Find the time taken for the mouse to travel to the point with position vector $(6\mathbf{i} - 3\mathbf{j})$ m.

4 A helicopter starts from the point with position vector $\begin{pmatrix}120\\-10\end{pmatrix}$ m relative to a fixed origin, and moves with constant velocity $\begin{pmatrix}-30\\40\end{pmatrix}\,\text{m s}^{-1}$. Find:

a the position vector of the helicopter t seconds later

b the time at which the helicopter is due north of the origin.

Hint When the helicopter is due north of the origin, the **i**-component of its position vector will be 0.

(P) **5** At time $t = 0$, the particle P is at the point with position vector $4\mathbf{i}$, and moving with constant velocity $\mathbf{i} + \mathbf{j}\,\text{m s}^{-1}$. A second particle Q is at the point with position vector $-3\mathbf{j}$ and moving with velocity $\mathbf{v}\,\text{m s}^{-1}$. After 8 seconds, the paths of P and Q meet. Find the speed of Q.

(P) **6** At noon, a ferry F is 400 m due north of an observation point O and is moving with a constant velocity of $(7\mathbf{i} + 7\mathbf{j})\,\text{m s}^{-1}$, and a speedboat S is 500 m due east of O, moving with a constant velocity of $(-3\mathbf{i} + 15\mathbf{j})\,\text{m s}^{-1}$.

a Write down the position vectors of F and S at time t seconds after noon.

b Show that F and S will collide, and find the position vector of the point of collision.

7 A particle starts at rest and moves with constant acceleration. After 5 seconds its velocity is $\begin{pmatrix}3\\4\end{pmatrix}\,\text{m s}^{-1}$.

a Find the acceleration of the particle.

b The displacement vector of the particle from its starting position after 5 seconds.

8 An object moves with constant acceleration so that its velocity changes from $(15\mathbf{i} + 4\mathbf{j})\,\text{m s}^{-1}$ to $(5\mathbf{i} - 3\mathbf{j})\,\text{m s}^{-1}$ in 4 seconds. Find:

a the acceleration of the particle

Given that the initial position vector of the particle relative to a fixed origin O is $10\mathbf{i} - 8\mathbf{j}$ m,

b find the position vector of the particle after t seconds.

9 A plane moves with constant acceleration $\begin{pmatrix}-1\\1.5\end{pmatrix}\,\text{m s}^{-2}$.

When $t = 0$, the velocity of the plane is $\begin{pmatrix}70\\-30\end{pmatrix}\,\text{m s}^{-1}$. Find:

a the velocity of the plane after 10 seconds

b the distance of the plane from its starting point after 10 seconds.

(P) **10** A model boat moves with constant acceleration $(0.2\mathbf{i} + 0.6\mathbf{j})$ m s^{-2}. After 20 seconds its velocity is $(4\mathbf{i} + 3\mathbf{j})$ m s^{-1}. Find the displacement vector of the boat from its starting position after 20 seconds.

(P) **11** A particle A starts at the point with position vector $12\mathbf{i} + 12\mathbf{j}$. The initial velocity of A is $(-\mathbf{i} + \mathbf{j})$ m s^{-1}, and it has constant acceleration $(2\mathbf{i} - 4\mathbf{j})$ m s^{-2}. Another particle, B, has initial velocity $\mathbf{i}$ m s^{-1} and constant acceleration $2\mathbf{j}$ m s^{-2}. After 3 seconds the two particles collide. Find:

a the speeds of the two particles when they collide

b the position vector of the point where the two particles collide

c the position vector of B's starting point.

(E/P) **12** A ship is moving such that at time 12:00 its position is O and its velocity is $(-4\mathbf{i} + 8\mathbf{j})$ km h^{-1}. At 14:00, the ship is travelling with velocity $(-2\mathbf{i} - 6\mathbf{j})$ km h^{-1}.

Relative to O, the ship has displacement $\mathbf{s}$ at time t hours after 12:00 where $t \geqslant 0$.

Modelling the ship as a particle with constant acceleration, find:

a the acceleration of the ship **(2 marks)**

b an expression for $\mathbf{s}$ in terms of t **(2 marks)**

c the time at which the ship is directly south-west of O. **(3 marks)**

At time t hours after 12:00, another ship has displacement $\mathbf{r} = (40 - 25t)\mathbf{j}$ relative to O.

d Find the position vector of the point where the two ships meet. **(4 marks)**

(E/P) **13** A particle moves so that its position vector, in metres, relative to a fixed origin O at time t seconds is $\mathbf{r} = (2t^2 - 3)\mathbf{i} + (7 - 4t)\mathbf{j}$, where $t \geqslant 0$.

a Show that the particle is north-east of O when $t^2 + 2t - 5 = 0$. **(2 marks)**

b Hence determine the distance of the particle from O when it is north-east of O, giving your answer correct to 3 significant figures. **(3 marks)**

A second particle moves with constant acceleration $(3a\mathbf{i} - 2a\mathbf{j})$ m s^{-2}. When $t = 0$ the velocity of the particle is $(5\mathbf{i} + 6\mathbf{j})$ m s^{-1} and its position vector relative to O is $5\mathbf{j}$ m. When $t = 2$ seconds the particle is travelling with velocity $(b\mathbf{i} + 2b\mathbf{j})$ m s^{-1}.

c Find the speed and direction of the particle when $t = 2$. **(6 marks)**

d Find the distance between the two particles at this time. **(4 marks)**

Challenge

During an air show, a stunt aeroplane passes over a control tower with velocity $(20\mathbf{i} - 100\mathbf{j})$ m s^{-1}, and flies in a horizontal plane with constant acceleration $6\mathbf{j}$ m s^{-2}. A second aeroplane passes over the same control tower at time t seconds later, where $t > 0$, travelling with velocity $(70\mathbf{i} + 40\mathbf{j})$ m s^{-1}. The second aeroplane is flying in a higher horizontal plane with constant acceleration $-8\mathbf{j}$ m s^{-2}.

Given that the two aeroplanes pass directly over one another in their subsequent motion, find the value of t.

8.2 Vector methods with projectiles

Projectile motion is motion in a vertical plane with constant acceleration. Hence you can analyse it using the vector equations of motion. When using vectors with projectile questions you should consider **i** and **j** to be the unit vectors horizontally and vertically, unless you are told otherwise.

Links You can also analyse projectile motion by considering the horizontal and vertical components of velocity separately. ← Chapter 6

Example 4

A ball is struck by a racket from a point A which has position vector $20\mathbf{j}$ m relative to a fixed origin O. Immediately after being struck, the ball has velocity $(5\mathbf{i} + 8\mathbf{j})$ m s^{-1}, where **i** and **j** are unit vectors horizontally and vertically respectively. After being struck, the ball travels freely under gravity until it strikes the ground at point B.

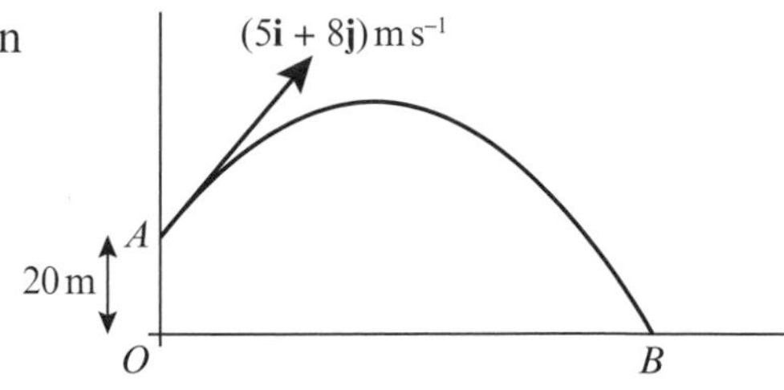

a Find the speed of the ball 1.5 seconds after being struck.

b Find an expression for the position vector, **r**, of the ball relative to O at time t seconds.

c Hence determine the distance OB.

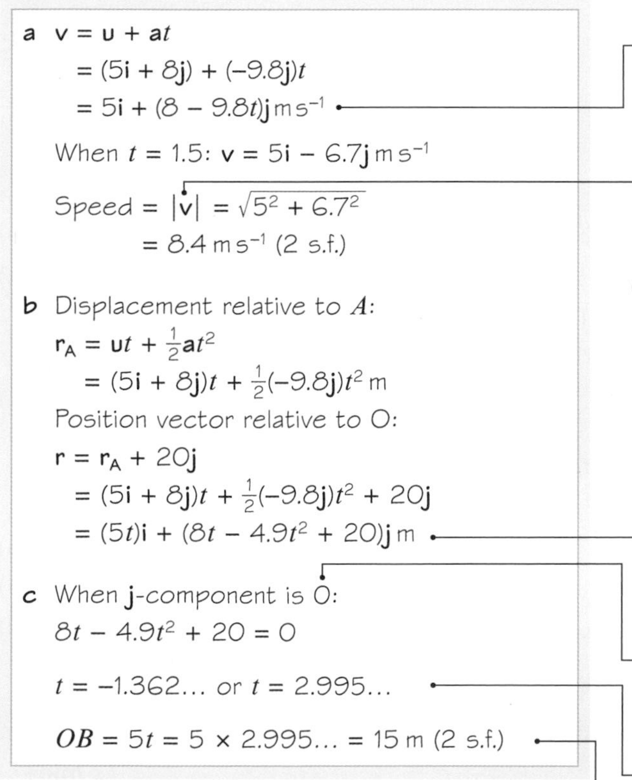

a $\mathbf{v} = \mathbf{u} + \mathbf{a}t$

$= (5\mathbf{i} + 8\mathbf{j}) + (-9.8\mathbf{j})t$

$= 5\mathbf{i} + (8 - 9.8t)\mathbf{j}$ m s^{-1}

When $t = 1.5$: $\mathbf{v} = 5\mathbf{i} - 6.7\mathbf{j}$ m s^{-1}

Speed $= |\mathbf{v}| = \sqrt{5^2 + 6.7^2}$

$= 8.4$ m s^{-1} (2 s.f.)

b Displacement relative to A:

$\mathbf{r}_A = \mathbf{u}t + \frac{1}{2}\mathbf{a}t^2$

$= (5\mathbf{i} + 8\mathbf{j})t + \frac{1}{2}(-9.8\mathbf{j})t^2$ m

Position vector relative to O:

$\mathbf{r} = \mathbf{r}_A + 20\mathbf{j}$

$= (5\mathbf{i} + 8\mathbf{j})t + \frac{1}{2}(-9.8\mathbf{j})t^2 + 20\mathbf{j}$

$= (5t)\mathbf{i} + (8t - 4.9t^2 + 20)\mathbf{j}$ m

c When **j**-component is 0:

$8t - 4.9t^2 + 20 = 0$

$t = -1.362...$ or $t = 2.995...$

$OB = 5t = 5 \times 2.995... = 15$ m (2 s.f.)

Acceleration due to gravity acts vertically downwards, and has vector $-9.8\mathbf{j}$ m s^{-2}.

Speed is the magnitude of the velocity vector.

Watch out The equation $\mathbf{r} = \mathbf{u}t + \frac{1}{2}\mathbf{a}t^2$ gives you the displacement relative to the starting position of the ball, A. To find the position vector of the ball relative to O you need to add on the position vector of A.

Rearrange so that you have separate expressions for the **i**-component and the **j**-component.

For all the points on the horizontal axes, the vertical component of the position vector is 0.

Use your calculator to solve the quadratic equation, then take the positive answer.

The horizontal component of the position vector tells you the horizontal distance from O.

Exercise 8B

For all questions in this exercise **i** and **j** are unit vectors horizontally and vertically respectively. Unless stated otherwise, take $g = 9.8\ \text{m s}^{-2}$.

1 A particle P is projected from the origin with velocity $(12\mathbf{i} + 24\mathbf{j})\ \text{m s}^{-1}$. The particle moves freely under gravity. Find:

a the position vector of P after 3 s

b the speed of P after 3 s.

2 In this question use $g = 10\ \text{m s}^{-2}$

A particle P is projected from the origin with velocity $(4\mathbf{i} + 5\mathbf{j})\ \text{m s}^{-1}$. The particle moves freely under gravity. Find:

a the position vector of P after t s

b the greatest height of the particle.

Hint When the particle is at its greatest height, the **j**-component of the velocity will be 0.

E/P **3** A ball is projected from a point A at the top of a cliff, with position vector $25\mathbf{j}$ m relative to the base of the cliff O. The base of the cliff is at sea level. The velocity of projection is $(3p\mathbf{i} + p\mathbf{j})\ \text{m s}^{-1}$, where p is a constant. After 2 seconds, the ball passes a point B with position vector $(q\mathbf{i} + 10\mathbf{j})$ m, where q is a constant, before hitting the sea at point C. The ball is modelled as a particle moving freely under gravity and the sea is modelled as a horizontal plane.

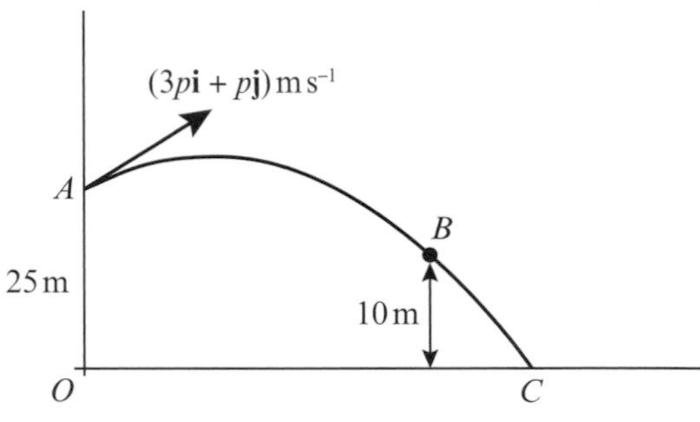

a Suggest, with reasons, which of these two modelling assumptions is most realistic. **(2 marks)**

b Find the velocity vector of the ball at point B. **(6 marks)**

A remote-control boat leaves O at the same time the ball is projected, and travels in a straight line towards C with constant acceleration. Given that the ball lands on the boat,

c find the acceleration of the boat. **(6 marks)**

E **4** A particle P is projected with velocity $(3u\mathbf{i} + 4u\mathbf{j})\ \text{m s}^{-1}$ from a fixed point O on horizontal ground. Given that P strikes the ground at a point 750 m from O,

a show that $u = 17.5$ **(6 marks)**

b calculate the greatest height above the ground reached by P **(3 marks)**

c find the angle the direction of motion of P makes with **i** when $t = 5$. **(4 marks)**

E **5** A particle is projected with velocity $(8\mathbf{i} + 10\mathbf{j})\ \text{m s}^{-1}$ from a point O at the top of a cliff and moves freely under gravity. Six seconds after projection, the particle strikes the sea at the point S. Calculate:

a the horizontal distance between O and S **(2 marks)**

b the vertical distance between O and S. **(3 marks)**

At time T seconds after projection, the particle is moving with velocity $(8\mathbf{i} - 14.5\mathbf{j})\ \text{m s}^{-1}$.

c Find the value of T and the position vector, relative to O, of the particle at this instant. **(6 marks)**

6 In this question use $g = 10\,\text{m s}^{-2}$

A body B is projected from a fixed point O on horizontal ground with velocity $a\mathbf{i} + b\mathbf{j}\,\text{m s}^{-1}$, where a and b are positive constants. The body moves freely under gravity until it hits the ground at the point P, where it immediately comes to rest.

The position vector of a point on the path of B relative to O is $(x\mathbf{i} + y\mathbf{j})\,\text{m}$.

a Show that $y = \dfrac{bx}{a} - \dfrac{5x^2}{a^2}$ **(5 marks)**

Given that $a = 8$, $OP = X\,\text{m}$ and the maximum vertical height of B above the ground is $Y\,\text{m}$,

b find, in terms of b,

i X **ii** Y **(6 marks)**

8.3 Variable acceleration in one dimension

The equations of motion for constant acceleration allow you to write velocity and displacement as functions of time.

When a body experiences **variable acceleration** you can model the acceleration as a function of time. You can use calculus to describe the relationship between displacement, velocity and acceleration.

Links Velocity, v, is the rate of change of displacement, s

$v = \dfrac{ds}{dt}$ $\quad s = \int v\,dt$

Acceleration, a, is the rate of change of velocity, v

$a = \dfrac{dv}{dt} = \dfrac{d^2s}{dt^2}$ $\quad v = \int a\,dt$ ← Year 1, Section 11.4

You need to be able to use any of the functions and techniques from your A level course to analyse motion in a straight line.

Example 5

A particle is moving in a straight line with acceleration at time t seconds given by

$a = \cos 2\pi t\,\text{m s}^{-2}$, where $t \geqslant 0$

The velocity of the particle at time $t = 0$ is $\dfrac{1}{2\pi}\,\text{m s}^{-1}$. Find:

a an expression for the velocity at time t seconds

b the maximum speed

c the distance travelled in the first 3 seconds.

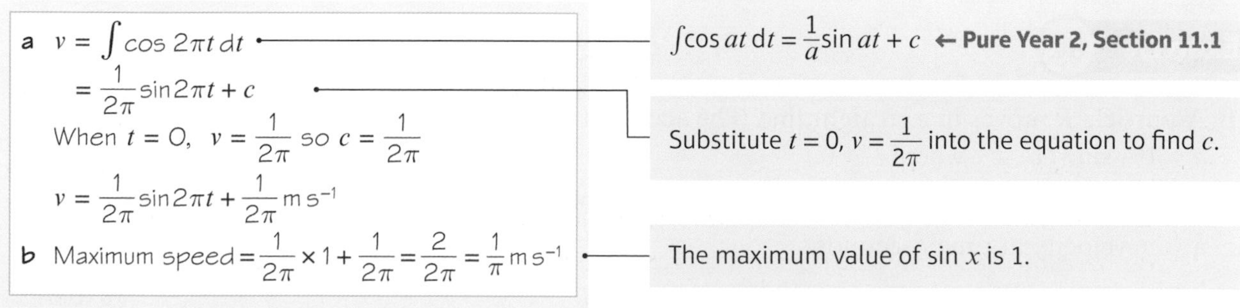

a $v = \int \cos 2\pi t\,dt$

$= \dfrac{1}{2\pi}\sin 2\pi t + c$

When $t = 0$, $v = \dfrac{1}{2\pi}$ so $c = \dfrac{1}{2\pi}$

$v = \dfrac{1}{2\pi}\sin 2\pi t + \dfrac{1}{2\pi}\,\text{m s}^{-1}$

b Maximum speed $= \dfrac{1}{2\pi} \times 1 + \dfrac{1}{2\pi} = \dfrac{2}{2\pi} = \dfrac{1}{\pi}\,\text{m s}^{-1}$

$\int \cos at\,dt = \dfrac{1}{a}\sin at + c$ ← Pure Year 2, Section 11.1

Substitute $t = 0$, $v = \dfrac{1}{2\pi}$ into the equation to find c.

The maximum value of $\sin x$ is 1.

c $s = \frac{1}{2\pi}\int_0^3 (\sin 2\pi t + 1)\,dt$

$= \frac{1}{2\pi}\left[-\frac{1}{2\pi}\cos 2\pi t + t\right]_0^3$

$= \frac{1}{2\pi}\left[\left(-\frac{1}{2\pi} + 3\right) - \left(-\frac{1}{2\pi}\right)\right]$

$= \frac{3}{2\pi}$ m or 0.477 m (3 s.f.)

To find the distance travelled in the first 3 seconds integrate v between $t = 0$ and $t = 3$: $\int_0^3 \left(\frac{1}{2\pi}\sin 2\pi t + \frac{1}{2\pi}\right)dt = \frac{1}{2\pi}\int_0^3 (\sin 2\pi t + 1)\,dt$

Example 6

A particle of mass 6 kg is moving on the positive x-axis. At time t seconds the displacement, s, of the particle from the origin is given by

$$s = 2t^{\frac{3}{2}} + \frac{e^{-2t}}{3}\text{ m, where } t \geqslant 0$$

a Find the velocity of the particle when $t = 1.5$.

Given that the particle is acted on by a single force of variable magnitude F N which acts in the direction of the positive x-axis,

b find the value of F when $t = 2$.

a $v = \frac{ds}{dt} = 3t^{\frac{1}{2}} - \frac{2e^{-2t}}{3}$ m s^{-1}

When $t = 1.5$ seconds:

$v = 3 \times 1.5^{0.5} - \frac{2e^{-3}}{3}$

$= 3.64$ m s^{-1} (3 s.f.)

b $a = \frac{dv}{dt} = 1.5t^{-0.5} + \frac{4e^{-2t}}{3}$ m s^{-2}

When $t = 2$ seconds:

$a = 1.5 \times 2^{-0.5} + \frac{4e^{-4}}{3}$

$= 1.0850\ldots$ m s^{-2}

$F = ma = 6 \times 1.0850\ldots = 6.51$ N (3 s.f.)

If $y = ae^{kt}$ then $\frac{dy}{dt} = kae^{kt}$

← Pure Year 2, Section 9.2

Differentiate to find an expression for v and substitute $t = 1.5$.

Problem-solving

You know the mass of the particle, so if you find the acceleration you can use $F = ma$ to find the magnitude of the force acting on it. Differentiate the velocity, then substitute $t = 2$ to find the acceleration when $t = 2$ seconds.

Exercise 8C

1 A particle P moves in a straight line. The acceleration, a, of P at time t seconds is given by $a = 1 - \sin\pi t$ m s^{-2}, where $t \geqslant 0$.

When $t = 0$, the velocity of P is 0 m s^{-1} and its displacement is 0 m. Find expressions for:

a the velocity at time t seconds

b the displacement at time t seconds.

2 A particle moving in a straight line has acceleration a, given by

$$a = \sin 3\pi t \text{ m s}^{-2},\ t \geqslant 0$$

At time t seconds the particle has velocity v m s^{-1} and displacement s m. Given that when $t = 0$, $v = \frac{1}{3\pi}$ and $s = 1$, find:

a an expression for v in terms of t

b the maximum speed of the particle

c an expression for s in terms of t.

(P) **3** An object moves in a straight line from a point O. At time t seconds the object has acceleration, a, where

$$a = -\cos 4\pi t \text{ m s}^{-2},\ 0 \leqslant t \leqslant 4$$

When $t = 0$, the velocity of the object is 0 m s^{-1} and its displacement is 0 m. Find:

a an expression for the velocity at time t seconds

b the maximum speed of the object

c an expression for the displacement of the object at time t seconds

d the maximum distance of the object from O

e the number of times the object changes direction during its motion.

Problem-solving

In part **e**, consider the number of times the velocity changes sign.

4 A body, M, of mass 5 kg moves along the positive x-axis. The displacement, s, of the body at time t seconds is given by $s = 3t^{\frac{2}{3}} + 2e^{-3t}$ m, where $t \geqslant 0$.

Find:

a the velocity of M when $t = 0.5$

b the acceleration of M when $t = 3$.

Given that M is acted on by a single force of variable magnitude F N which acts in the direction of the x-axis,

c find the value of F when $t = 3$ seconds.

(P) **5** A particle P moves in a straight line so that, at time t seconds, its displacement, s m, from a fixed point O on the line is given by

$$s = \begin{cases} \frac{1}{2}t, & 0 \leqslant t \leqslant 6 \\ \sqrt{t+3}, & t > 6 \end{cases}$$

Find:

a the velocity of P when $t = 4$

b the velocity of P when $t = 22$.

(P) **6** A particle P moves in a straight line so that, at time t seconds, its displacement from a fixed point O on the line is given by

$$s = \begin{cases} 3^t + 3t, & 0 \leqslant t \leqslant 3 \\ 24t - 36, & 3 < t \leqslant 6 \\ -252 + 96t - 6t^2, & t > 6 \end{cases}$$

Find:

a the velocity of P when $t = 2$

b the velocity of P when $t = 10$

c the greatest positive displacement of P from O

d the values of s when the speed of P is $18\,\text{m}\,\text{s}^{-1}$.

(P) **7** A particle moves in a straight line. At time t seconds after it begins its motion, the acceleration of the particle is $3\sqrt{t}\,\text{m}\,\text{s}^{-2}$ where $t > 0$.

Given that after 1 second the particle is moving with velocity $2\,\text{m}\,\text{s}^{-1}$, find the time taken for the particle to travel 16 m.

(E/P) **8** A runner takes part in a race in which competitors have to sprint 200 m in a straight line. At time t seconds after starting, her displacement, s, from the starting position is modelled as:

$$s = k\sqrt{t},\ 0 \leqslant t \leqslant T$$

Given that the runner completes the race in 25 seconds,

a find the value of k and the value of T **(2 marks)**

b find the speed of the runner when she crosses the finish line **(3 marks)**

c criticise this model for small values of t. **(2 marks)**

(E/P) **9** A particle is moving in a straight line. At time t seconds, where $t \geqslant 0$, the acceleration of P is $a\,\text{m}\,\text{s}^{-2}$ and the velocity $v\,\text{m}\,\text{s}^{-1}$ of P is given by

$$v = 2 + 8\sin kt$$

where k is a constant.

The initial acceleration of P is $4\,\text{m}\,\text{s}^{-2}$.

a Find the value of k. **(3 marks)**

Using the value of k found in part **a**,

b find, in terms of π, the values of t in the interval $0 \leqslant t \leqslant 4\pi$ for which $a = 0$ **(2 marks)**

c show that $4a^2 = 64 - (v - 2)^2$ **(5 marks)**

d find the maximum velocity and the maximum acceleration. **(2 marks)**

(E/P) **10** A particle P moves on the x-axis. At time t seconds the velocity of P is $v\,\text{m}\,\text{s}^{-1}$ in the direction of x increasing, where v is given by

$$v = \begin{cases} 10t - 2t^{\frac{3}{2}}, & 0 \leqslant t \leqslant 4 \\ 24 - \left(\dfrac{t-4}{2}\right)^4, & t > 4 \end{cases}$$

When $t = 0$, P is at the origin O.

Find:

a the greatest speed of P in the interval $0 \leqslant t \leqslant 4$ **(4 marks)**

b the distance of P from O when $t = 4$ **(3 marks)**

c the time at which P is instantaneously at rest for $t > 4$ **(1 mark)**

d the total distance travelled by P in the first 10 seconds of its motion. **(7 marks)**

8.4 Differentiating vectors

You can use calculus with vectors to solve problems involving motion in two dimensions with variable acceleration.

To differentiate a vector quantity in the form $f(t)\mathbf{i} + g(t)\mathbf{j}$ you differentiate each function of time separately.

- **If $\mathbf{r} = x\mathbf{i} + y\mathbf{j}$, then** $\mathbf{v} = \frac{d\mathbf{r}}{dt} = \dot{\mathbf{r}} = \dot{x}\mathbf{i} + \dot{y}\mathbf{j}$

  and $\mathbf{a} = \frac{d\mathbf{v}}{dt} = \frac{d^2\mathbf{r}}{dt^2} = \ddot{\mathbf{r}} = \ddot{x}\mathbf{i} + \ddot{y}\mathbf{j}$

Notation Dot notation is a short-hand for differentiation with respect to time:

$\dot{x} = \frac{dx}{dt}$ and $\dot{y} = \frac{dy}{dt}$

$\ddot{x} = \frac{d^2x}{dt^2}$ and $\ddot{y} = \frac{d^2y}{dt^2}$

Example 7

A particle P of mass 0.8 kg is acted on by a single force $\mathbf{F}$ N. Relative to a fixed origin O, the position vector of P at time t seconds is $\mathbf{r}$ metres, where

$$\mathbf{r} = 2t^3\mathbf{i} + 50t^{-\frac{1}{2}}\mathbf{j},\ t \geqslant 0$$

Find:

a the speed of P when $t = 4$

b the acceleration of P as a vector when $t = 2$

c $\mathbf{F}$ when $t = 2$.

a $\mathbf{v} = \dot{\mathbf{r}} = 6t^2\mathbf{i} - 25t^{-\frac{3}{2}}\mathbf{j}\,\text{m s}^{-1}$

Differentiate $2t^3$ and $5t^{-\frac{1}{2}}$ separately to find the **i**- and **j**-components of the velocity.

When $t = 4$: $\mathbf{v} = \left(96\mathbf{i} - \frac{25}{8}\mathbf{j}\right)\text{m s}^{-1}$

Speed $= \sqrt{96^2 + \left(\frac{25}{8}\right)^2} = 96.1\,\text{m s}^{-1}$ (3 s.f.)

The speed is the magnitude of $\mathbf{v}$.

b $\mathbf{a} = \ddot{\mathbf{r}} = \left(12t\mathbf{i} + \frac{75}{2}t^{-\frac{5}{2}}\mathbf{j}\right)\text{m s}^{-2}$

When $t = 2$: $\mathbf{a} = 24\mathbf{i} + 6.6291\ldots\mathbf{j}\,\text{m s}^{-2}$

$\mathbf{a} = \ddot{\mathbf{r}} = \frac{d^2\mathbf{r}}{dt^2}$ or alternatively, $\mathbf{a} = \dot{\mathbf{v}} = \frac{d\mathbf{v}}{dt}$

c $\mathbf{F} = m\mathbf{a} = 0.8(24\mathbf{i} + 6.6291\ldots\mathbf{j})$

$= (19.2\mathbf{i} + 5.30\mathbf{j})$ N (3 s.f.)

Use $\mathbf{F} = m\mathbf{a}$ and round each coefficient to 3 significant figures.

Exercise 8D

1 At time t seconds, a particle P has position vector $\mathbf{r}$ m with respect to a fixed origin O, where

$$\mathbf{r} = (3t - 4)\mathbf{i} + (t^3 - 4t)\mathbf{j},\ t \geqslant 0$$

Find:

a the velocity of P when $t = 3$

b the acceleration of P when $t = 3$.

2 A particle P of mass 3 grams moving in a plane is acted on by a force $\mathbf{F}$ N. Its velocity at time t seconds is given by $\mathbf{v} = (t^2\mathbf{i} + (2t - 3)\mathbf{j})\,\text{m s}^{-1}$, $t \geqslant 0$.

Find $\mathbf{F}$ when $t = 4$.

(P) **3** In this question **i** and **j** are the unit vectors east and north respectively.

A particle P is moving in a plane. At time t seconds, the position vector of P, $\mathbf{r}$ m, relative to a fixed origin O is given by $\mathbf{r} = 5e^{-3t}\mathbf{i} + 2\mathbf{j}$, $t \geqslant 0$.

a Find the time at which the particle is directly north-east of O.

b Find the speed of the particle at this time.

c Explain why the particle is always moving directly west.

(E) **4** At time t seconds, a particle P has position vector $\mathbf{r}$ m with respect to a fixed origin O, where

$$\mathbf{r} = 4t^2\mathbf{i} + (24t - 3t^2)\mathbf{j},\ t \geqslant 0$$

a Find the speed of P when $t = 2$. **(3 marks)**

b Show that the acceleration of P is a constant and find the magnitude of this acceleration. **(3 marks)**

(E) **5** A particle P is initially at a fixed origin O. At time $t = 0$, P is projected from O and moves so that, at time t seconds after projection, its position vector $\mathbf{r}$ m relative to O is given by

$$\mathbf{r} = (t^3 - 12t)\mathbf{i} + (4t^2 - 6t)\mathbf{j},\ t \geqslant 0$$

Find:

a the speed of projection of P **(5 marks)**

b the value of t at the instant when P is moving parallel to **j** **(3 marks)**

c the position vector of P at the instant when P is moving parallel to **j**. **(3 marks)**

The motion of the particle is due to it being acted on by a single variable force, **F** N.

d Given that the mass of the particle is 0.5 kg, find the magnitude of **F** when $t = 5$ s. **(4 marks)**

(E/P) **6** A particle P is moving in a plane. At time t seconds, the position vector of P, $\mathbf{r}$ m, is given by $\mathbf{r} = (3t^2 - 6t + 4)\mathbf{i} + (t^3 + kt^2)\mathbf{j}$, where k is a constant.

When $t = 3$, the speed of P is $12\sqrt{5}\,\mathrm{m\,s^{-1}}$.

a Find the two possible values of k. **(6 marks)**

b For each of these values of k, find the magnitude of the acceleration of P when $t = 1.5$. **(4 marks)**

(E) **7** Relative to a fixed origin O, the position vector of a particle P at time t seconds is **r** metres, where

$$\mathbf{r} = 6t^2\mathbf{i} + t^{\frac{5}{2}}\mathbf{j},\ t \geqslant 0$$

At the instant when $t = 4$, find:

a the speed of P **(5 marks)**

b the acceleration of P, giving your answer as a vector. **(2 marks)**

(E/P) **8** A particle P moves in a horizontal plane. At time t seconds, the position vector of P is **r** metres relative to a fixed origin O where **r** is given by

$$\mathbf{r} = (18t - 4t^3)\mathbf{i} + ct^2\mathbf{j},\ t \geqslant 0,$$

where c is a positive constant. When $t = 1.5$, the speed of P is $15\,\mathrm{m\,s^{-1}}$. Find:

a the value of c **(6 marks)**

b the acceleration of P when $t = 1.5$. **(3 marks)**

E 9 At time t seconds, a particle P has position vector $\mathbf{r}$ metres relative to a fixed origin O, where

$$\mathbf{r} = (2t^2 - 3t)\mathbf{i} + (5t + t^2)\mathbf{j},\ t \geqslant 0$$

Show that the acceleration of P is constant and find its magnitude. **(5 marks)**

E/P 10 A particle P moves in a horizontal plane. At time t seconds, the position vector of P is $\mathbf{r}$ metres relative to a fixed origin O, and $\mathbf{r}$ is given by $\mathbf{r} = (20t - 2t^3)\mathbf{i} + kt^2\mathbf{j}$, $t \geqslant 0$, where k is a positive constant. When $t = 2$, the speed of P is $16\,\text{m s}^{-1}$. Find:

a the value of k **(6 marks)**

b the acceleration of P at the instant when it is moving parallel to $\mathbf{j}$. **(4 marks)**

8.5 Integrating vectors

You can integrate vectors in the form $\text{f}(t)\mathbf{i} + \text{g}(t)\mathbf{j}$ by integrating each function of time separately.

- $\mathbf{v} = \int \mathbf{a}\,\text{d}t$ **and** $\mathbf{r} = \int \mathbf{v}\,\text{d}t$

Watch out When you integrate a vector, the constant of integration will also be a vector. Write it in the form $\mathbf{c} = p\mathbf{i} + q\mathbf{j}$.

Example 8

A particle P is moving in a plane. At time t seconds, its velocity $\mathbf{v}\,\text{m s}^{-1}$ is given by

$$\mathbf{v} = 3t\mathbf{i} + \tfrac{1}{2}t^2\mathbf{j},\ t \geqslant 0$$

When $t = 0$, the position vector of P with respect to a fixed origin O is $(2\mathbf{i} - 3\mathbf{j})$ m. Find the position vector of P at time t seconds.

$$\mathbf{r} = \int \mathbf{v}\,\text{d}t = \int \left(3t\mathbf{i} + \tfrac{1}{2}t^2\mathbf{j}\right)\text{d}t$$

$$= \frac{3t^2}{2}\mathbf{i} + \frac{t^3}{6}\mathbf{j} + \mathbf{c}$$

You integrate $3t$ and $\frac{1}{2}t^2$ in the usual way, using $\int t^n\,\text{d}t = \frac{t^{n+1}}{n+1}$. You must include the constant of integration, which is a vector, $\mathbf{c}$.

When $t = 0$, $\mathbf{r} = 2\mathbf{i} - 3\mathbf{j}$:

$$2\mathbf{i} - 3\mathbf{j} = 0\mathbf{i} + 0\mathbf{j} + \mathbf{c}$$

$$\mathbf{c} = 2\mathbf{i} - 3\mathbf{j}$$

You are given an **initial condition** (or **boundary condition**) which allows you to find $\mathbf{c}$. Substitute $t = 0$ and $\mathbf{r} = 2\mathbf{i} - 3\mathbf{j}$ into the integrated expression and solve to find $\mathbf{c}$.

← Pure Year 1, Section 13.3

Hence

$$\mathbf{r} = \frac{3t^2}{2}\mathbf{i} + \frac{t^3}{6}\mathbf{j} + 2\mathbf{i} - 3\mathbf{j} = \left(\frac{3t^2}{2} + 2\right)\mathbf{i} + \left(\frac{t^3}{6} - 3\right)\mathbf{j}$$

The position vector of P at time t seconds is

$$\left(\left(\frac{3t^2}{2} + 2\right)\mathbf{i} + \left(\frac{t^3}{6} - 3\right)\mathbf{j}\right)\text{m}.$$

Collect together the terms in $\mathbf{i}$ and $\mathbf{j}$ to complete your answer.

Example 9

A particle P is moving in a plane so that, at time t seconds, its acceleration is $(4\mathbf{i} - 2t\mathbf{j})\,\text{m s}^{-2}$. When $t = 3$, the velocity of P is $6\mathbf{i}\,\text{m s}^{-1}$ and the position vector of P is $(20\mathbf{i} + 3\mathbf{j})$ m with respect to a fixed origin O. Find:

a the angle between the direction of motion of P and $\mathbf{i}$ when $t = 2$

b the distance of P from O when $t = 0$.

a $\mathbf{v} = \int \mathbf{a}\,dt = \int (4\mathbf{i} - 2t\mathbf{j})\,dt$

$= 4t\mathbf{i} - t^2\mathbf{j} + \mathbf{c}$

When $t = 3$, $\mathbf{v} = 6\mathbf{i}$:

$6\mathbf{i} = 12\mathbf{i} - 9\mathbf{j} + \mathbf{c}$

$\mathbf{c} = -6\mathbf{i} + 9\mathbf{j}$

Hence

$\mathbf{v} = 4t\mathbf{i} - t^2\mathbf{j} - 6\mathbf{i} + 9\mathbf{j}$

$= ((4t - 6)\mathbf{i} + (9 - t^2)\mathbf{j})\,\text{m s}^{-1}$

When $t = 2$:

$\mathbf{v} = (8 - 6)\mathbf{i} + (9 - 4)\mathbf{j} = 2\mathbf{i} + 5\mathbf{j}\,\text{m s}^{-1}$

The angle $\mathbf{v}$ makes with $\mathbf{i}$ is given by

$\tan\theta = \frac{5}{2} \Rightarrow \theta \approx 68.2°$.

When $t = 2$, the angle between the direction of motion of P and $\mathbf{i}$ is 68.2° (1 d.p.)

b $\mathbf{r} = \int \mathbf{v}\,dt = \int ((4t - 6)\mathbf{i} + (9 - t^2)\mathbf{j})\,dt$

$= (2t^2 - 6t)\mathbf{i} + \left(9t - \frac{t^3}{3}\right)\mathbf{j} + \mathbf{d}$

When $t = 3$, $\mathbf{r} = 20\mathbf{i} + 3\mathbf{j}$:

$20\mathbf{i} + 3\mathbf{j} = (18 - 18)\mathbf{i} + (27 - 9)\mathbf{j} + \mathbf{d}$

$= 18\mathbf{j} + \mathbf{d}$

$\mathbf{d} = 20\mathbf{i} - 15\mathbf{j}$

Hence

$\mathbf{r} = ((2t^2 - 6t)\mathbf{i} + \left(9t - \frac{t^3}{3}\right)\mathbf{j} + 20\mathbf{i} - 15\mathbf{j})\,\text{m}$

When $t = 0$, $\mathbf{r} = (20\mathbf{i} - 15\mathbf{j})\,\text{m}$:

$OP = |20\mathbf{i} - 15\mathbf{j}| = \sqrt{20^2 + 15^2} = 25\,\text{m}$

When $t = 0$, the distance of P from O is 25 m.

The direction of motion of P is the direction of the velocity vector of P. Your first step is to find the velocity by integrating the acceleration.

You then use the fact that the velocity is $6\mathbf{i}\,\text{m s}^{-1}$ when $t = 3$ to find the constant of integration.

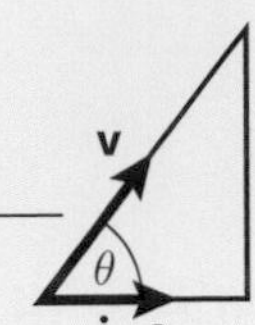

You find the angle the velocity vector makes with **i** using trigonometry.

You find the position vector by integrating the velocity vector. Remember to include the constant of integration.

The constant of integration is a vector. This constant is different from the constant in part **a** so you should give it a different letter.

Watch out Read the question carefully to work out whether you need to find a vector or a scalar quantity. The **distance** from O is the magnitude of the displacement vector, so use Pythagoras' Theorem.

Example 10

The velocity of a particle P at time t seconds is $((3t^2 - 8)\mathbf{i} + 5\mathbf{j})\,\text{m s}^{-1}$. When $t = 0$, the position vector of P with respect to a fixed origin O is $(2\mathbf{i} - 4\mathbf{j})\,\text{m}$.

a Find the position vector of P after t seconds.

A second particle Q moves with constant velocity $(8\mathbf{i} + 4\mathbf{j})\,\text{m s}^{-1}$. When $t = 0$, the position vector of Q with respect to the fixed origin O is $2\mathbf{i}\,\text{m}$.

b Prove that P and Q collide.

a Let the position vector of P after t seconds be $\mathbf{p}$ metres.

$\mathbf{p} = \int \mathbf{v}\,dt = \int((3t^2 - 8)\mathbf{i} + 5\mathbf{j})\,dt$

$= (t^3 - 8t)\mathbf{i} + 5t\mathbf{j} + \mathbf{c}$

When $t = 0$, $\mathbf{p} = 2\mathbf{i} - 4\mathbf{j}$:

$2\mathbf{i} - 4\mathbf{j} = 0\mathbf{i} + 0\mathbf{j} + \mathbf{c} \Rightarrow \mathbf{c} = 2\mathbf{i} - 4\mathbf{j}$

Hence

$\mathbf{p} = (t^3 - 8t)\mathbf{i} + 5t\mathbf{j} + 2\mathbf{i} - 4\mathbf{j}$

$= (t^3 - 8t + 2)\mathbf{i} + (5t - 4)\mathbf{j}$

The position vector of P after t seconds is $((t^3 - 8t + 2)\mathbf{i} + (5t - 4)\mathbf{j})$ m.

b Let the position vector of Q after t seconds be $\mathbf{q}$ m.

$\mathbf{r} = \mathbf{r}_0 + \mathbf{v}t$

$\mathbf{q} = 2\mathbf{i} + (8\mathbf{i} + 4\mathbf{j})t = (8t + 2)\mathbf{i} + 4t\mathbf{j}$

Equating the position vectors of P and Q:

$(t^3 - 8t + 2)\mathbf{i} + (5t - 4)\mathbf{j} = (8t + 2)\mathbf{i} + 4t\mathbf{j}$

Equate coefficients of $\mathbf{j}$: $5t - 4 = 4t$

$\Rightarrow t = 4$

Check with coefficients of $\mathbf{i}$:

When $t = 4$, $t^3 - 8t + 2 = 4^3 - 8(4) + 2$

$= 34$

and $8t + 2 = 8(4) + 2 = 34$

So the particles will collide when $t = 4$ seconds.

There are two position vectors in this question and to write them both as **r** m would be confusing. It is sensible to write the position vector of P as **p** m and the position vector of Q as **q** m.

Use the equation for the position vector of a particle moving with constant velocity. You could also integrate 8**i** + 4**j** with the boundary condition **q** = 2**i** when $t = 0$.

Problem-solving

Equate the position vectors for each particle. If they collide there will be a single value of t for which **p** = **q**. This means that the coefficients of **i** will be equal **and** the coefficients of **j** will be equal.

The coefficient of **i** involves a t^3 term so it is easier to start by equating the **j** components.

Now check **i** as well, as the particles only collide if **both** coefficients match.

Exercise 8E

(E) **1** A particle P starts from rest at a fixed origin O. The acceleration of P at time t seconds (where $t \geqslant 0$) is $(6t^2\mathbf{i} + (8 - 4t^3)\mathbf{j})\,\text{m s}^{-2}$. Find:

a the velocity of P when $t = 2$ **(3 marks)**

b the position vector of P when $t = 4$. **(3 marks)**

(E) **2** A particle P is moving in a plane with velocity $\mathbf{v}\,\text{m s}^{-1}$ at time t seconds where

$$\mathbf{v} = (3t^2 + 2)\mathbf{i} + (6t - 4)\mathbf{j},\ t \geqslant 0$$

When $t = 2$, P has position vector $9\mathbf{j}$ m with respect to a fixed origin O. Find:

a the distance of P from O when $t = 0$ **(4 marks)**

b the acceleration of P at the instant when it is moving parallel to the vector **i**. **(4 marks)**

(E) **3** At time t seconds, where $t \geqslant 0$, the particle P is moving in a plane with velocity $\mathbf{v}\,\text{m s}^{-1}$ and acceleration $\mathbf{a}\,\text{m s}^{-2}$, where $\mathbf{a} = (2t - 4)\mathbf{i} + 6\sin t\mathbf{j}$.

Given that P is instantaneously at rest when $t = \frac{\pi}{2}$ seconds, find:

a $\mathbf{v}$ in terms of π and t **(5 marks)**

b the exact speed of P when $t = \frac{3\pi}{2}$ **(3 marks)**

E/P **4** At time t seconds (where $t \geqslant 0$), the particle P is moving in a plane with acceleration $\mathbf{a}\,\text{m}\,\text{s}^{-2}$, where

$$\mathbf{a} = (5t - 3)\mathbf{i} + (8 - t)\mathbf{j}$$

When $t = 0$, the velocity of P is $(2\mathbf{i} - 5\mathbf{j})\,\text{m}\,\text{s}^{-1}$. Find:

a the velocity of P after t seconds **(3 marks)**

b the value of t for which P is moving parallel to $\mathbf{i} - \mathbf{j}$ **(4 marks)**

c the speed of P when it is moving parallel to $\mathbf{i} - \mathbf{j}$. **(3 marks)**

E/P **5** At time t seconds (where $t \geqslant 0$), a particle P is moving in a plane with acceleration $(2\mathbf{i} - 2t\mathbf{j})\,\text{m}\,\text{s}^{-2}$. When $t = 0$, the velocity of P is $2\mathbf{j}\,\text{m}\,\text{s}^{-1}$ and the position vector of P is $6\mathbf{i}\,\text{m}$ with respect to a fixed origin P.

a Find the position vector of P at time t seconds. **(5 marks)**

At time t seconds (where $t \geqslant 0$), a second particle Q is moving in the plane with velocity $((3t^2 - 4)\mathbf{i} - 2t\mathbf{j})\,\text{m}\,\text{s}^{-1}$. The particles collide when $t = 3$.

b Find the position vector of Q at time $t = 0$. **(4 marks)**

E **6** At time $t = 0$ a particle P is at rest at a point with position vector $(4\mathbf{i} - 6\mathbf{j})\,\text{m}$ with respect to a fixed origin O. The acceleration of P at time t seconds (where $t \geqslant 0$) is $((4t - 3)\mathbf{i} - 6t^2\mathbf{j})\,\text{m}\,\text{s}^{-2}$. Find:

a the velocity of P when $t = \frac{1}{2}$ **(5 marks)**

b the position vector of P when $t = 6$. **(5 marks)**

E **7** At time t seconds (where $t \geqslant 0$) the particle P is moving in a plane with acceleration $\mathbf{a}\,\text{m}\,\text{s}^{-2}$, where $\mathbf{a} = (8t^3 - 6t)\mathbf{i} + (8t - 3)\mathbf{j}$.

When $t = 2$, the velocity of P is $(16\mathbf{i} + 3\mathbf{j})\,\text{m}\,\text{s}^{-1}$. Find:

a the velocity of P after t seconds **(4 marks)**

b the value of t when P is moving parallel to $\mathbf{i}$. **(3 marks)**

E/P **8** At time t seconds the velocity of a particle P is $((4t - 3)\mathbf{i} + 4\mathbf{j})\,\text{m}\,\text{s}^{-1}$. When $t = 0$, the position vector of P is $(\mathbf{i} + 2\mathbf{j})$ m, relative to a fixed origin O.

a Find an expression for the position vector of P at time t seconds. **(4 marks)**

A second particle Q moves with constant velocity $(5\mathbf{i} + k\mathbf{j})\,\text{m}\,\text{s}^{-1}$.
When $t = 0$, the position vector of Q is $(11\mathbf{i} + 5\mathbf{j})\,\text{m}$.

b Given that the particles P and Q collide, find:

i the value of k

ii the position vector of the point of collision. **(6 marks)**

Challenge

A particle P is moving in a plane. At time t seconds, P is moving with velocity $\mathbf{v}$ m s^{-1}, where $\mathbf{v} = 3t\cos t\mathbf{i} + 5t\mathbf{j}$. Given that P is initially at the point with position vector $4\mathbf{i} + \mathbf{j}$ m relative to a fixed origin O, find the position vector of P when $t = \frac{\pi}{2}$.

Mixed exercise 8

1 A constant force $\mathbf{F}$ N acts on a particle of mass 4 kg for 5 seconds. The particle was initially at rest, and after 5 seconds it has velocity $6\mathbf{i} - 8\mathbf{j}$ m s^{-1}. Find $\mathbf{F}$.

(P) 2 A force $2\mathbf{i} - \mathbf{j}$ N acts on a particle of mass 2 kg. If the initial velocity of the particle is $\mathbf{i} + 3\mathbf{j}$ m s^{-1}, find the distance of the particle from its initial position after 3 seconds.

3 In this question $\mathbf{i}$ and $\mathbf{j}$ are the unit vectors due east and north respectively. At 2 pm the coastguard spots a rowing dinghy 500 m due south of a fixed observation point. The dinghy has constant velocity $(2\mathbf{i} + 3\mathbf{j})$ m s^{-1}.

a Find, in terms of t, the displacement vector of the dinghy relative to the observation point t seconds after 2 pm.

b Find the distance of the dinghy from the observation point at 2.05 pm.

(E/P) 4 In this question $\mathbf{i}$ and $\mathbf{j}$ are the unit vectors due east and north respectively. At 8 am two ships A and B have position vectors $\mathbf{r}_A = (\mathbf{i} + 3\mathbf{j})$ km and $\mathbf{r}_B = (5\mathbf{i} - 2\mathbf{j})$ km relative to a fixed origin, O. Their velocities are $\mathbf{v}_A = (2\mathbf{i} - \mathbf{j})$ km h^{-1} and $\mathbf{v}_B = (-\mathbf{i} + 4\mathbf{j})$ km h^{-1} respectively.

a Write down the position vectors of A and B t hours later. **(3 marks)**

b Show that t hours after 8 am the displacement vector of B relative to A is given by

$((4 - 3t)\mathbf{i} + (-5 + 5t)\mathbf{j})$ km **(2 marks)**

c Show that the two ships do not collide. **(3 marks)**

d Find the distance between A and B at 10 am. **(3 marks)**

5 A particle is projected with velocity $(8\mathbf{i} + 10\mathbf{j})$ m s^{-1}, where $\mathbf{i}$ and $\mathbf{j}$ are unit vectors horizontally and vertically respectively, from a point O at the top of a cliff and moves freely under gravity. Six seconds after projection, the particle strikes the sea at the point S. Calculate:

a the horizontal distance between O and S

b the vertical distance between O and S.

 **6**

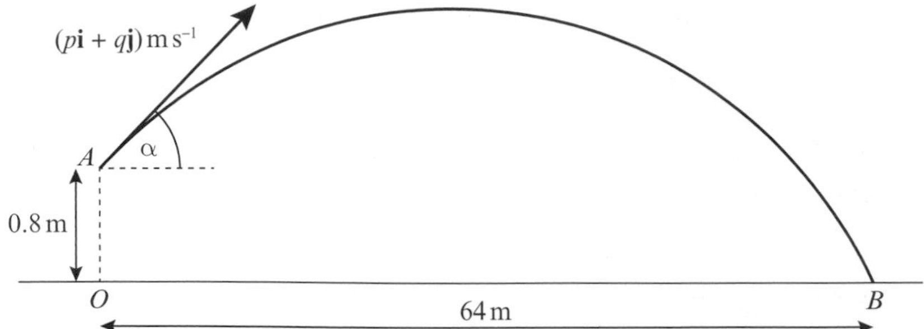

A cricket ball is hit from a point A with velocity of $(p\mathbf{i} + q\mathbf{j})\,\text{m s}^{-1}$, at an angle α above the horizontal. $\mathbf{i}$ and $\mathbf{j}$ are the unit vectors horizontally and vertically upwards respectively. The point A is 0.8 m vertically above the point O, which is on horizontal ground.

The ball takes 4 seconds to travel from A to B, where B is on the ground and $OB = 64\,\text{m}$, as shown in the diagram. By modelling the motion of the ball as that of a particle moving freely under gravity,

a find the value of p and the value of q **(5 marks)**

b find the initial speed of the ball **(2 marks)**

c find the exact value of $\tan\alpha$ **(1 mark)**

d find the length of time for which the cricket ball is at least 5 m above the ground. **(6 marks)**

e State an additional physical factor which may be taken into account in a refinement of the above model to make it more realistic. **(1 mark)**

E **7** A particle P moves in a straight line in such a way that, at time t seconds, its velocity, $\mathbf{v}\,\text{m s}^{-1}$, is given by

$$\mathbf{v} = \begin{cases} t\sqrt{14 + 2t^2}, & 0 \leqslant t \leqslant 5 \\ \dfrac{1000}{t^2}, & t > 5 \end{cases}$$

When $t = 0$, P is at the point O. Calculate the displacement of P from O:

a when $t = 5$ **(3 marks)**

b when $t = 6$. **(3 marks)**

E/P **8** A particle P of mass 0.4 kg is moving in a straight line under the action of a single variable force of magnitude F N. At time t seconds (where $t \geqslant 0$) the displacement x m of P from a fixed point O is given by $x = 2t + \frac{k}{t+1}$, where k is a constant. Given that when $t = 0$, the velocity of P is $6\,\text{m s}^{-1}$, find:

a the value of k **(5 marks)**

b the distance of P from O when $t = 0$ **(1 mark)**

c the value of F when $t = 3$. **(4 marks)**

E **9** A ball, attached to the end of an elastic string, is moving in a vertical line. The motion of the ball is modelled as a particle B moving along a vertical line so that its displacement, x m, from a fixed point O on the line at time t seconds is given by $x = 0.6\cos\left(\frac{\pi t}{3}\right)$. Find:

a the distance of B from O when $t = \frac{1}{2}$ **(2 marks)**

b the smallest positive value of t for which B is instantaneously at rest **(4 marks)**

c the magnitude of the acceleration of B when $t = 1$. Give your answer to 3 significant figures. **(3 marks)**

(E) **10** A light spot S moves along a straight line on a screen. At time $t = 0$, S is at a point O. At time t seconds (where $t \geqslant 0$) the distance, x cm, of S from O is given by $x = 4te^{-0.5t}$. Find:

a the acceleration of S when $t = \ln 4$ **(5 marks)**

b the greatest distance of S from O. **(2 marks)**

(E/P) **11** Two particles P and Q move in a plane so that at time t seconds, where $t \geqslant 0$, P and Q have position vectors $\mathbf{r}_P$ metres and $\mathbf{r}_Q$ metres respectively, relative to a fixed origin O, where

$$\mathbf{r}_P = (3t^2 + 4)\mathbf{i} + \left(2t - \tfrac{1}{2}\right)\mathbf{j}$$

$$\mathbf{r}_Q = (t + 6)\mathbf{i} + \frac{3t^2}{2}\mathbf{j}$$

Find:

a the velocity vectors of P and Q at time t seconds **(5 marks)**

b the speed of P when $t = 2$ **(2 marks)**

c the value of t at the instant when the particles are moving parallel to one another. **(4 marks)**

d Show that the particles collide and find the position vector of their point of collision. **(6 marks)**

12 At time t seconds, a particle P has position vector $\mathbf{r}$ m with respect to a fixed origin O, where

$$\mathbf{r} = (3t^2 - 4)\mathbf{i} + (8 - 4t^2)\mathbf{j}$$

a Show that the acceleration of P is a constant.

b Find the magnitude of the acceleration of P and the size of the angle which the acceleration makes with $\mathbf{j}$.

(E) **13** At time t seconds, a particle P has position vector $\mathbf{r}$ m with respect to a fixed origin O, where

$$\mathbf{r} = 2\cos 3t\mathbf{i} - 2\sin 3t\mathbf{j}$$

a Find the velocity of P when $t = \frac{\pi}{6}$ **(5 marks)**

b Show that the magnitude of the acceleration of P is constant. **(4 marks)**

(E/P) **14** A particle of mass 0.5 kg is acted upon by a variable force $\mathbf{F}$. At time t seconds, the velocity $\mathbf{v}$ m s^{-1} is given by $\mathbf{v} = (4ct - 6)\mathbf{i} + (7 - c)t^2\mathbf{j}$, where c is a constant.

a Show that $\mathbf{F} = (2c\mathbf{i} + (7 - c)t\mathbf{j})$ N. **(4 marks)**

b Given that when $t = 5$ the magnitude of $\mathbf{F}$ is 17 N, find the possible values of c. **(5 marks)**

(E) **15** At time t seconds (where $t \geqslant 0$) the particle P is moving in a plane with acceleration $\mathbf{a}$ m s^{-2}, where $\mathbf{a} = (8t^3 - 6t)\mathbf{i} + (8t - 3)\mathbf{j}$.

When $t = 2$, the velocity of P is $(16\mathbf{i} + 3\mathbf{j})$ m s^{-1}. Find:

a the velocity of P after t seconds **(3 marks)**

b the value of t when P is moving parallel to $\mathbf{i}$. **(4 marks)**

(E) **16** A particle P moves so that its acceleration $\mathbf{a}$ m s^{-2} at time t seconds, where $t \geqslant 0$, is given by

$$\mathbf{a} = 4t\mathbf{i} + 5t^{-\frac{1}{2}}\mathbf{j}$$

When $t = 0$, the velocity of P is $10\mathbf{i}$ m s^{-1}.

Find the speed of P when $t = 5$. **(6 marks)**

(E/P) **17** In this question $\mathbf{i}$ and $\mathbf{j}$ are horizontal unit vectors due east and due north respectively.

A clockwork train is moving on a flat, horizontal floor. At time $t = 0$, the train is at a fixed point O and is moving with velocity $3\mathbf{i} + 13\mathbf{j}$ m s^{-1}. The velocity of the train at time t seconds is $\mathbf{v}$ m s^{-1}, and its acceleration, $\mathbf{a}$ m s^{-2}, is given by $\mathbf{a} = 2t\mathbf{i} + 3\mathbf{j}$.

a Find $\mathbf{v}$ in terms of t. **(3 marks)**

b Find the value of t when the train is moving in a north-east direction. **(3 marks)**

Challenge

1 A particle moves on the positive x-axis such that its displacement, s m, from O at time t seconds is given by

$$s = (20 - t^2)\sqrt{t + 1},\ t \geqslant 0$$

a State the initial displacement of the particle.

b Show that the particle changes direction exactly once and determine the time at which this occurs.

c Find the exact speed of the particle when it crosses O.

2 Relative to a fixed origin O, the particle R has position vector $\mathbf{r}$ metres at time t seconds, where

$$\mathbf{r} = (6\sin \omega t)\mathbf{i} + (4\cos \omega t)\mathbf{j}$$

and ω is a positive constant.

a Find $\dot{\mathbf{r}}$ and hence show that $v^2 = 2\omega^2(13 + 5\cos 2\omega t)$, where v m s^{-1} is the speed of R at time t seconds.

b Deduce that $4\omega \leqslant v \leqslant 6\omega$.

c At the instant when $t = \frac{\pi}{3\omega}$, find the angle between $\mathbf{r}$ and $\dot{\mathbf{r}}$, giving your answer in degrees to one decimal place.

Summary of key points

1 If a particle starts from the point with position vector $\mathbf{r}_0$ and moves with constant velocity $\mathbf{v}$, then its displacement from its initial position at time t is $\mathbf{v}t$ and its position vector $\mathbf{r}$ is given by $\mathbf{r} = \mathbf{r}_0 + \mathbf{v}t$.

2 For an object moving in a plane with constant acceleration:

- $\mathbf{v} = \mathbf{u} + \mathbf{a}t$
- $\mathbf{r} = \mathbf{u}t + \frac{1}{2}\mathbf{a}t^2$

where

- $\mathbf{u}$ is the initial velocity
- $\mathbf{a}$ is the acceleration
- $\mathbf{v}$ is the velocity at time t
- $\mathbf{r}$ is the displacement at time t.

3 If $\mathbf{r} = x\mathbf{i} + y\mathbf{j}$, then $\mathbf{v} = \frac{d\mathbf{r}}{dt} = \dot{\mathbf{r}} = \dot{x}\mathbf{i} + \dot{y}\mathbf{j}$

and $\mathbf{a} = \frac{d\mathbf{v}}{dt} = \frac{d^2\mathbf{r}}{dt^2} = \ddot{\mathbf{r}} = \ddot{x}\mathbf{i} + \ddot{y}\mathbf{j}$

4 $\mathbf{v} = \int \mathbf{a}\,dt$ and $\mathbf{r} = \int \mathbf{v}\,dt$

2 Review exercise

(E) **1**

A uniform plank AB of length 5 m and weight 200 N, rests in a horizontal position on supports at C and D, where $AC = 0.5$ m and $BD = 0.75$ m. A builder of weight 800 N stands on the plank at M where $AM = 2$ m, as shown in the diagram. The builder is modelled as a particle and the plank is modelled as a rod. Calculate:

a the magnitude of the reaction at C **(3)**

b the magnitude of the reaction at D. **(3)**

c State how you have used the modelling assumption that the builder is a particle. **(1)**

← Section 4.3

(E/P) **2**

A uniform plank AC of length $5l$ m and mass m kg, rests in a horizontal position on supports at B and C, where $AB = l$ m and $BC = 4l$ m. The plank is modelled as a rod. Show that:

a the magnitude of the reaction at B is $\frac{5}{8}mg$ **(3)**

b the magnitude of the reaction at C is $\frac{3}{8}mg$. **(3)**

c State how you have used the modelling assumption that the plank is:

i uniform **(1)**

ii a rod. **(1)**

← Section 4.3

(E) **3**

A uniform rod AD of length 10 m and weight 500 N, rests in a horizontal position on supports at B and C, where $AB = 2$ m and $BC = 4$ m.

a Calculate the largest weight that can be placed at D before the rod starts to tip. **(3)**

b Calculate the largest weight that can be placed at A before the rod starts to tip. **(3)**

← Sections 4.3, 4.5

4

200 N　　2000 N

A lever consists of a uniform steel rod AB of weight 200 N and length 3 m, which rests on a pivot at C. A 2000 N weight is placed at B, and is supported by a force of 200 N applied vertically downwards at A. Given that the lever is in equilibrium, calculate the length CB.

← Sections 4.3, 4.5

5 A particle of mass 3 kg is moving up a rough slope that is inclined at an angle α to the horizontal where $\tan\alpha = \frac{5}{12}$. A force of magnitude P N acts horizontally on the particle towards the plane. Given that the coefficient of friction between the particle and the slope is 0.2 and that the particle is moving at a constant velocity, calculate the value of P.

← Sections 5.2, 5.3

(P) **6** A particle of mass 2 kg sits on a smooth slope that is inclined at 45° to the horizontal. A force of F N acts at an angle of 30° to the plane on the particle causing it to accelerate up the hill at 2 m s^{-2}.

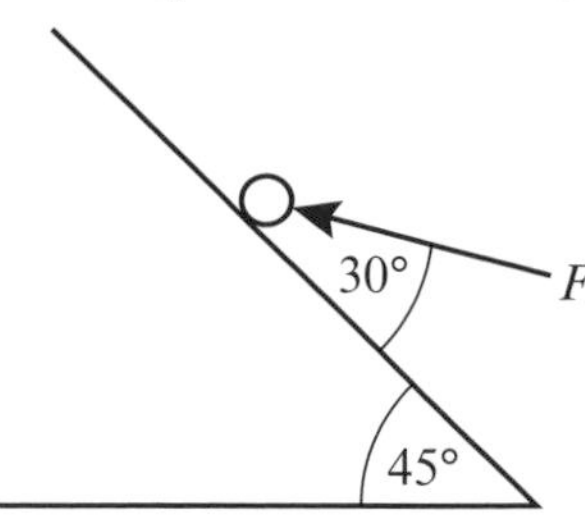

Show that $F = \frac{2}{\sqrt{3}}(4 + \sqrt{2}g)$ N.

← Sections 5.2, 5.3

(E/P) **7** A shipping container of mass 15000 kg is being pulled by a winch up a rough slope that is inclined at 10° to the horizontal. The winch line imparts a constant force of 42000 N, which acts parallel to and up the slope, causing the shipping container to accelerate at a constant rate of 0.1 m s^{-2}. Calculate:

a the reaction between the shipping container and the slope **(2)**

b the coefficient of friction, μ, between the shipping container and the slope. **(3)**

When the shipping container is travelling at 2 m s^{-1} the engine is turned off.

c Find the time taken for the shipping container to come to rest. **(3)**

d Determine whether the shipping container will remain at rest, justifying your answer carefully. **(2)**

← Sections 5.2, 5.3

(E) **8** A ball is projected horizontally from a tabletop at a height of 0.8 m above level ground. Given that the initial velocity of the ball is 2 m s^{-1}, find:

a the time taken for the ball to reach the ground **(3)**

b the horizontal distance between the table edge and the point where the ball lands. **(2)**

← Section 6.1

(E/P) **9** A football is kicked horizontally off a 20 m platform and lands a distance of 40.0 m from the edge of the platform.

a Find the initial horizontal velocity of the football. **(5)**

b State two assumptions you have made in your calculations, and comment on the validity of each assumption. **(2)**

← Section 6.1

(E) **10** A projectile is launched from a point on horizontal ground with speed 150 m s^{-1} at an angle of 10° above the horizontal. Find:

a the time the projectile takes to reach its highest point above the ground **(4)**

b the range of the projectile. **(4)**

← Sections 6.2, 6.3

(E/P) **11**

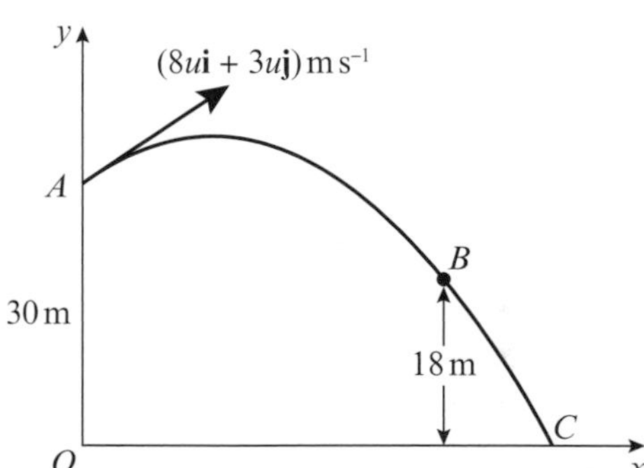

In this question, the unit vectors **i** and **j** are in a vertical plane, **i** being horizontal and **j** being vertical.

A particle P is projected from a point A with position vector $30\mathbf{j}$ m with respect to a fixed origin O. The velocity of projection is $(8u\mathbf{i} + 3u\mathbf{j})\text{ m s}^{-1}$. The particle moves freely under gravity, passing through a point B, which has position vector $(k\mathbf{i} + 18\mathbf{j})$ m, where k is a constant, before reaching the point C on the x-axis, as shown in the figure above. The particle takes 3 s to move from A to B.

Find:

a the value of u **(4)**

b the value of k **(2)**

c the angle the velocity of P makes with the x-axis as it reaches C. **(6)**

← Sections 6.2, 6.3

(E) **12** A particle P is projected from the origin with velocity $(12\mathbf{i} + 24\mathbf{j})$ m s^{-1}, where **i** and **j** are horizontal and vertical unit vectors respectively. The particle moves freely under gravity. Find:

a the position vector of P after 3 s **(4)**

b the speed of P after 3 s. **(4)**

← Sections 6.2, 6.3

(P) **13** A projectile is launched from a point on a horizontal plane with initial speed u m s^{-1} at an angle of elevation α. The particle moves freely under gravity until it strikes the plane. The range of the projectile is R m.

a Show that the time of flight of the particle is $\frac{2u\sin\alpha}{g}$ seconds.

b Show that $R = \frac{u^2\sin 2\alpha}{g}$.

c Deduce that, for a fixed u, the greatest possible range is when $\alpha = 45°$.

d Given that $R = \frac{2u^2}{5g}$, find the two possible values of the angle of elevation at which the projectile could have been launched. ← Section 6.4

(E/P) **14** A smooth bead B of mass 1 kg is threaded on a light inextensible string. The ends of the string are attached to two fixed points A and C where A is vertically above C. The bead is held in equilibrium by a horizontal force F. AB and BC make angles of 30° and 60° respectively with the vertical, as shown in diagram.

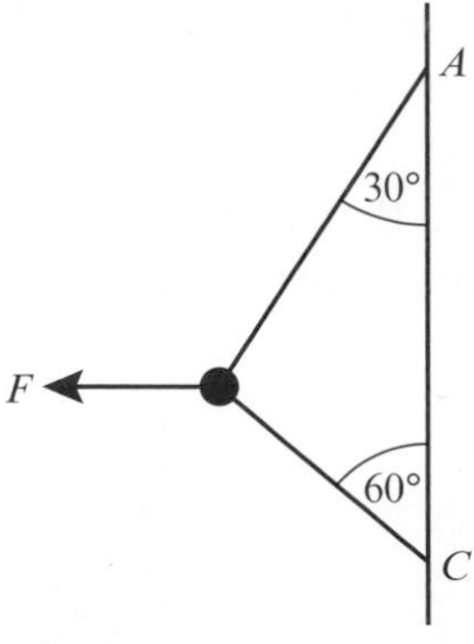

a Show that the tension in the string is $\frac{2g}{\sqrt{3}-1}$ N. **(3)**

b Calculate the magnitude of F. **(3)**

c State how you have used the fact that the bead is smooth in your calculations. **(1)**

← Section 7.2

(E) **15** A crate of mass 500 kg sits on a hill which is inclined at an angle α to the horizontal where $\tan\alpha = \frac{7}{24}$. The coefficient of friction between the hill and the crate is 0.15 , and the crate is held at rest by a force of magnitude F N which acts parallel to and up the line of greatest slope of the hill.

By modelling the crate as a particle,

a show that the normal reaction of the hill on the crate is $480g$ N **(3)**

b work out the minimum value of F. **(3)**

← Section 7.3

(E/P) **16** A ladder PQ of mass 25 kg and length 6 metres, rests with its base, P, on rough horizontal ground and its top, Q, leaning against a smooth vertical wall. The coefficient of friction between the ladder and the ground is 0.25. The ladder lies in a vertical plane perpendicular to the wall and the ground, and is inclined at an angle 60° to the horizontal.

A builder of mass 75 kg climbs up the ladder. Modelling the builder as a particle and the ladder as a uniform rod, find the maximum distance up the ladder the builder can climb before the ladder begins to slip. **(10)**

← Section 7.4

E/P **17** A uniform ladder PQ of mass m kg and length l metres, rests with one end P on rough horizontal ground and the other end Q against a smooth vertical wall. The coefficient of friction between the ladder and the ground is μ. The ladder lies in a vertical plane perpendicular to the wall and the ground, and is inclined at an angle α to the horizontal. Given that the ladder is on the point of slipping, find an expression for μ in terms of α. **(10)**

← Section 7.4

E/P **18** A non-uniform ladder AB of weight 240 N and length 6 m rests with its end A on smooth horizontal ground and its end B against a rough vertical wall. The coefficient of friction between the ladder and the wall is 0.3.The centre of mass of the ladder is 2 m from A. The ladder lies in a vertical plane perpendicular to the wall and the ground, and is inclined at an angle α to the horizontal, where $\tan\alpha = \frac{3}{2}$. The ladder can be prevented from sliding down the wall by applying a horizontal force of magnitude P N to the bottom of the ladder. By modelling the ladder as a non-uniform rod determine the minimum value of P. **(10)**

← Section 7.4

E/P **19** A sled of mass 5 kg is released from rest on a hill that is angled at α to the horizontal where $\tan\alpha = \frac{1}{5}$. The coefficient of friction between the sled and the hill is 0.15. By modelling the sled as a particle work out how long it takes the sled to travel 200 m. **(6)**

← Section 7.5

E/P **20** At 10 am two aeroplanes P and Q have position vectors $\mathbf{r}_P = (400\mathbf{i} + 200\mathbf{j})$ km and $\mathbf{r}_Q = (500\mathbf{i} - 100\mathbf{j})$ km relative to a fixed origin O. Their velocities are $\mathbf{v}_P = (300\mathbf{i} + 250\mathbf{j})$ km h^{-1} and $\mathbf{v}_Q = (600\mathbf{i} - 200\mathbf{j})$ km h^{-1}.

a Write down expressions for the position vectors of P and Q after a time t hours. **(4)**

b Find the displacement vector of Q relative to P at 10 am. **(2)**

c Work out the distance between P and Q at noon. **(4)**

← Section 8.1

E/P **21** A particle P of mass 2 kg moves in a straight line under the action of a variable force F N. At time t ($t \geqslant 0$), the displacement x m of P from a fixed point O is given by $x = 3t - \frac{2k}{2t-1}$, where k is a constant. When $t = 0$, the velocity of P is 10 m s^{-1}.

a Show that $k = \frac{7}{4}$ **(4)**

b Find the distance of P from O when $t = 2$ s. **(2)**

← Sections 8.3, 8.4

E **22** A particle P moves in a plane such that at time t seconds, where $t \geqslant 0$, it has position vector

$$\mathbf{r} = \left(\left(\tfrac{1}{3}t^3 + 2t\right)\mathbf{i} + \left(\tfrac{1}{2}t^2 - 1\right)\mathbf{j}\right)\text{ m}$$

Find:

a the velocity vector of P at time t seconds **(2)**

b the speed of P when $t = 5$ s **(3)**

c the magnitude and direction of the acceleration of P when $t = 2$ s. **(4)**

← Sections 8.3, 8.4

E **23** A particle is acted upon by a variable force F. At time t seconds the displacement of the particle in metres relative to a fixed origin O is given by

$$\mathbf{r} = (4t^2 + 1)\mathbf{i} + (2t^2 - 3)\mathbf{j}$$

a Find the velocity of the particle when $t = 3$ s. **(3)**

b Show that the acceleration of the particle is constant. **(2)**

← Sections 8.3, 8.4

E/P **24** A particle P moves so that its velocity $\mathbf{v}$ m s^{-1} at time t seconds, where $t \geqslant 0$, is given by $\mathbf{v} = -2t\mathbf{i} + 3\sqrt{t}\mathbf{j}$. When $t = 0$, the displacement of P relative to a fixed origin is $2\mathbf{j}$ m.

Find the distance of P from O when $t = 4$ s. **(7)**

← Sections 8.3, 8.5

E **25** A particle moves in a plane with acceleration $\mathbf{a}$ m s^{-2} where

$$\mathbf{a} = t(2 - 3t^2)\mathbf{i} - 4(2t + 1)\mathbf{j},\ t \geqslant 0$$

When $t = 0$, the velocity of P is $(3\mathbf{i} + \mathbf{j})$ m s^{-1}. Find:

a the velocity of P after t s **(4)**

b the time at which P is moving in the direction of $\mathbf{i}$. **(2)**

← Sections 8.3, 8.5

E **26** In this question $\mathbf{i}$ and $\mathbf{j}$ are horizontal unit vectors due east and due north respectively.

A wind surfer is surfing on a lake. The acceleration of the wind surfer at time t s is given by $\mathbf{a} = (-4t\mathbf{i} - 2\mathbf{j})$ m s^{-2}. At time $t = 0$ s the windsurfer is moving directly east at a speed of 8 m s^{-1}.

a Find $\mathbf{v}$ in terms of t. **(4)**

b Find the value of t when the windsurfer is moving in a southerly direction. **(3)**

← Sections 8.3, 8.5

Challenge

1

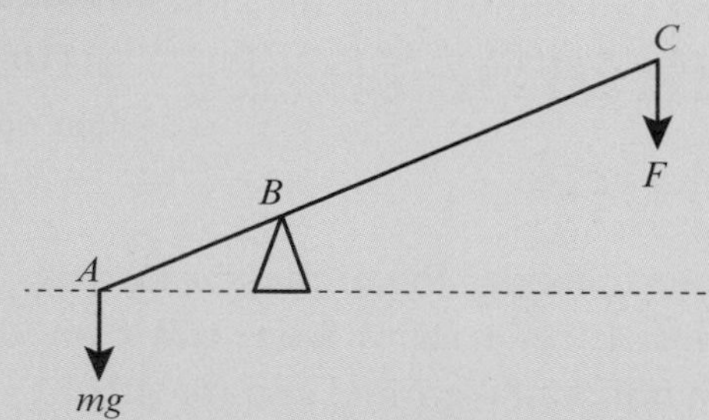

A lever consists of a uniform steel rod AC of weight 100 N and length $2k$ m, which rests on a pivot at B that has a height of $0.3k$ m. $AB = 0.5k$ m. A mass m kg is attached to the lever at A. The mass is lifted by means of a force of magnitude F N that is applied vertically downwards at C. Show that $F > \frac{1}{3}(mg - 100)$.

← Section 4.3, 4.5

2 A particle P travels in a straight line such that its velocity, v m s^{-1} at time t seconds, is given by

$$v = 3\sin kt + \cos kt,\ t \geqslant 0$$

where k is a constant and angles are measured in degrees. At time $t = 0$, the particle is at a fixed origin, O, and has acceleration 1.5 m s^{-2}.

Work out the maximum distance of the particle from the origin in its subsequent motion, and the first time at which this occurs.

← Section 8.1

3 A straight hill slopes upwards at an angle of θ to the horizontal, where $0 \leqslant \theta < 90°$. A projectile is launched perpendicular to the plane of the hill, with an initial velocity of u m s^{-1}, and lands a distance d m down the hill.

Show that $d = \frac{2u^2}{g}\tan\theta\sec\theta$.

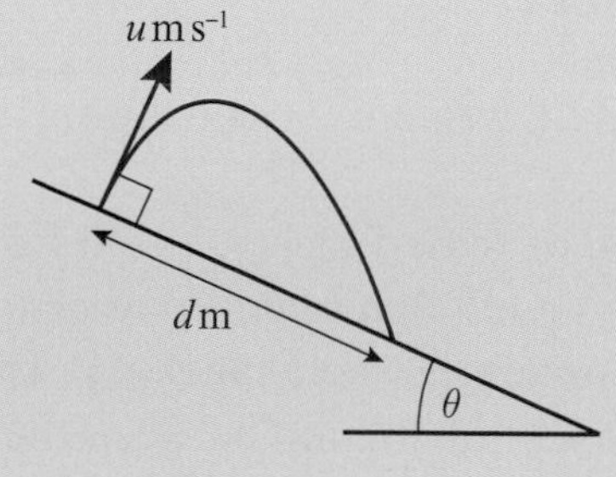

← Section 6.4

Exam-style practice

Mathematics
A Level
Paper 3: Statistics and Mechanics

Time: 2 hours
You must have: Mathematical Formulae and Statistical Tables, Calculator

---

SECTION A: STATISTICS

1 An electrical engineer makes components for computer systems. She claims that the components last longer than 500 hours on 52% of occasions.

In a random sample of 40 of the components, X last longer than 500 hours.

a Find $P(X \geqslant 22)$. **(1)**

b Write down two conditions under which the normal approximation may be used as an approximation to the binomial distribution. **(2)**

A random sample of 250 components was taken and 120 lasted longer than 500 hours.

c Assuming the engineer's claim to be correct, use a normal approximation to find the probability that 120 or fewer components last longer than 500 hours. **(3)**

d Using your answer to part **c**, comment on the engineer's claim. **(1)**

2 $P(A) = 0.4$, $P(B) = 0.55$ and $P(C) = 0.26$.

Given that $P(A \cap B) = 0.2$, that events A and C are mutually exclusive and that events B and C are statistically independent,

a Draw a Venn diagram to illustrate events A, B and C. **(5)**

b Show that events A and B are not statistically independent. **(2)**

c Find $P(A|B')$. **(2)**

d Find $P(C|(A \cap B)')$. **(2)**

3 The daily mean air temperature, t °C, is recorded in Perth for the month of October.

t	$12 \leqslant t < 15$	$15 \leqslant t < 18$	$18 \leqslant t < 20$	$20 \leqslant t < 22$	$22 \leqslant t < 26$
f	2	6	11	7	5

a State, with a reason, whether t is a discrete or continuous variable. **(1)**

b Use your calculator to find estimates for the mean and standard deviation of the temperatures. **(2)**

c Give two reasons why a histogram could be used to display this data. **(2)**

d Use linear interpolation to find the 10th to 90th interpercentile range. **(3)**

A meteorologist believes that there is a positive correlation between the daily mean air temperature and the number of hours of sunshine. She takes a random sample of 8 days from the data set above and finds that the product moment correlation coefficient is 0.612.

e Stating your hypotheses clearly, test at the 5% level of significance, whether or not the product moment correlation coefficient for the population is greater than zero. **(3)**

4 An industrial chemical process produces an amount of a substance, q grams, dependent on the temperature, t °C applied. The table below shows the outcomes of five experiments.

t	10	20	32	41	57
q	160	700	2000	3300	6400

A chemist believes that the relationship between the variables can be modelled by an equation of the form $q = kt^n$, where k and n are constants to be determined. The data are coded using $x = \log t$ and $y = \log q$. The product moment correlation coefficient between x and y is found to be 0.9998.

a State with a reason whether this value supports the suggested model. **(1)**

b Given that the equation of the regression line of y on x is $y = 0.0761 + 2.1317x$, find the value of k and the value of n. **(3)**

c Explain, giving a reason, whether it would be sensible to use this model to predict the amount of substance produced when $t = 85$ °C. **(1)**

5 The weights of cats in a particular town are normally distributed. A cat that weighs between 3.5 kg and 4.6 kg is said to be of 'standard' weight. Given that 2.5% of cats weigh less than 3.416 kg and 5% of cats weigh greater than 4.858 kg,

a find the proportion of cats that are of standard weight. **(6)**

15 cats are chosen at random.

b Find the probability that at least 10 of these cats are of standard weight. **(2)**

In a second town, the weights of cats are also normally distributed with standard deviation 0.51 kg. A random sample of 12 cats was taken and the sample mean was 4.73 kg.

c Test, at the 10% level of significance, whether or not the mean weight of all the cats in the town is different from 4.5 kg. State your hypotheses clearly. **(4)**

6 Jemima plays two games of tennis. The probability that she wins the first game is 0.62. If she wins the first game, the probability that she wins the second is 0.75. If she loses the first game, the probability that she wins the second is 0.45. Find the probability that she wins both games given that she wins the second game. **(4)**

SECTION B: MECHANICS

7 At time t seconds, where $t \geqslant 0$, a particle P moves such that its velocity, $\mathbf{v}$ m s^{-1}, is given by

$$\mathbf{v} = (2 - 6t^2)\mathbf{i} - t\mathbf{j}$$

When $t = 1$ the displacement of the particle from a fixed origin O is $5\mathbf{i}$ m.

Find the distance of the particle from O when $t = 3$ seconds, giving your answer to 3 significant figures. **(7)**

8 An arrow is fired from horizontal ground with an initial speed of $100\,\text{m s}^{1}$ at an angle of 30° above the horizontal.

By modelling the arrow as a particle work out:

a the time taken for the arrow to hit the ground **(3)**

b the maximum height of the arrow **(3)**

c the speed of the arrow after 3 seconds. **(6)**

9 In this question **i** and **j** are the unit vectors due east and north respectively.

A cyclist makes a journey between two points A and B. At time $t = 0$ s the cyclist is moving due east at $2\,\text{m s}^{-1}$.

The cyclist is modelled as a particle.

Relative to A, the position vector of the cyclist at time t seconds is **r** metres.

Given that the acceleration of the cyclist is constant and equal to $0.2\mathbf{i} - 0.8\mathbf{j}\,\text{m s}^{-2}$, find:

a the position vector of the cyclist after 10 seconds **(2)**

b the distance of the cyclist from A after 10 seconds. **(2)**

After 10 seconds the cyclist stops accelerating and heads due east at a constant speed of $5\,\text{m s}^{-1}$.

c Find the value of t when the cyclist is directly south-east of A. **(2)**

After a further 30 s the cyclist reached point B.

d Work out the bearing of B from A to the nearest degree. **(2)**

10

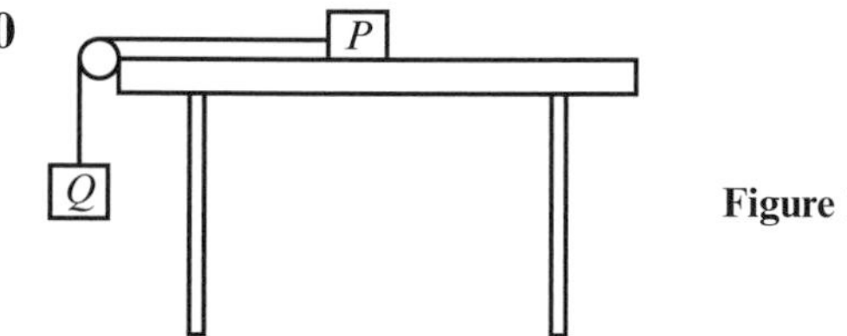

Figure 1

Two particles P and Q, of masses 3 kg and 2 kg respectively, are attached to the ends of a light inextensible string. P lies on a rough horizontal table. The string passes over a small smooth pulley fixed on the edge of the table. Q hangs freely below the pulley, as shown in Figure 1. The coefficient of friction between A and the table is μ. The particles are released from rest with the string taut. Immediately after release, P accelerates at a rate of $0.5\,\text{m s}^{-2}$.

a Find the tension in the string immediately after the particles begin to move. **(3)**

b Show that $\mu = 0.54$ (2 s.f.) **(3)**

After two seconds the string breaks.

c Assuming that P remains on the table, calculate how long it takes P to come to rest. **(6)**

d State how you have used the information that the string is inextensible in your calculations. **(1)**

11

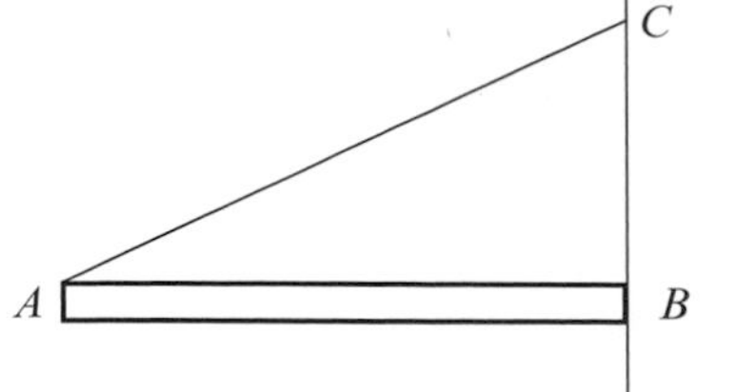

Figure 2

A uniform rod AB of length l m and mass m kg rests with one end touching a rough vertical wall at B.

The rod is kept horizontal by a light inextensible string AC where C lies on the wall directly above B.

The plane ABC is perpendicular to the wall and $\angle BAC$ is α, where $\tan \alpha = \frac{5}{12}$.

a Show that the tension in the string is $\frac{13}{10}mg$ N. **(4)**

b Calculate the coefficient of friction between the rod and the wall. **(6)**

Percentage points of the normal distribution

The values z in the table are those which a random variable $Z \sim N(0, 1)$ exceeds with probability p; that is, $P(Z > z) = 1\ \Phi - (z) = p$.

p	z	p	z
0.5000	0.0000	0.0500	1.6449
0.4000	0.2533	0.0250	1.9600
0.3000	0.5244	0.0100	2.3263
0.2000	0.8416	0.0050	2.5758
0.1500	1.0364	0.0010	3.0902
0.1000	1.2816	0.0005	3.2905

Critical values for correlation coefficients

This table concerns tests of the hypothesis that a population correlation coefficient ρ is 0. The values in the table are the minimum value which need to be reached by a sample correlation coefficient in order to be significant at the level shown, on a one-tailed test.

Product moment coefficient					
		Level			**Sample**
0.10	**0.05**	**0.025**	**0.01**	**0.005**	**Level**
0.8000	0.9000	0.9500	0.9800	0.9900	4
0.6870	0.8054	0.8783	0.9343	0.9587	5
0.6084	0.7293	0.8114	0.8822	0.9172	6
0.5509	0.6694	0.7545	0.8329	0.8745	7
0.5067	0.6215	0.7067	0.7887	0.8343	8
0.4716	0.5822	0.6664	0.7498	0.7977	9
0.4428	0.5494	0.6319	0.7155	0.7646	10
0.4187	0.5214	0.6021	0.6851	0.7348	11
0.3981	0.4973	0.5760	0.6581	0.7079	12
0.3802	0.4762	0.5529	0.6339	0.6835	13
0.3646	0.4575	0.5324	0.6120	0.6614	14
0.3507	0.4409	0.5140	0.5923	0.6411	15
0.3383	0.4259	0.4973	0.5742	0.6226	16
0.3271	0.4124	0.4821	0.5577	0.6055	17
0.3170	0.4000	0.4683	0.5425	0.5897	18
0.3077	0.3887	0.4555	0.5285	0.5751	19
0.2992	0.3783	0.4438	0.5155	0.5614	20
0.2914	0.3687	0.4329	0.5034	0.5487	21
0.2841	0.3598	0.4227	0.4921	0.5368	22
0.2774	0.3515	0.4133	0.4815	0.5256	23
0.2711	0.3438	0.4044	0.4716	0.5151	24
0.2653	0.3365	0.3961	0.4622	0.5052	25
0.2598	0.3297	0.3882	0.4534	0.4958	26
0.2546	0.3233	0.3809	0.4451	0.4869	27
0.2497	0.3172	0.3739	0.4372	0.4785	28
0.2451	0.3115	0.3673	0.4297	0.4705	29
0.2407	0.3061	0.3610	0.4226	0.4629	30
0.2070	0.2638	0.3120	0.3665	0.4026	40
0.1843	0.2353	0.2787	0.3281	0.3610	50
0.1678	0.2144	0.2542	0.2997	0.3301	60
0.1550	0.1982	0.2352	0.2776	0.3060	70
0.1448	0.1852	0.2199	0.2597	0.2864	80
0.1364	0.1745	0.2072	0.2449	0.2702	90
0.1292	0.1654	0.1966	0.2324	0.2565	100

Answers

Prior knowledge 1

1 a $A = \log 3, B = \log 2$
b Gradient = log 2, y-intercept = log 3
2 For each 1 cm increase in handspan, the height increases by approximately 11.3 cm.
3 $P(X \geqslant 32) = 0.0061 < 0.01$
H_0 can be rejected.

Exercise 1A

1 a $y = ax^n$ b $a = 15.8$ (3 s.f.), $n = 0.4$
2 a $y = kb^x$ b $k = 2.51, b = 39.8$ (3 s.f.)
3 $a = 1 \times 10^{172}, n = -2.739$ (3 d.p.)
4 a

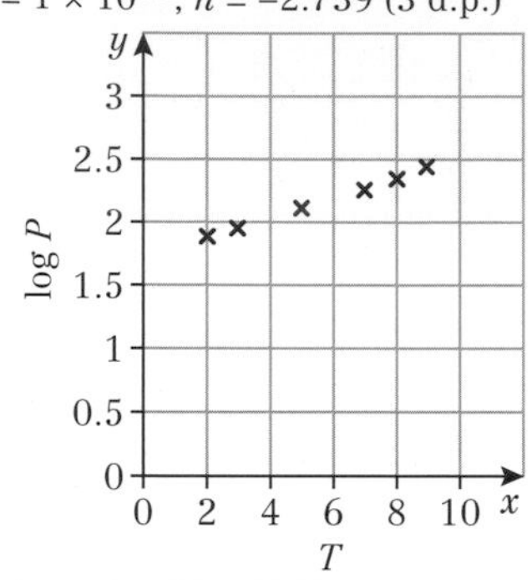

T	2	3	5	7	8	9
$\log P$	1.86	1.93	2.10	2.25	2.33	2.41

b Strong positive correlation
c Yes – the variables show a linear relationship when $\log P$ is plotted against T.
d $a = 50.1$ (3 s.f.), $b = 1.2$
e For every month that passes, the population of moles increases by 20%.
5 a $t = a + bn$ would show a linear relationship. This graph is not a straight line.
b $a = 0.5, k = 0.6$
6 $r = 0.389c^{1.31}$
7 $a = 1.0, n = 1.8$
8 a $a = 1.23$ (3 s.f.), $b = 1.12$ (3 s.f.)
b b is the rate of change of g per degree.
c 35 °C is outside the range of the data (extrapolation).

Challenge

a A graph of $\log T$ against $\log E$ shows a straight line.
b $\log E = 1.09 - 1.96(\log T)$, $a = 12.3$ (3 s.f.), $b = -1.96$ (3 s.f.)
c log 0 is undefined.

Exercise 1B

1 Answers close to
a 0.9 b −0.7 c −0.3
2 a The type and strength of linear correlation between v and m.
b 0.870
3 a −0.854
b There is negative correlation. The relatively older young people took less time to reach the required level.
4 a

Time, t	1	3	4	5	7
Atoms, n	231	41	17	7	2
$\log n$	2.36	1.61	1.23	0.845	0.301

b −0.980 (3 s.f.)
c There is an almost perfect negative correlation with data in the form $\log n$ against t, which suggests an exponential decay curve.
d $a = 307$ (3 s.f.), $b = 0.479$ (3 s.f.)
5 a

Width, w	3	4	6	8	11
Mass, m	23	40	80	147	265
$\log w$	0.477	0.602	0.778	0.903	1.04
$\log m$	1.36	1.60	1.90	2.17	2.42

b 0.9995
c A graph of $\log w$ against $\log m$ is close to a straight line as the value of r is close to 1, therefore $m = kw^n$ is a good model for this data.
d $n = 1.88$ or 1.89 (3 s.f.), $k = 2.91$ (3 s.f.)
6 a −0.833
b −0.833 is close to −1 so the data values show a strong to moderate negative correlation. A linear regression model is suitable for these data.
7 a A 'trace or tr' of rain is an amount less than 0.05mm.
b −0.473 (3 s.f.), treating 'tr' values as 0.
c The data shows a weak negative correlation so a linear model may not be best, there may be other variables affecting the relationship or a different model might be a better fit.

Challenge

r for x and y: 0.999 (3 s.f.)
r for $\log x$ and $\log y$: 1.00 (3 s.f.)
r for x and $\log y$: 0.985 (3 s.f.)
Therefore the most suitable model would be in the form $y = ax^n$

Exercise 1C

1 a $H_0: \rho = 0, H_1: \rho \neq 0$, critical value = ± 0.3120. Reject H_0: there is reason to believe at the 5% level of significance that there is a correlation between the scores.
b $H_0: \rho = 0, H_1: \rho \neq 0$, critical value = ± 0.3665. Accept H_0: there is no evidence of correlation between the two scores at the 2% level of significance.
2 a −0.960 (3 s.f.)
b $H_0: \rho = 0, H_1: \rho \neq 0$, critical value = ± 0.8745. Reject H_0: there is reason to believe at the 1% level of significance that there is a correlation between the scores.
3 a The type and strength of linear correlation between two variables.
b 0.935 (3 s.f.)
c $H_0: \rho = 0, H_1: \rho > 0$, critical value = 0.4973. Reject H_0: there is reason to believe that students who do well in theoretical biology are likely to do well in practical biology.
d There is a probability of 0.05 that the null hypothesis is true.
4 a 0.686 (3 s.f.)
b $H_0: \rho = 0, H_1: \rho > 0$, critical value = 0.6215. Reject H_0: there is reason to believe that there is a linear correlation between the English and Mathematics marks.
5 $H_0: \rho = 0, H_1: \rho > 0$, $r = 0.793$ (3 s.f.), critical value = 0.8822. Accept H_0. There is evidence that the company is incorrect to believe that profits increase with sales.

Online Full worked solutions are available in SolutionBank.

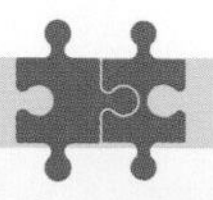

6 H_0: $\rho = 0$, H_1: $\rho < 0$, critical value = −0.4409. Accept H_0. There is evidence that the researcher is incorrect to believe that there is negative correlation between the amount of solvent and the rate of the reaction.
7 2.5%
8 8
9 a −0.846 (3 s.f.)
b H_0: $\rho = 0$, H_1: $\rho < 0$, critical value = −0.8822. Accept H_0. There is evidence that the employee is incorrect to believe that there is negative correlation between humidity and visibility.

Mixed exercise 1

1 a 0.9998
b r is close to 1, so a graph of log t against log x shows a straight line, suggesting that the relationship is in the form $t = ax^n$.
c $n = 1.38$, $a = 0.617$ (3 s.f.)
2 a $a = 0.232$ (3 s.f.), $b = 1.08$ (3 s.f.)
b 151 °C is outside the range of the data (extrapolation).
3 As a person's age increases, their score on the memory test decreases.
4 a Each cow should be given 7 units. The yield levels off at this point.
b 0.952 (3 s.f.)
c It would be less than 0.952. The yield of the last 3 cows is no greater than that of the 7th cow.
5 a −0.972
b There is strong negative correlation. As c increases, f decreases.
6 a 0.340 (3 d.p.)
b H_0: $\rho = 0$, H_1: $\rho \neq 0$, critical value = ±0.6319. Accept H_0. There is not enough evidence that there is a correlation between age and salary.
7 a 0.937 (3 s.f.)
b H_0: $\rho = 0$, H_1: $\rho \neq 0$, critical value = ±0.6319. Reject H_0. There is evidence that there is a correlation between the age of a machine and its maintenance costs.
8 H_0: $\rho = 0$, H_1: $\rho < 0$, critical value = −0.5822. Reject H_0. There is evidence that the greater the altitude, the lower the temperature.
9 H_0: $\rho = 0$, H_1: $\rho > 0$, critical value 0.5822, 0.972 > 0.5822. Reject H_0. There is evidence that age and weight are positively correlated.
10 a 0.940
b H_0: $\rho = 0$, H_1: $\rho > 0$, critical value 0.7293. Reject H_0. There is evidence that sunshine hours and ice cream sales are positively correlated.
11 $r = 0.843$ (3 s.f.), H_0: $\rho = 0$, H_1: $\rho > 0$, critical value 0.8054. Reject H_0. There is evidence that mean windspeed and daily maximum gust are positively correlated.
12 $r = -0.793$ (3 s.f.), H_0: $\rho = 0$, H_1: $\rho < 0$, critical value −0.7545. Reject H_0. There is evidence that temperature and pressure are negatively correlated.

Large data set
Student's own answers

Prior knowledge 2

1 a 0.75
b 0
c 0.25
2 a 0.12
b

C, P, 𝓔
0.08, 0.12, 0.48, 0.32

c 0.32
3 a $\frac{2}{7}$ b $\frac{6}{7}$

Exercise 2A

1 a $A \cap B'$ b $A' \cup B$
c $(A \cap B) \cup (A' \cap B')$ d $A \cap B \cap C$
e $A \cup B \cup C$ f $(A \cup B) \cap C'$
2 a b c

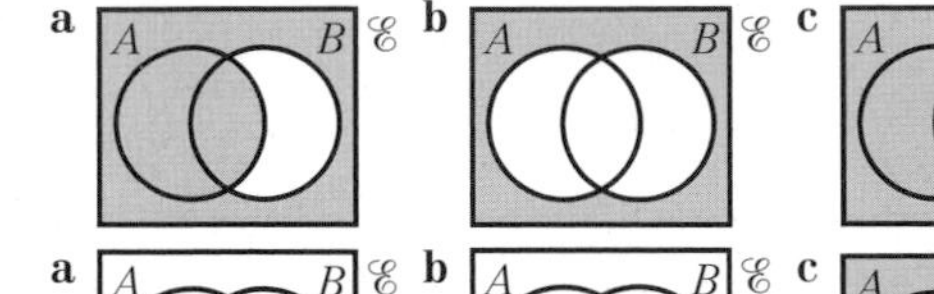

3 a b c

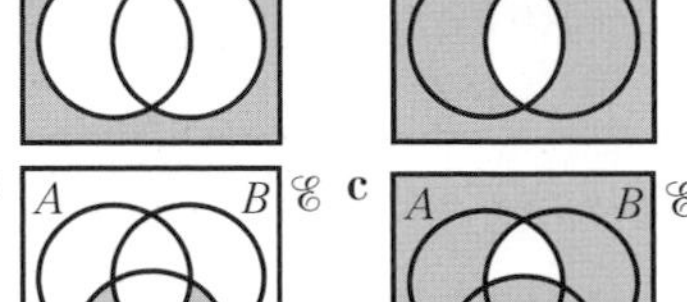

4 a 0.0769 b 0.25 c 0.0192
d 0.308 e 0.75 f 0.231
5 a 0.6 b 0.8 c 0.4 d 0.9
6 a 0.25 b 0.5 c 0.65 d 0.1
7 a

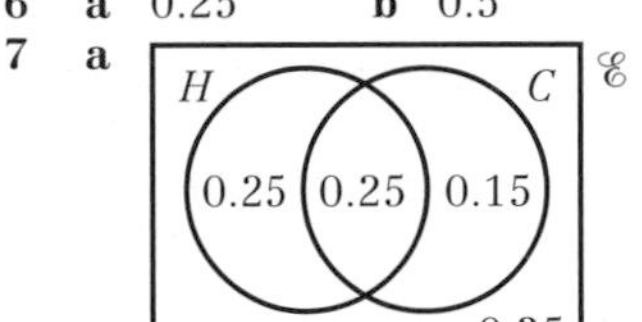

b i 0.65 ii 0.15 iii 0.85
8 a

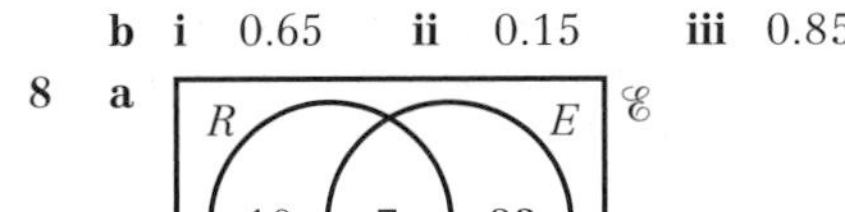

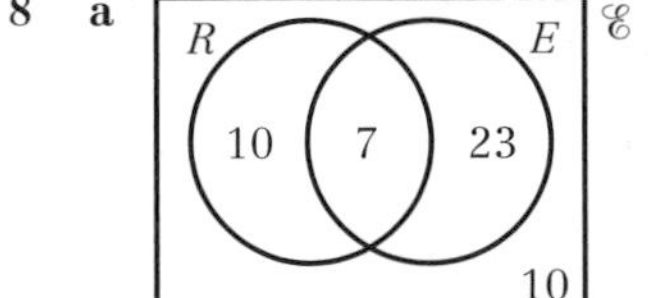

b i 7 ii $\frac{1}{5}$ iii $\frac{43}{50}$
9 a

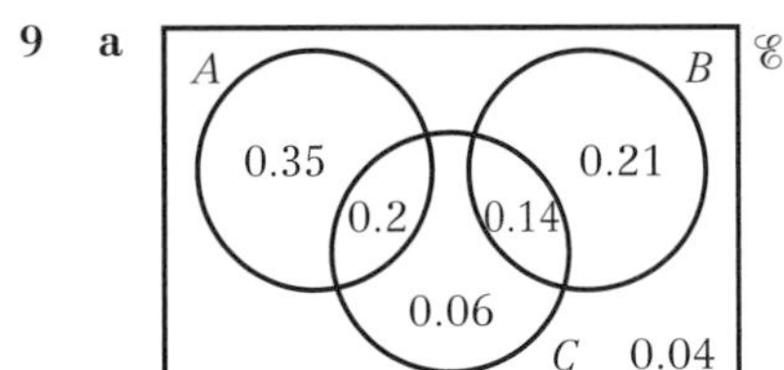

b i 0.1 ii 0.76 iii 1
10 a

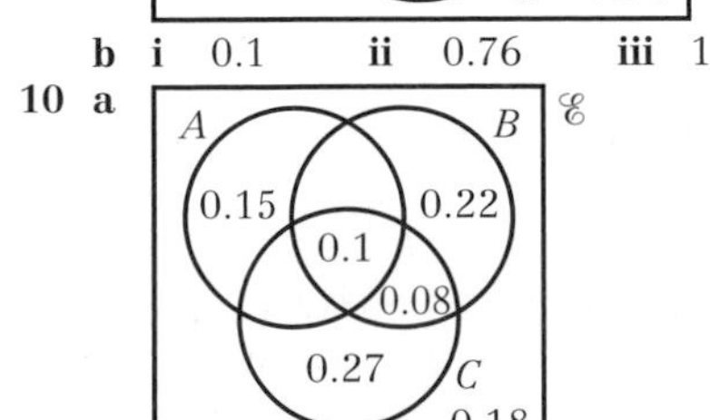

b i 0.53 ii 0.18
c Not independent.
$P(A' \cap C) = 0.35$, $P(A') \times P(C) = 0.75 \times 0.45 = 0.3375$

11 a 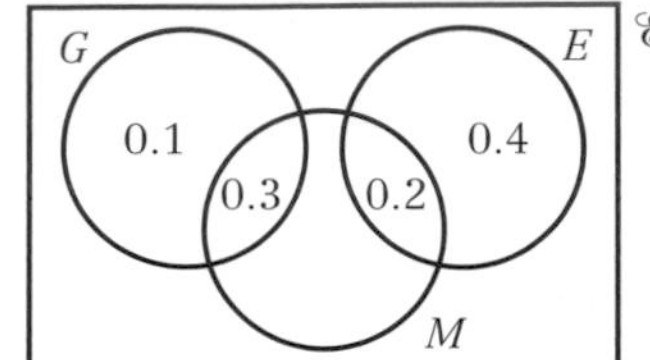

b **i** 0.6 **ii** 0.5

c Not independent.
$P(G' \cap M) = 0.2$, $P(G') \times P(M) = 0.6 \times 0.5 = 0.3$

12 a xy **b** $x + y + xy$ **c** $1 - y + xy$

Challenge

a xyz

b $x + y + z + xyz - xy - yz - xz$

c $z - yz + xyz$

Exercise 2B

1 a $\frac{29}{60}$ **b** $\frac{18}{29}$ **c** $\frac{18}{35}$ **d** $\frac{14}{31}$

2 a

	Badminton	Squash	Total
Male	21	22	43
Female	15	17	32
Total	36	39	75

b **i** $\frac{22}{39}$ **ii** $\frac{15}{36}$ or $\frac{5}{12}$ **iii** $\frac{17}{32}$

3 a

	Girls	Boys	Total
Vanilla	13	2	15
Chocolate	12	10	22
Strawberry	20	23	43
Total	45	35	80

b **i** $\frac{23}{43}$ **ii** $\frac{13}{15}$ **iii** $\frac{10}{35}$ or $\frac{2}{7}$

4 a

Red spinner / Blue spinner	1	2	3	4
1	2	3	4	5
2	3	4	5	6
3	4	5	6	7
4	5	6	7	8

b **i** $\frac{1}{4}$ **ii** $\frac{1}{4}$ **iii** $\frac{1}{4}$

5 a

Dice 2 / Dice 1	1	2	3	4	5	6
1	1	2	3	4	5	6
2	2	4	6	8	10	12
3	3	6	9	12	15	18
4	4	8	12	16	20	24
5	5	10	15	20	25	30
6	6	12	18	24	30	36

b $\frac{1}{6}$ **c** $\frac{1}{4}$

d All outcomes are equally likely.

6 0.0769 (3 s.f.) or $\frac{1}{13}$

7 a 0.333 **b** 0.667

c Assume that the coins are not biased.

8 a

	D	D′	Total
S	18	38	56
S′	59	5	64
Total	77	43	120

b **i** $\frac{43}{120}$ **ii** $\frac{5}{120}$ **iii** $\frac{18}{77}$ **iv** $\frac{38}{56}$

9 a

	Women	Men	Total
Stick	26	18	44
No stick	37	29	66
Total	63	47	110

b **i** $\frac{44}{110}$ or $\frac{2}{5}$ **ii** $\frac{26}{63}$ **iii** $\frac{18}{44}$ or $\frac{9}{22}$

10 a $\frac{6}{25}$ **b** $\frac{13}{30}$ **c** $\frac{29}{64}$ **d** $\frac{31}{90}$

Exercise 2C

1 a 0.7 **b** 0.3

c 0.483 (3 s.f.) **d** 0.571 (3 s.f.)

2 a 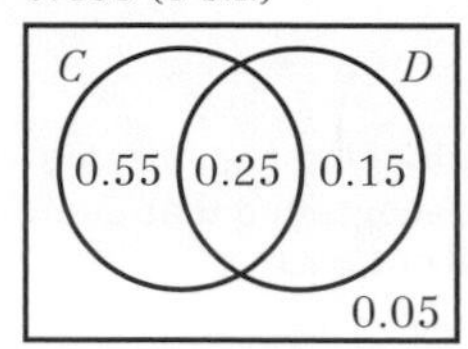

b **i** 0.95 **ii** 0.625 **iii** 0.313 (3 s.f.) **iv** 0.25

3 a 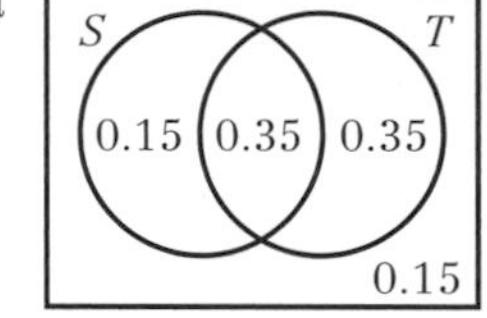

b **i** 0.35 **ii** 0.5 **iii** 0.7 **iv** 0.231 (3 s.f.)

4 a $\frac{3}{8}$ **b** $\frac{2}{5}$ **c** $\frac{6}{11}$ **d** $\frac{13}{19}$

5 a $\frac{9}{80}$ **b** $\frac{9}{32}$ **c** $\frac{1}{5}$ **d** $\frac{12}{35}$

6 a 0.6 **b** 0.4

c 0.299 (3 s.f.) **d** 0.329 (3 s.f.)

7 a $\frac{9}{23}$ **b** $\frac{3}{23}$

c $P(B|C) = 0.111... \neq P(B) = 0.345...$ So B and C are not independent

8 a 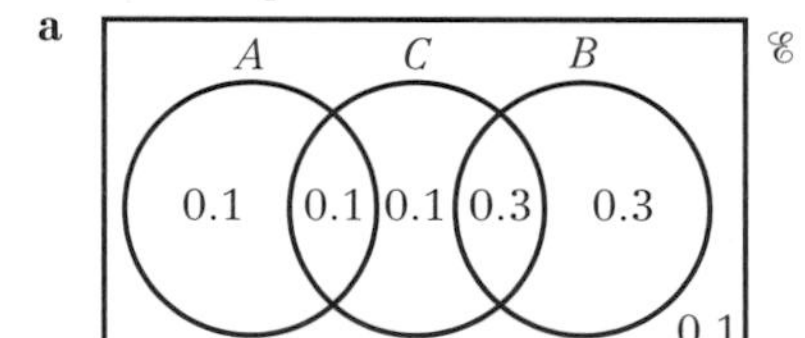

b **i** 0.2 **ii** 0.6 **iii** 0.5

9 a 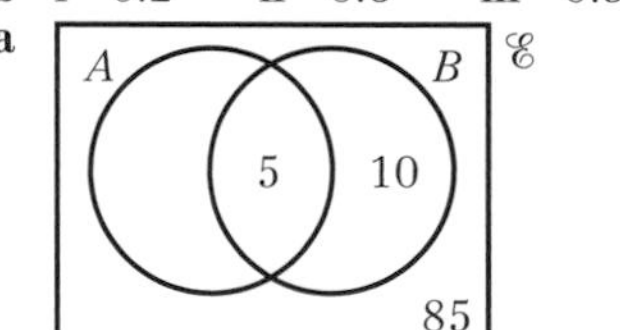

b $\frac{1}{3}$

c No one who doesn't have the disease would be given a false negative result. However, only $\frac{1}{3}$ of the people who have a positive result would have the disease.

10 a 0.7 **b** 0.7 **c** They are independent.

11 $x = 0.21$, $y = 0.49$

12 $c = \frac{7}{30}$, $d = \frac{4}{15}$

Exercise 2D

1 a 0.3 **b** 0.6 **c** 0.8 **d** 0.9

2 a 0.8

b **i** 0.2 **ii** 0.615 (3 s.f.) **iii** 0.429 (3 s.f.)

c $P(C \cap D) \neq P(C) \times P(D)$

3 **a** 0.9
b **i** 0.8 **ii** 0.2 **iii** 0.5
4 **a** 0.15 **b** 0.45 **c** 0.55 **d** 0.25 **e** 0.3
5 0.1
6 **a** 0.5 **b** 0.3 **c** 0.3
7 **a** 0.3 **b** 0.35 **c** 0.4
8 **a** 0.0833 (3 s.f.) **b** 0.15
c 0.233 (3 s.f.) **d** 0.357 (3 s.f.)
e 0.643 (3 s.f.) **f** 0.783 (3 s.f.)
9 **a** 0.67 **b** 0.476 (3 s.f.) **c** 0.126
d

$\mathscr{E}$
A: 0.174
C: 0.174
B: 0.25
0.126 (A and C)
0.12 (A and B)
0.156

e 0.294

10 **a** 0.28 **b** 0.7
c 0.333 (3 s.f.) **d** 0.467 (3 s.f.)
11 **a** 0.1 **b** 0.143 (3 s.f.)
c $P(A) \times P(B) = 0.3 \times 0.7 = 0.21$, $P(A \cap B) = 0.15$
This suggests that the events are not independent. If Anna is late, Bella is *less* likely to be late and vice versa.
12 **a** 0.5 **b** 0.333 (3 s.f.) **c** 0.833 (3 s.f.)
d $P(K|J) = 0.833... \neq P(K) = 0.7$. So J and K are not independent

Challenge
a $\frac{1}{15}$ **b** $\frac{5}{12}$ **c** $\frac{2}{3}$

Exercise 2E

1 **a**

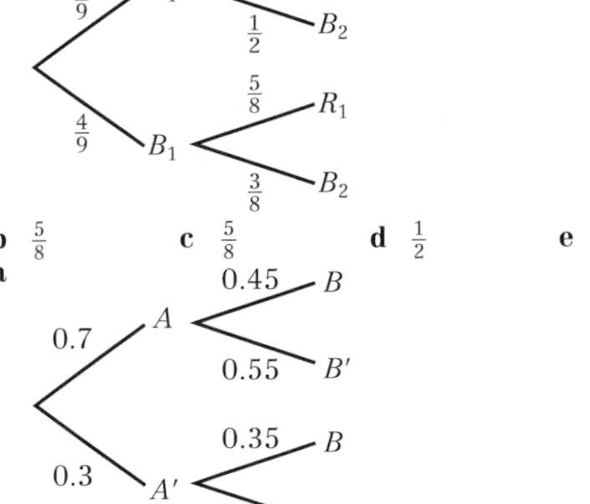

b $\frac{5}{8}$ **c** $\frac{5}{8}$ **d** $\frac{1}{2}$ **e** $\frac{1}{2}$
2 **a**

b **i** 0.315 **ii** 0.195 **iii** 0.75
3 **a**

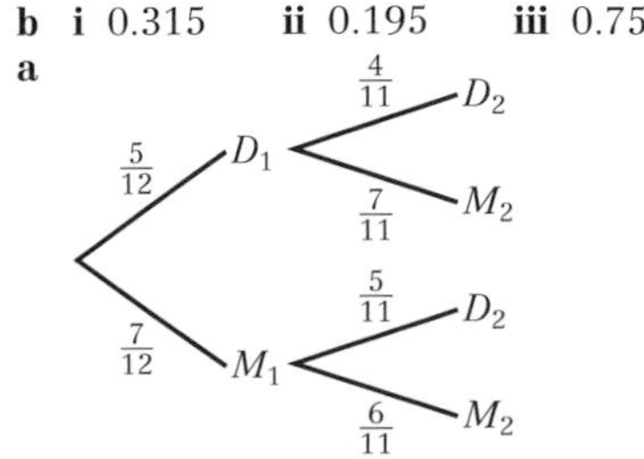

b 0.152 (3 s.f.) **c** 0.530 (3 s.f.) **d** 0.222 (3 s.f.)
4 0.36
5 **a** 0.25 **b** 0.333

6 **a**

4/11 B_1: 3/10 B_2, 7/10 G_2
7/11 G_1: 4/10 B_2, 6/10 G_2

b $\frac{7}{11}$ **c** $\frac{3}{5}$

7 **a**

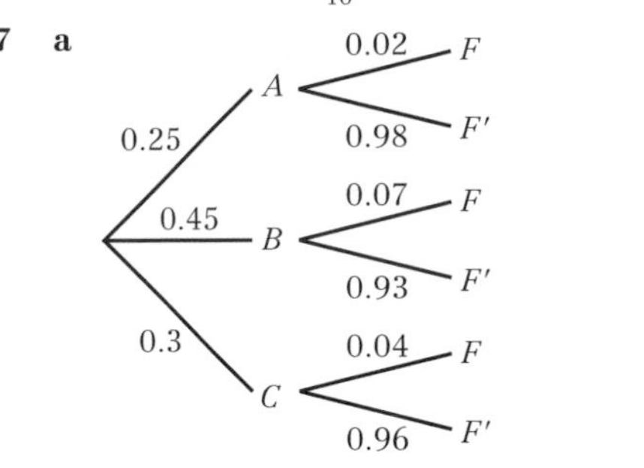

b **i** 0.0315 **ii** 0.0485 **c** 0.103 (3 s.f.)
8 **a**

0.04 C: 0.9 Positive, 0.1 Negative
0.96 C': 0.02 Positive, 0.98 Negative

b 0.945 (3 s.f.) **c** 0.00423
d The probability that a positive result is a false positive (positive result for someone without the condition) = $P(-|+) = 0.348$. Over one third of positive results are false positives and 10% of people with the condition give negative results.
9 **a**

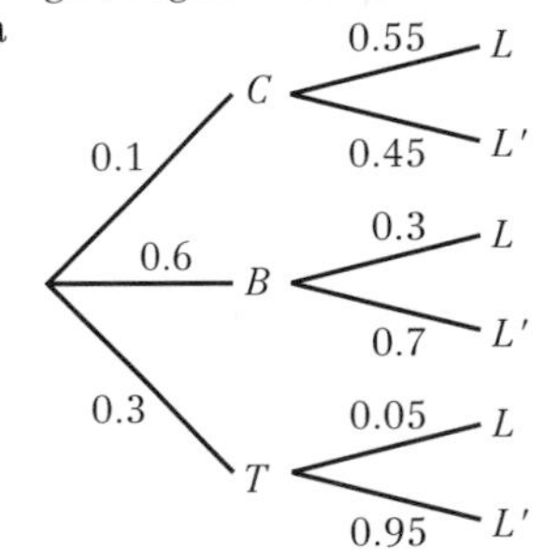

b **i** 0.015 **ii** 0.25 **c** 0.78
10 **a**

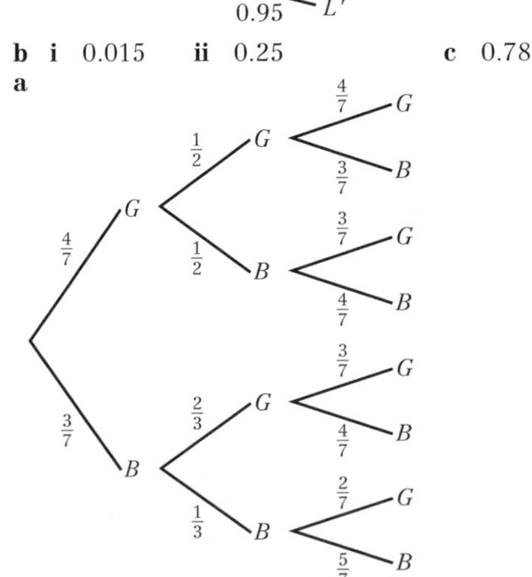

b $\frac{3}{7}$

c Adding together the probabilities on the 4 branches of the tree diagram where the counter from box B is blue: $\frac{12}{98}+\frac{16}{98}+\frac{24}{147}+\frac{15}{157}=\frac{27}{49}$

d Adding together the probabilities on the 2 branches of the tree diagram where events C and D both occur. $\frac{12}{98}+\frac{15}{147}=\frac{11}{49}$

e $\frac{37}{49}$ f $\frac{8}{13}$

11 She has not taken into account the fact that the jelly bean is eaten after being selected. The correct answer is 0.5.

Mixed exercise 2

1 a 0.55 b 0.45 c 0.5 d 0.429 (3 s.f.)

2 a

J K L $\mathscr{E}$
0.15 0.1 0.0675 0.0825 0.2825 0.3175

b i 0.6 ii 0.6 iii 0.222 (3 s.f.) iv 0.471 (3 s.f.)

3 a 0.433 (3 s.f.) b 0.6 c 0.72
d 0.25 e 0.577 (3 s.f.)

4 a

$\frac{6}{15}$ R: $\frac{5}{14}$ R, $\frac{9}{14}$ G
$\frac{9}{15}$ G: $\frac{6}{14}$ R, $\frac{8}{14}$ G

b i $\frac{12}{35}$ ii $\frac{18}{35}$

c $\frac{2}{5}$ d $\frac{6}{65}$

5 a 0.74 b 0.757 (3 s.f.) c 0.703

6 a $\frac{4}{15}$ b $\frac{15}{41}$
c 0.117 (3 s.f.) d 0.146 (3 s.f.)

7 a 0.3 b 0.42

c

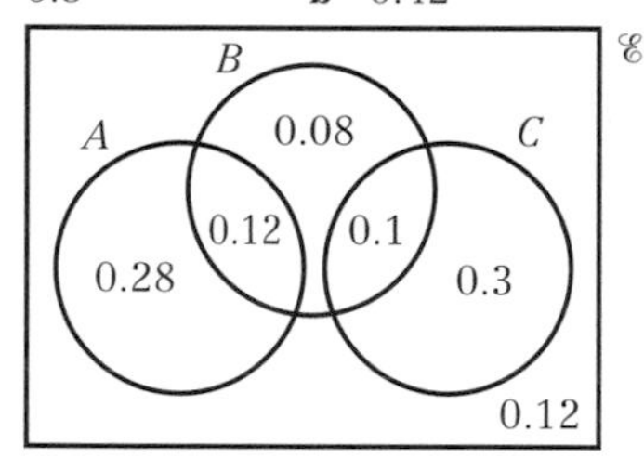

d i 0.25 ii 0.28

8 a In some football matches, neither team scores.
b 0.12 c 0.179 (3 s.f.)

Challenge

a $0.4 \leqslant p \leqslant 0.6$ b $0.2 \leqslant q \leqslant 0.5$

Prior knowledge 3

1 a 0.540 (3 s.f.) b 0.390 (3 s.f.)
2 a 0.124 (3 s.f.) b 0.584 (3 s.f.) c 0.869 (3 s.f.)
3 a 0.211 (3 s.f.) b 0.599 (3 s.f.)

For Chapter 3, student answers may differ slightly from those shown here when calculators are used rather than table values.

Exercise 3A

1 a Continuous – lengths can take any value
b Discrete – scores can only take certain values
c Continuous – masses can take any value
d Discrete – show sizes can only take certain values

2

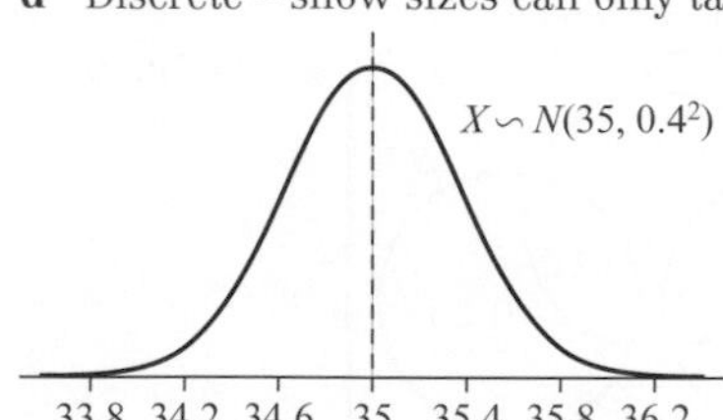

3 The distribution is not symmetrical.
4 a 0.68 b 0.95
5 49
6 60 g
7 $\mu = 56.7$ (3 s.f.), $\sigma^2 = 4.69^2$ (3 s.f.)
8 a 0.5 b 0.683 (3 s.f.) c 0.954 (3 s.f.)
d Incorrect: although $P(X > 100) > 0$, it is very small since 100 is more than 3 standard deviations away from the mean, so the model as a whole is still reasonable.
9 a 36 b Between 2 and 3

Exercise 3B

1 a 0.9332 b 0.9772 c 0.2119
2 a 0.0478 b 0.2525 c 0.2782
3 a 0.1587 b 0.4985 c 0.5948
4 a 0.2635 b 0.1714 c 0.0373
5 a i 0.7475 ii 0.2525
b Sum is 1, combined probabilities include every possible value.
6 a 0.3176 b 0.6824
7 a 0.1814 b 0.4295
8 a i 0.1056 ii 0.1056 b 0.0012
9 a i 0.3605 ii 0.2375 b 0.3380
10 a 0.0766 b 0.1906 c 0.3296
11 a 0.0228 b 0.7345
12 a 0.4013 b 0.0001986

Exercise 3C

1 a 27.38 b 33.37 c 31.27 d 35.30
2 a 8.16 b 10.85 c 12.02 d 11.45
3 a i 19.1 ii 18.3
b 0.0915
4 a i 70.6 ii 80.8 b 0.075
5 a i 81.0 ii 80.6 b 0.0364
6 a 4.095 (3 d.p.) b 5.005 (3 d.p.)
c 4.5 is the mean, so 50% of badgers will have a mass less than 4.5.
7 a 73.52 (2 d.p.) b 8.09 (2 d.p.)
8 a 61.68 (2 d.p.) b 5.13 (2 d.p.)
c Tom is correct in this case; the normal distribution is symmetric about the mean, so 50% of bars will have mass less than the mean.
9 a Short: Up to 165 cm,
Regular: Between 165 cm and 178 cm
Long: Over 178 cm
b That the population follows the normal distribution over the whole range of values i.e. that there are no extreme outliers.

Exercise 3D

1 **a** 0.9830 **b** 0.9131 **c** 0.2005 **d** 0.3520
e 0.4893 **f** 0.0516 **g** 0.1823 **h** 0.8836
2 **a** 1.33 **b** 1.86 **c** 1.0364 **d** −1.6449
e 1.06 **f** 2.55 **g** 1.2816 **h** 0.5244
3 **a** 0 **b** −0.16 **c** 0.2 **d** 0.74
4 **a** $\Phi(0)$ **b** $\Phi(0.5)$ **c** $1 - \Phi(-0.25)$
d $\Phi(0.0833) - \Phi(-1.17)$
5 **a** 1.96 **b** 87.8 (3 s.f.)
6 **a** −1.0364 **b** 54.9 cm
7 **a** $-1.2816 < z < 1.2816$ **b** 1103–1247 hours

Exercise 3E

1 11.5
2 3.87
3 31.6
4 25
5 $\mu = 13.1, \sigma = 4.32$.
6 $\mu = 28.3, \sigma = 2.59$.
7 $\mu = 12, \sigma = 3.56$.
8 $\mu = 35, \sigma = 14.8$ or $\sigma = 14.9$.
9 4.75
10 $\sigma = 1.99, a = 2.18$.
11 **a** 203.37 mm **b** 0.1504 **c** 0.0516
12 **a** 0.1299 mm **b** 0.5587 **c** 0.0644
13 **a**

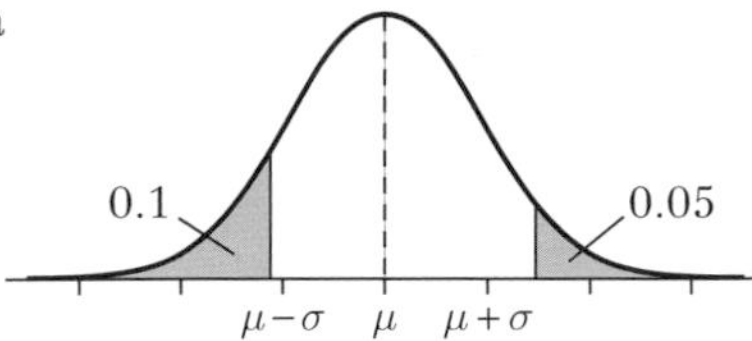

b $\mu = 23.26, \sigma = 4.100$ **c** 0.4469
14 **a** $\mu = 16.79, \sigma = 0.9421$ **b** 1.27

Challenge

a Let z be such that $\Phi(z) = 0.75$,
then upper quartile = $\mu + z\sigma$ and lower quartile = $\mu - z\sigma$,
so $q = (\mu + z\sigma) - (\mu - z\sigma) = 2z\sigma$.
Calculate that $z = 0.674$, then $q = 1.348\sigma$ and thus $\sigma = 0.742q$ (3 s.f.).
b Since $q = (\mu + z\sigma) - (\mu - z\sigma) = 2z\sigma$ (i.e. the μs cancel), q is not dependent on μ and vice versa, and it is not possible to write μ in terms of q.

Exercise 3F

1 **a i** Yes, n is large (> 50) and p is close to 0.5.
ii $X \sim N(72, 5.37^2)$
b i No, n is not large enough (< 50).
c i Yes, n is large (> 50) and p is close to 0.5.
ii $X \sim N(130, 7.90^2)$
d i No, p is too far from 0.5.
e i Yes, n is large (> 50) and p is close to 0.5.
ii $X \sim N(192, 9.99^2)$
f i Yes, n is large (> 50) and p is close to 0.5.
ii $X \sim N(580, 15.6^2)$
2 **a** 0.1253 **b** 0.0946 **c** 0.6723
3 **a** 0.0097 **b** 0.5596 **c** 0.0559
4 **a** 0.6203 **b** 0.4540 **c** 0.0102
5 0.006
6 0.3767
7 **a** n large, p close to 0.5. **b** 0.1593
c 0.5772 **d** 115
8 **a** 0.6277 **b** 0.8457
9 **a** 0.0786 **b** 0.26%

Exercise 3G

1 **a** Not significant. Accept H_0.
b Significant. Reject H_0.
c Not significant. Accept H_0.
d Significant. Reject H_0.
e Not significant. Accept H_0.
2 **a** $\overline{X} < 119.39...$ or 119 (3 s.f.)
b $\overline{X} > 13.2$
c $\overline{X} < 84.3$
d $\overline{X} > 0.877$ or $\overline{X} < -0.877$
e $\overline{X} > -7.31$ or $\overline{X} < -8.69$
3 Result is significant so reject H_0. There is evidence that the new formula is an improvement.
4 **a** $\overline{X} \geqslant 103.29$
b 102.5 < 103.29, so there is not enough evidence to reject the null hypothesis
5 Insufficient evidence; accept H_0.
6 **a** 0.9256 **b** 0.9455
c $H_0: \mu = 5.7$, $H_1: \mu < 5.7$. There is sufficient evidence to suggest mean diameter less than 5.7 mm.
7 **a** 136.48 g **b** 0.01289
c $H_0: \mu = 860$, $H_1: \mu \neq 860$. Insufficient evidence to suggest mean mass is different to 860 g.
8 $H_0: \mu = 9.5$, $H_1: \mu > 9.5$. Critical region is $\overline{X} \geqslant 10.715$. $\overline{x} = 12.2 > 10.715$, so reject H_0 and conclude that the mean daily windspeed is greater than 9.5 knots.

Mixed exercise 3

1 **a** 0.0401 **b** 0.3307 **c** 188 cm
2 **a** 12.7% or 12.8% **b** 51.1% or 51.2%
3 **a** 0.0668 **b** 0.0521 **c** 0.9314
4 **a** 3.65 **b** 0.1357 **c** 32.5
5 **a** 8.60 ml **b** 0.123 **c** 109 ml
6 **a** $\mu = 30, \sigma = 14.8$ or $\sigma = 14.9$ **b** 38.03
7 Mean 10.2 cm, standard deviation 3.76 cm
8 **a** 0.3085
b 0.370 or 0.371
c The first score was better, since fewer of the students got this score or more.
9 **a** 4.25 or 4.26 **b** 0.050 (2 d.p.) **c** 0.8729
10 **a** 8.54 minutes **b** 0.1758
11 Mean 6.12 mm, standard deviation 0.398 mm
12 0.0778
13 **a** n is large and p is close to 0.5.
b $\mu = 40, \sigma^2 = 24$
c 0.0262
14 **a** 0.0147
b n is large and p is close to 0.5; $\mu = 55.2, \sigma = 5.46$
c 0.68%
15 **a** n is large and p is close to 0.5.
b 0.5232 **c** 166
16 0.6339
17 **a** 0.5914 **b** 0.0197
c Assuming the claim is correct, there would be a less than 2% chance that 95 seedlings produce apples within 3 years. Therefore it is unlikely that the claim is correct.
18 **a** 0.5801 **b** 0.0594
c Assuming the claim is true, there is a less than 6% chance that 170 or more people would be cured out of 300, so it is likely that the herbalist has under-stated the actual cure rate.
19 $\overline{X} > 7.66$ (3 s.f.)
20 Test statistic = −1.5491... > −1.6449
Not significant so accept H_0. There is insufficient evidence to suggest that the mean contents of a bottle is lower than the manufacturer's claim.

21 a Accept $0.032 \sim 0.034$ b Accept < 0.069
c Test statistic = $1.4815... < 1.6449$
Not significant so accept H_0. Insufficient evidence of an increase in the mean breaking strength of climbing rope.

22 a $Z = \pm 1.96 \rightarrow 1000 + 1.96\sigma = 1010$
$1.96\sigma = 10 \rightarrow \sigma^2 = 26.03$ (2 d.p)
b Test statistic = $2.0165... < 2.3263$
Not significant so accept H_0. There is insufficient evidence of a deviation in mean from 1010. So we can assume condition **i** is being met.

23 a Accept $0.845 \sim 0.846$
b Test statistic = $3.0145... > 2.5758$
Significant so reject H_0. There is evidence that the mean length of eggs from this island is different from elsewhere.

24 a $X \sim N\left(\mu_1, \frac{\sigma^2}{n}\right)$ b Need $n = 28$ or more

Challenge
a 0.2510 b $\bar{X} \leqslant 105, \bar{X} \geqslant 135$
c 102 is in the critical region, so at the 5% significance level there is evidence to reject the manager's claim. It is probable that less than 48% of people support the manager.

Review exercise 1

Student answers may differ slightly from those shown here when calculators are used rather than table values.

1 a 0.9991
b Very close to 1 suggesting a strong correlation between the two variables
c $a = 0.1244$, $n = 1.4437$

2 a $a = 0.611$, $b = 1.04$ b Not in range – extrapolation

3 a 0.93494...
b $0.935 > 0.7155$ so reject H_0; levels of serum and disease are positively correlated.

4 $r = -0.4063$, critical value for $n = 6$ is -0.6084 so no evidence.

5 a $H_0: \rho = 0$, $H_1: \rho < 0$. $r = -0.9313$, critical value for $n = 5$ is -0.8783 so there is evidence. Reject H_0.
b Would lead to conclusion that there is no evidence (critical value = 0.9343). Accept H_0.

6 a 0.338 (3 s.f.) b 0.46
c 0.743 (3 s.f.) d 0.218 (3 s.f.)

7 a
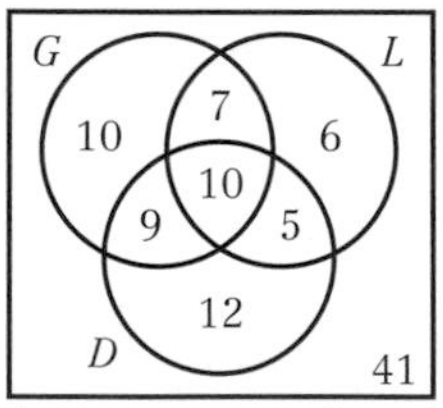

b 0.1
c 0.41 d 0.21 e 0.667 (3 s.f.)

8 a 0.1 b
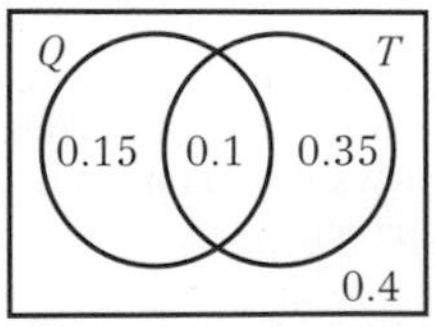

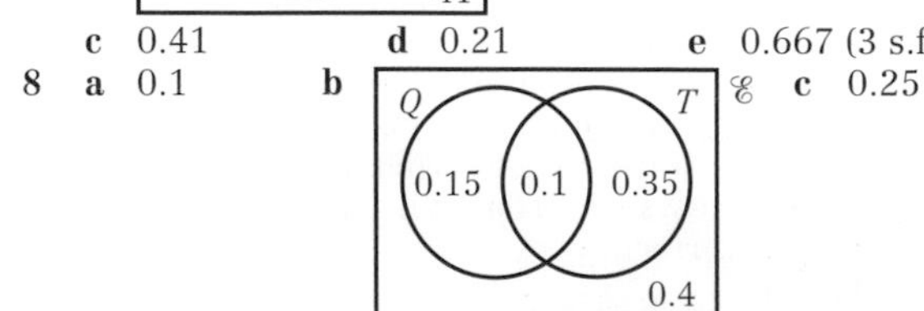

c 0.25

9 a
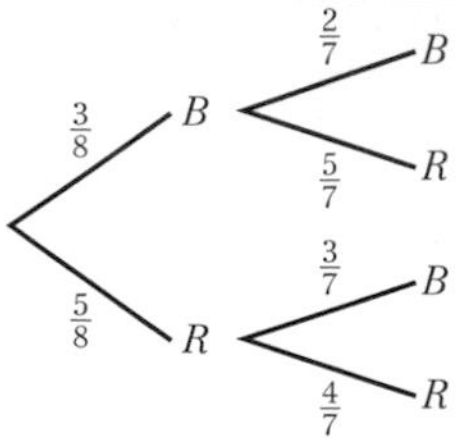

b i $\frac{21}{56}$ or $\frac{3}{8}$ ii $\frac{2}{7}$

10 a
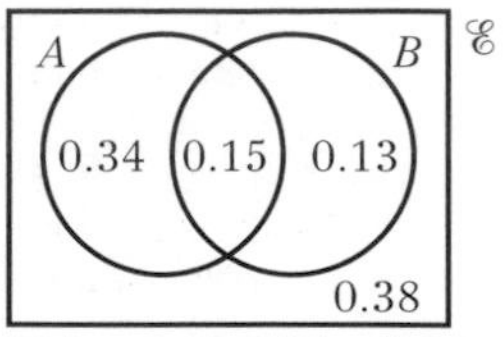

b $P(A) = 0.49$, $P(B) = 0.28$
c $\frac{17}{36}$ or 0.472 (3 s.f.)
d No: $P(A) \times P(B) \neq P(A \cap B)$

11 a 0.5 b 0.35
c
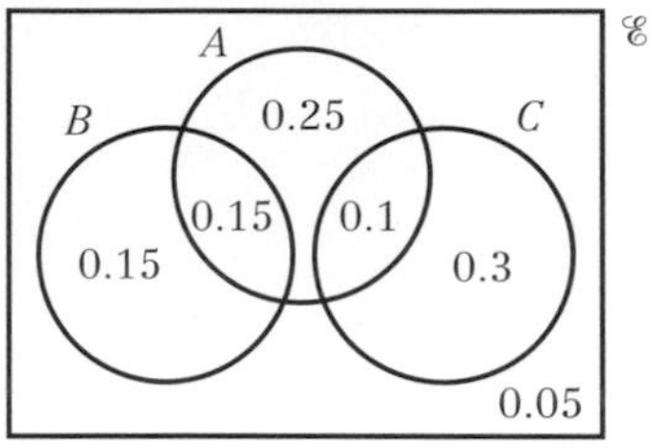

d i 0.25 ii 0.4 iii $\frac{2}{3}$

12 a 0.198 (3 s.f.) b $\frac{45}{79}$ or 0.570 (3 s.f.)

13 a 0.2743 b 12

14 a 0.0618 b 0.9545
c 0.00281 d This is a bad assumption.

15 a Mean 1.700 m, standard deviation 0.095 m
b 0.337

16 a 1.07 (3 s.f.) b 0.505 (3 s.f.) c 0.981 (3 s.f.)

17 a 0.2225 b 0.2607
c 0.2567 d 0.0946

18 a 0.06703 (5 d.p.) b 0.34% (2 s.f.)

19 $H_0: \mu = 18$, $H_1: \mu < 18$, $0.0262 < 0.05$ so reject H_0 *or* in critical region. There is evidence that the (mean) time to complete the puzzles has reduced.

20 a 2.36 metres (3 s.f.) b 0.524 (3 s.f.)
c $H_0: \mu = 1.9$, $H_1: \mu \neq 1.9$. Less than 1.769 m and greater than 2.031 m
d 2.09 m is in the critical region so there is evidence to suggest the mean length is not 1.9 m.

21 $H_0: \mu = 12$, $H_1: \mu < 12$. Given the mean = 12 °C, $P(X < 11.1) < 0.05$ therefore there is evidence to suggest the mean is not 12 °C.

Challenge
1 a $0.4 \leqslant p \leqslant 0.7$ b $0 \leqslant q \leqslant 0.25$
2 a CVs are 144.78 and 173.22 so CRs are $\leqslant 144$ and $\geqslant 174$
b 173 is not in the CR so there is evidence to support the politician's claim.

Prior knowledge 4

1 13.1 cm, 12.0 cm
2 a 10.8 N b 21.6 N c 7.8 N (2 s.f.)

Exercise 4A

1 a 6 Nm clockwise b 10.5 Nm clockwise
c 13 Nm anticlockwise d 0 Nm
2 a 10 Nm anticlockwise b 30.5 Nm anticlockwise
c 13.3 Nm clockwise d 33.8 Nm anticlockwise
3 a i 313.6 Nm clockwise ii 156.8 Nm anticlockwise
b Sign is a particle.
4 a 0 Nm b 0 Nm
c 36 Nm anticlockwise d 36 Nm anticlockwise
5 2.5 N

Online Full worked solutions are available in SolutionBank.

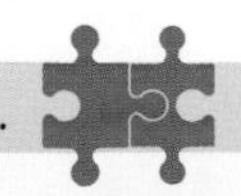

Exercise 4B

1 **a** 5 Nm anticlockwise **b** 13 Nm clockwise
c 19 Nm anticlockwise **d** 11 Nm anticlockwise
e 4 Nm clockwise **f** 7 Nm anticlockwise
2 **a** 16 Nm clockwise **b** 1 Nm anticlockwise
c 10 Nm clockwise **d** 7 Nm clockwise
e 0.5 Nm anticlockwise **f** 9.59 Nm anticlockwise
3 6 m
4 1.6
5 528 448 Nm anticlockwise
6 $6000 \times x\sin\theta > 8000 \times \frac{1}{2}x\cos\theta$
$6000\sin\theta > 4000\cos\theta$
$\frac{\sin\theta}{\cos\theta} > \frac{4000}{6000}$
$\tan\theta > \frac{2}{3}$

Exercise 4C

1 **a** 10 N, 10 N **b** 15 N, 5 N
c 12 N, 8 N **d** 12.6 N, 7.4 N
2 **a** 7.5, 17.5 **b** 30, 35 **c** 245, $2\frac{2}{3}$
3 0.5 m
4 59 N
5 31 cm from the broomhead
6 **a** 16.25 N, 13.75 N **b** 3.2 m
7 **a** 784 N **b** 0.625 m
8 **a** 122.5 N **b** 1.17 m
9 **a** $\frac{9}{2}T_C = 4W + 8 \times 30$
$\frac{9}{2}T_C = 4W + 240$
$9T_C = 8W + 480$
$T_C = \frac{8}{9}W + \frac{160}{3}$
b $T_A = \frac{W}{9} - \frac{70}{3}$ **c** 750 N
10 14.1 N
11 **a** $R_A = 60$ N **b** 33.6 kg

Challenge
3 kg, 5 kg, 1 kg, 2 kg, 4 kg (from left to right)

Exercise 4D

1 $R_A = 2.4$ N, $R_B = 3.6$ N
2 **a** 10g N **b** 3.5 m from A
3 $\frac{1}{3}$m from A
4 **a** 29.4 N, 118 N **b** 2.11 m
5 **a** 160 N **b** 2.77 m
6 **a** 3 m
b Centre of mass lies at the midpoint of the seesaw.
c 2 m towards Sophia.
7 $R_C = 5R_D$
$R_C + R_D = 80 + W$
$R_D = \frac{80 + W}{6}$
Taking moments about $A: 6R_C + 20R_D = 80 \times 10 + xW$
$50R_D = 800 + xW$
$25W - 3xW = 400$
$W = \frac{400}{25 - 3x}$
8 1.6 m

Challenge
1.61 m

Exercise 4E

1 5
2 $\frac{2}{3}$m
3 2.05 m
4 **a** $C = 15$ N, $D = 5$ N **b** $2 \times 12 \neq 20 \times 0.5$
c $2.14 \leqslant x \leqslant 4.78$ m
5 2.5 m
6 **a** Taking moments about N:
$mg \times ON = \frac{3}{4}mg \times 2a$ so $ON = \frac{3}{2}a$
b $\frac{23}{20}mg$ N
7 40 N

Mixed exercise 4

1 **a** 105 N **b** 140 N
c 1.03 m to the right of D
2 **a** $(1 \times 150) + W(x - 1) = 1.5\left(\frac{150 + W}{2}\right)$
$150 + Wx - W = 112.5 + 0.75W$
$37.5 = 1.75W - Wx$
$150 = 7W - Wx$
$W = \frac{150}{7 - 4x}$
b $0 \leqslant x < \frac{7}{4}$
3 **a** 40 g **b** $x = \frac{1}{2}$
c **i** The weight acts at the centre of the plank.
ii The plank remains straight.
iii The man's weight acts at a single point.
4 **a** $2.5 \times 100 = 3.5W + 150(3.5 - x)$
$250 = 3.5W + 525 - 150x$
$150x = 3.5W + 275$
$300x = 7W + 550$
b $W = 790 - 300x$ **c** $x = 2.53$, $W = 30$
5 **a** 200 N **b** 21 cm
6 **a** 36 kg **b** 2.2 m
7 **a** 19.6 N **b** 5
8 $\frac{2}{3}$m
9 **a** 125 N **b** 1.8 m
10 4.88 m
11 2.39 m
12 **a** $\frac{10000}{x}$ **b** $500\text{ kg} \leqslant M \leqslant 2000\text{ kg}$
c This model has the crane only able to lift weights of 500 kg at full extension, not very practical.

Challenge
1 3.28 m
2 **a** 69.1 N **b** 163 N

Prior knowledge 5

1 $(\mathbf{i} + 2\mathbf{j})\,\text{ms}^{-2}$
2 **a** 16.55 **b** 25.02°

Exercise 5A

1 **a** **i** 11.3 N (3 s.f.) **ii** 4.10 N (3 s.f.)
iii $(11.3\mathbf{i} + 4.10\mathbf{j})$ N
b **i** 0 N **ii** −5 N **iii** $-5\mathbf{j}$ N
c **i** −5.14 N (3 s.f.) **ii** 6.13 N (3 s.f.)
iii $(-5.14\mathbf{i} + 6.13\mathbf{j})$ N
d **i** −3.86 N (3 s.f.) **ii** −4.60 N (3 s.f.)
iii $(-3.86\mathbf{i} - 4.60\mathbf{j})$ N
2 **a** **i** −2 N **ii** 6.93 N (3 s.f.)
b **i** 8.13 N (3 s.f.) **ii** 10.3 N (3 s.f.)
c **i** $(P\cos\alpha + Q - R\sin\beta)$ N
ii $(P\sin\alpha - R\cos\beta)$ N
3 **a** 39.3 N (3 s.f.) at an angle of 68.8° above the horizontal
b 27.9 N (3 s.f.) at an angle of 16.2° above the horizontal
c 3.01 N (3 s.f.) at an angle of 53.3° above the horizontal
4 **a** $B = 30.4$ N, $\theta = 4.72°$
b $B = 28.5$ N, $\theta = 29.8°$
c $B = 13.9$ N, $\theta = 7.52°$

5 a $\frac{\sqrt{3}}{5}$m s^{-2} b 48 N
6 $20\sqrt{2}$N
7 36.3 kg (3 s.f.)
8 $2T\sin 60 + 20g = 80g$
$T = \frac{60g}{2\sin 60} = 20\sqrt{3}g$
9 $\mathbf{F}_1 = 12\sqrt{3}$N, $\mathbf{F}_2 = 20$ N

Challenge
$\mathbf{F}_1 = 6\sqrt{2} - \sqrt{6}$N, $\mathbf{F}_2 = 2\sqrt{3} - 2$

Exercise 5B

1 a

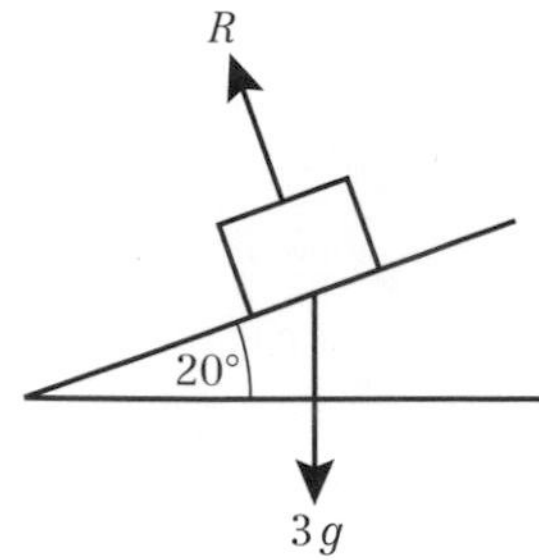

b 27.6 N (3 s.f.) c 3.35 m s^{-2}

2 a

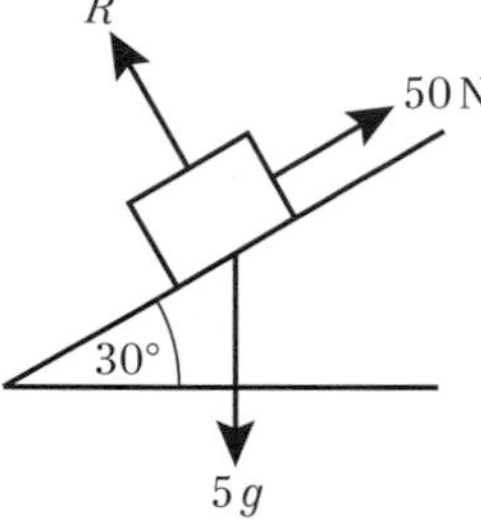

b 42.4 N (3 s.f.) c 5.1 m s^{-2}
3 a 3.92 N (3 s.f.) b 5.88 m s^{-2} (3 s.f.)
4 a

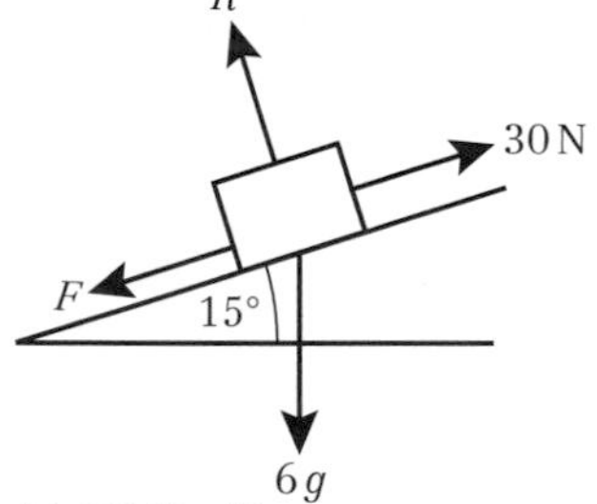

b 14.8 N (3 s.f.)
5 a 0.589 kg (3 s.f.) b 4.9 m s^{-2}
6 0.296 ms^{-2} (3 s.f.)
7 15.0 N (3 s.f.)
8 R(↗): $26\cos 45 - mg\sin\alpha - 12 = m \times 1$
$13\sqrt{2} - 12 = m + \frac{1}{2}mg$
$m = \frac{13\sqrt{2} - 12}{1 + \frac{g}{2}}$
$m = 1.08$ kg (3 s.f.)

Challenge
a $mg\sin\theta = ma$ and $mg\sin(\theta + 60) = 4ma$
$4\sin\theta = \sin(\theta + 60)$
$4\sin\theta = \sin\theta\cos 60 + \cos\theta\sin 60$
$4\sin\theta = \frac{1}{2}\sin\theta + \frac{\sqrt{3}}{2}\cos\theta$
$\frac{7}{2}\sin\theta = \frac{\sqrt{3}}{2}\cos\theta$
$\tan\theta = \frac{\sqrt{3}}{7}$
b 13.9°

Exercise 5C

1 a i 3 N ii $F = 3$ N and body remains at rest
b i 7 N ii $F = 7$ N and body remains at rest
c i 7 N ii $F = 7$ N and body accelerates
iii 1 m s^{-2}
d i 6 N ii $F = 6$ N and body remains at rest
e i 9 N
ii $F = 9$ N and body remains at rest in limiting equilibrium
f i 9 N
ii $F = 9$ N and body accelerates
iii 0.6 m s^{-2}
g i 3 N ii $F = 3$ N and body remains at rest
h i 5 N
ii $F = 5$ N and body remains at rest in limiting equilibrium
i i 5 N ii $F = 5$ N and body accelerates
iii 0.2 m s^{-2}
j i 6 N ii $F = 6$ N and body accelerates
iii 1.22 m s^{-2} (3 s.f.)
k i 5 N ii $F = 5$ N and body accelerates
iii 3.85 m s^{-2} (3 s.f.)
l i 12.7 N (3 s.f.)
ii The body accelerates.
iii 5.39 m s^{-2} (3 s.f.)
2 a $R = 88$ N, $\mu = 0.083$ (2 s.f.)
b $R = 80.679$ N, $\mu = 0.062$ (2 s.f.)
c $R = 118$ N, $\mu = 0.13$ (2 s.f.)
3 0.242 (3 s.f.)
4 0.778 N (3 s.f.)
5 56.1 N (3 s.f.)
6 16.5 N (3 s.f.)
7 a Use $v = u + at$ to find $a = -\frac{2}{3}$ms^{-2}
R(→): $-\mu mg = -\frac{2}{3}m$
$\mu = \frac{2}{3g}$
b The coefficient of friction remains unchanged. The air resistance has no effect on the coefficient of friction, which is dependent on the properties of the wheels and the rails.

Challenge
R(↙): $mg\sin\alpha - \mu mg\cos\alpha = ma$
$g\sin\alpha - \mu g\cos\alpha = a$

Mixed exercise 5

1 a 32.0 N (3 s.f.) b 0.5 m s^{-2}
2 $F_1 = 27.8$ N, $F_2 = 24.2$ N (3 s.f.)
3 a

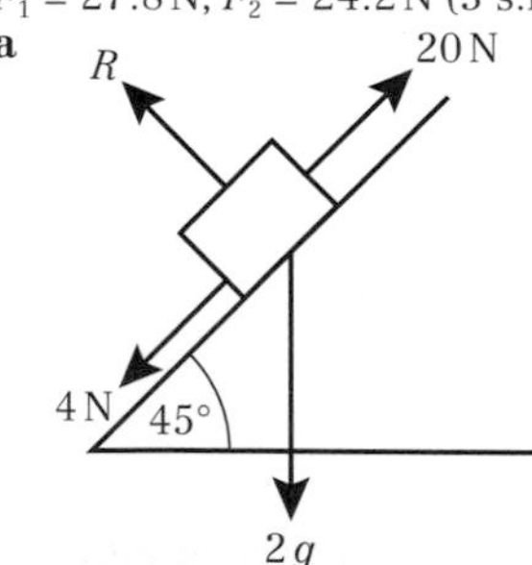

b 13.9 N (3 s.f.)
c Res (↗): $16 - 2g\sin 45 = 2a$
$a = \frac{16 - 2g\sin 45}{2} = 1.1$ ms^{-2} (2 s.f)
4 2.06 ms^{-2} (3 s.f.)
5 R(→): $F = 150\cos 45 + 100\cos 30$
$= \frac{150\sqrt{2}}{2} + \frac{100\sqrt{3}}{2}$
$= 25(3\sqrt{2} + 2\sqrt{3})$N
6 $\mu = \frac{5\sqrt{3}}{93}$

7 11.4 N
8 3.41 m s^{-2} (3 s.f.)
9 a 4400 N (2 s.f.)
b 0.59 (2 s.f.)
c i e.g. The force due to air resistance will not remain constant in the subsequent motion of the car.
ii e.g. Whilst skidding the car is unlikely to travel in a straight line.

Challenge
$F_{MAX} = 0.2 \times 400g \cos 15° = 760$ N (2 s.f.)
Component of weight that acts down the slipway:
$400g \sin 15° = 1000$ N (2 s.f.)
1000 N > 760 N so boat will come to momentary rest then accelerate back down the slope.
The boat will take 6.9 seconds to reach the water.

Prior knowledge 6

1 a 11.5 m
b 3.1 s
2 $x = v\cos\theta$, $y = v\sin\theta$
3 a i $\cos\theta = \frac{12}{13}$ ii $\tan\theta = \frac{5}{12}$
b i $\sin\theta = \frac{8}{17}$ ii $\cos\theta = \frac{15}{17}$

Exercise 6A

1 a 122.5 m b 100 m
2 a $x = 36$ m, $y = 19.6$ m b 41 m
3 $u = 16.6$ m s^{-1}
4 77.5 m s^{-1}
5 0.59 s
6 1.9 m
7 a 0.5 s b 0.42 c 3.5 m

Exercise 6B

1 a $u_x = 19.2$ m s^{-1}, $u_y = 16.1$ m s^{-1}
b $(19.2\mathbf{i} + 16.1\mathbf{j})$ m s^{-1}
2 a $u_x = 16.9$ m s^{-1}, $u_y = -6.2$ m s^{-1}
b $(16.9\mathbf{i} - 6.2\mathbf{j})$ m s^{-1}
3 a $u_x = 32.3$ m s^{-1}, $u_y = 13.5$ m s^{-1}
b $(32.3\mathbf{i} + 13.5\mathbf{j})$ m s^{-1}
4 a $u_x = 26.9$ m s^{-1}, $u_y = -7.8$ m s^{-1}
b $(26.9\mathbf{i} - 7.8\mathbf{j})$ m s^{-1}
5 10.8 m s^{-1}, 56.3°
6 6.4 m s^{-1}, 51.3° below the horizontal
7 a 33.7° b $k = 3$ or $k = -3$

Exercise 6C

1 3.1 (2 s.f.)
2 8.5 m (2 s.f.)
3 a 44 m (2 s.f.) b 79 m
4 a 2.7 s (2 s.f.) b 790 m (2 s.f.)
5 a 10 m (2 s.f.) b 41 m (2 s.f.)
6 a 3.9 s (2 s.f.) b 56 m (2 s.f.)
7 55° (nearest degree)
8 a $(36\mathbf{i} + 27.9\mathbf{j})$ m b 13 m s^{-1} (2 s.f.)
9 a 22° (2 s.f.) b 97 m (2 s.f.)
10 a 16 (2 s.f.) b 1.6 s (2 s.f.)
11 a 4.4 b 88 c 50° (2 s.f.)
12 a 1.1 s (2 s.f.) b 34 m (2 s.f.)
13 $\alpha = 40.6°$ (nearest 0.1°)
$U = 44$ (2 s.f.)
14 a 15.6 m s^{-1} b 2.92 s c 22.3 m s^{-1}
15 a $k = 7.35$
b i 13.6 m s^{-1} ii 72.9°
16 a 10.7 m s^{-1} b e.g. weight of the ball; air resistance

Challenge
R(→): $s = 12t$ and $s = (20\cos\alpha)t$
so $\cos\alpha = 0.6$ and $\sin\alpha = 0.8$
R(↑): $s = -4.9t^2 + 40$ and $s = (20\sin\alpha)t - 4.9t^2$
So $t = \frac{40}{20\sin\alpha} = \frac{40}{16} = 2.5$ seconds

Exercise 6D

1 R(↑): $v^2 = U^2\sin^2\alpha - 2gh$
At maximum height, $v = 0$ so $0 = U^2\sin^2\alpha - 2gh$
Rearrange to give $h = \frac{U^2\sin^2\alpha}{2g}$
2 a R(→): $x = 21\cos\alpha \times t$, so $t = \frac{x}{21\cos\alpha}$
R(↑): $y = 21\sin\alpha \times \frac{x}{21\cos\alpha} - \frac{1}{2}g\left(\frac{x}{21\cos\alpha}\right)^2$
$y = x\tan\alpha - \frac{x^2}{90\cos^2\alpha}$
b $\tan\alpha = 1.25$
3 a R(↑): $s = U\sin\alpha t - \frac{g}{2}t^2$
When particle strikes plane, $s = 0 = t(U\sin\alpha - \frac{g}{2}t)$
So $t = 0$ or $t = \frac{2U\sin\alpha}{g}$
b R(→): $s = ut = U\cos\alpha\left(\frac{2U\sin\alpha}{g}\right) = \frac{U^2\sin 2\alpha}{g}$
c Range $s = \frac{U^2\sin 2\alpha}{g}$ is greatest when $\sin 2\alpha = 1$
Occurs when $2\alpha = 90° \Rightarrow \alpha = 45°$
d 12° and 78°
4 Using $v = u + at$, at max height $t = \frac{v}{g}$
So time taken to return to the ground $= \frac{v}{g}$
Using $s = ut + \frac{1}{2}at^2$, distance travelled by one part =
$2v\left(\frac{v}{g}\right) = \frac{2v^2}{g}$
So two parts of firework are $\frac{2v^2}{g} + \frac{2v^2}{g} = \frac{4v^2}{g}$ apart.
5 a R(→): $x = U\cos\alpha \times t$, so $t = \frac{x}{U\cos\alpha}$
R(↑): $y = U\sin\alpha \times t - \frac{1}{2}gt^2$
Substitute for $t \Rightarrow y = U\sin\alpha\left(\frac{x}{U\cos\alpha}\right) - \frac{1}{2}g\left(\frac{x}{U\cos\alpha}\right)^2$
Use $\tan\alpha = \frac{\sin\alpha}{\cos\alpha}$ and rearrange to give
$y = x\tan\alpha - \frac{gx^2}{2U^2\cos^2\alpha}$.
b 13.7 m
6 a R(→): $x = U\cos\alpha \times t$, so $t = \frac{x}{U\cos\alpha}$
R(↑): $y = U\sin\alpha \times t - \frac{1}{2}gt^2$
Substitute for $t \Rightarrow y = U\sin\alpha\left(\frac{x}{U\cos\alpha}\right) - \frac{1}{2}g\left(\frac{x}{U\cos\alpha}\right)^2$
Use $\tan\alpha = \frac{\sin\alpha}{\cos\alpha}$ and $\frac{1}{\cos\alpha} = \sec\alpha$, and rearrange to give
$y = x\tan\alpha - \frac{gx^2}{2U^2}\sec^2\alpha$.
Use $\sec^2\alpha = 1 + \tan^2\alpha$, and rearrange to give
$y = x\tan\alpha - \frac{gx^2}{2U^2}(1 + \tan^2\alpha)$.
b 93.8 m
c 4.4 s
7 a R(→): $x = 9 = U\cos\alpha \times t$, so $t = \frac{9}{U\cos\alpha}$
R(↑): $y = U\sin\alpha \times t - \frac{1}{2}gt^2$
Substitute for $t \Rightarrow y = U\sin\alpha\left(\frac{9}{U\cos\alpha}\right) - \frac{1}{2}g\left(\frac{9}{U\cos\alpha}\right)^2$
Use $\tan\alpha = \frac{\sin\alpha}{\cos\alpha}$ and $y = 0.9$. Rearrange to give
$0.9 = 9\tan\alpha - \frac{81g}{2U^2\cos^2\alpha}$.
b 10.3 m s^{-1}
8 a R(→): $x = kt$, so $t = \frac{x}{k}$
R(↑): $y = 2kt - \frac{gt^2}{2}$
Substitute for $t \Rightarrow y = 2x - \frac{gx^2}{2k^2}$
b i $\frac{4k^2}{g}$ m ii $\frac{2k^2}{g}$ m

Challenge

For the projectile: $y = x\tan\alpha - \frac{gx^2}{2U^2\cos^2\alpha}$ so for $a = 45°$

$y = x - \frac{gx^2}{U^2}$

For the slope: $y = -x$

Projectile intersects the slope when $-x = x - \frac{gx^2}{U^2} \Rightarrow x = \frac{2U^2}{g}$, $y = -\frac{2U^2}{g}$

Distance $= \sqrt{\left(\frac{2U^2}{g}\right)^2 + \left(\frac{2U^2}{g}\right)^2} = \sqrt{8\left(\frac{U^2}{g}\right)^2} = \frac{2\sqrt{2}U^2}{g}$

Mixed exercise 6

1 a 45 m b 6.1 s
2 $h = 35$ (2 s.f.)
3 a 36 m b 30 m (2 s.f.)
4 a 140 m (2 s.f.) b 36 ms^{-1} (2 s.f.)
5 a R(↑): $s = U\sin\theta\, t - \frac{g}{2}t^2$
When particle strikes plane, $s = 0 = t(U\sin\theta - \frac{g}{2}t)$
So $t = 0$ or $t = \frac{2U\sin\theta}{g}$
R(→): $s = Ut = U\cos\theta\left(\frac{2U\sin\theta}{g}\right) = \frac{U^2\sin 2\theta}{g}$
b $\frac{U^2}{g}$ c 20.9°, 69.1° (nearest 0.1°)
6 a 2.0 s (2 s.f.) b 3.1 s (2 s.f.) c 36 m s^{-1} (2 s.f.)
7 a 2 m b 5.77° or 84.2°
8 a 0.65 s b 1.5 m c 23.8 m s^{-1}
9 a Particle P: $x = 18t$, Particle Q: $x = 30\cos\alpha\, t$
When particles collide: $18t = 30\cos\alpha\, t \Rightarrow \cos\alpha = \frac{3}{5}$
b $\frac{4}{3}$ s

Challenge

62 m s^{-1}

Prior knowledge 7

1 0.278 (3 s.f.)
2 12.25 N

Exercise 7A

1 a i $Q - 5\cos 30° = 0$ ii $P - 5\sin 30° = 0$
iii $Q = 4.33$ N $P = 2.5$ N
b i $P\cos\theta + 8\sin 40° - 7\cos 35° = 0$
ii $P\sin\theta + 7\sin 35° - 8\cos 40° = 0$
iii $\theta = 74.4°$ (allow 74.3°) $P = 2.20$ N (allow 2.19)
c i $9 - P\cos 30° = 0$
ii $Q + P\sin 30° - 8 = 0$
iii $Q = 2.80$ N $P = 10.4$ N
d i $Q\cos 60° + 6\cos 45° - P = 0$
ii $Q\sin 60° - 6\sin 45° = 0$
iii $Q = 4.90$ N $P = 6.69$ N
e i $6\cos 45° - 2\cos 60° - P\sin\theta = 0$
ii $6\sin 45° + 2\sin 60° - P\cos\theta - 4 = 0$
iii $\theta = 58.7°$ $P = 3.80$ N
f i $9\cos 40° + 3 - P\cos\theta - 8\sin 20° = 0$
ii $P\sin\theta + 9\sin 40° - 8\cos 20° = 0$
iii $\theta = 13.6°$ $P = 7.36$ N
2 a i ii $Q = 4$ N, $P = 8$ N

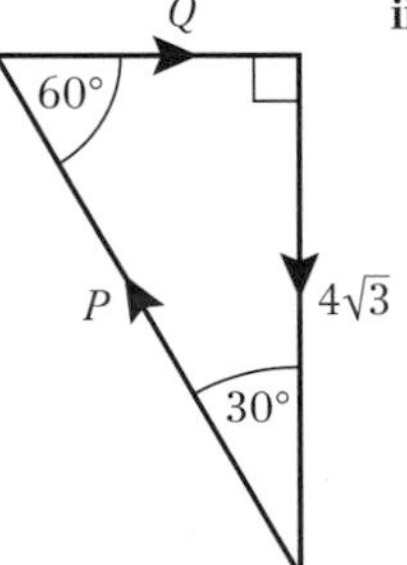

b i

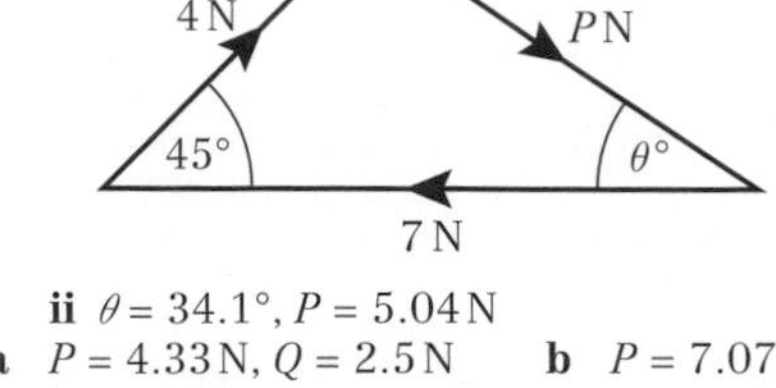

ii $\theta = 34.1°$, $P = 5.04$ N
3 a $P = 4.33$ N, $Q = 2.5$ N b $P = 7.07$ N, $Q = 7.07$ N
c $P = 4.73$ N, $Q = 4.20$ N d $P = 3.00$ N, $Q = 0.657$ N
e $P = 9.24$ N, $Q = 4.62$ N

Challenge

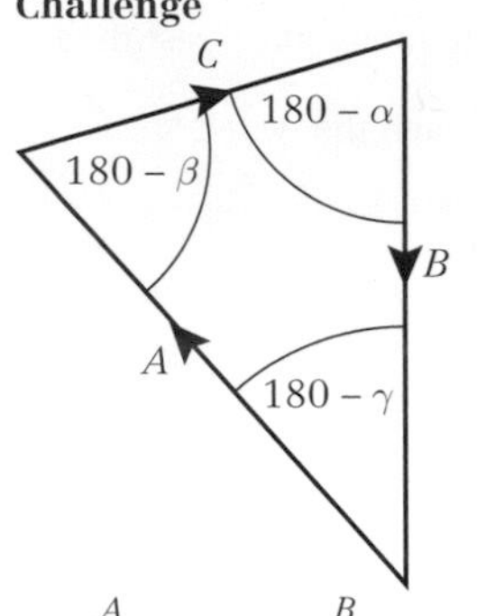

$\frac{A}{\sin(180 - \alpha)} = \frac{B}{\sin(180 - \beta)} = \frac{C}{\sin(180 - \gamma)}$

$\frac{A}{\sin\alpha} = \frac{B}{\sin\beta} = \frac{C}{\sin\gamma}$

Exercise 7B

1 35 N (2 s.f.)
2 a 20 N b 1.77
3 a 33.7° b 14.4
4 30 N and 43 N (2 s.f.)
5 a 5.46 N b 0.76 kg
c Assumption that there is no friction between the string and the bead.
6 a 1.46 N b 55 g
7 a 2.6 b 4.4 N
8 a $F = 19.6m$, $R = 9.8m$ b $F' = 17m$ (2 s.f.), $R' = 0$
9 13.9 N
10 39.2 N
11 a 15.7 N (3 s.f.)
b 37.2 N (3 s.f.)
c Assumption that there is no friction between the string and the pulley.
12 $R = 0.40$ N (2 s.f.)

Exercise 7C

1 0.446
2 0.123
3 a 1.5 N b Not limiting
4 a 40 kg
b The assumption is that the crate and books may be modelled as a particle.
5 a 11.9 N b 6.40 N
6 0.601 (accept 0.6)
7 a 13.3 N b $X = 10.7$ N
8 a 22.7 N
b 9.97 N down the plane
c $\mu \geqslant 0.439$
9 a $X = 44.8$ (accept 44.7) b $R = 51.3$ N
10 $T = 102$ (3 s.f.)
11 $2.75 \leqslant T \leqslant 3.87$ N
12 0.758
13 a 0.75 b 67.2 N

Online Full worked solutions are available in SolutionBank.

Exercise 7D

1 34.6 N, 50 N, 17.3 N, 0.35
2 a 22.8 N b 98 N, 22.8 N c 0.233
d The weight of a uniform ladder passes through its midpoint.
3 a 41.6° b 24.0°
c No friction at the wall.
4 a $5\frac{1}{3}$ m
b i The ladder may not be uniform.
ii There would be friction between the ladder and the vertical wall.
5 a 1.99 N b 0.526
c There is no friction between the rail and the pole.
6 R(↑): $R = mg$, R(→): $N = \mu mg$
Let the length of the ladder be $2l$
Taking moments about A,
$mgl\cos\theta = 2\mu mgl\sin\theta$
Cancelling and rearranging gives
$2\mu\tan\theta = 1$
7 104 N, 64.5 N, 0.620
8 R(↑): $R + \mu_2 N = mg$, R(→): $N = \mu_1 R$
Taking moments about the base of the ladder
$2N\tan\theta + 2\mu_2 N = mg$
$2\tan\theta + 2\mu_2 = \frac{1}{\mu_1} + \mu_2$
$2\tan\theta = \frac{1}{\mu_1} - \mu_2$
$\tan\theta = \frac{1 - \mu_1\mu_2}{2\mu_1}$
9 a $\frac{W\sqrt{3}}{6}$
b $F = P = \frac{W\sqrt{3}}{6}$, $R = W$, $F \leqslant \mu R$, $\mu \geqslant \frac{\sqrt{3}}{6}$
c $\frac{W}{4}$
10 a 6.50 N, 10.8 N b $\frac{1}{\sqrt{3}}$
11 a $\frac{19}{2\sqrt{3}}W$ b $\left(\frac{19}{2\sqrt{3}} - 2\right)W \leqslant P \leqslant \left(\frac{19}{2\sqrt{3}} + 2\right)W$
c It allows us to assume that the weight of the ladder acts through its midpoint.
d i The magnitude of the reaction at Y will get smaller.
ii The range of values for P will get smaller.

Exercise 7E

1 3.35 m s^{-2} (3 s.f.)
2 a 27.7 N (3 s.f.) b 2.12 m s^{-2}
3 a 2.43 m s^{-2} (3 s.f.) b 4.93 m s^{-1} (3 s.f.)
4 0.165 (3 s.f.)
5 0.20 (2 s.f.)
6 0.15 (2 s.f.)
7 a 88.8 N (3 s.f.) b 0.24 (2 s.f.)
8 a $\frac{13g}{15}$ b 23.5 m (3 s.f.)
c 2.35 s (3 s.f.) d 12.4 m s^{-1} (3 s.f.)
9 0.180 (3 s.f.)
10 R(↖): $R = mg\cos\alpha$, R(↗): $\mu R - mg\sin\alpha = ma$
$\mu mg\cos\alpha - mg\sin\alpha = ma$
$\mu g\cos\alpha - g\sin\alpha = a$
11 a 0.155
b The particle will slide down the hill. The component of weight down the slope $5g\sin 10 = 8.51$ N is greater than the friction $0.155 \times 5g\cos 10 = 7.48$ N.

Exercise 7F

1 2.8 m s^{-1}
2 a 1.12 m s^{-2} b 4100 N
c The resistances are unlikely to be constant as the resistance will increase as the speed increases.
3 a 21.9 N b 0.418 (3 s.f.) c 38 N (2 s.f.)
4 a 18 N (2 s.f.) b 2 c $\frac{2}{7}$ s
5 a For particle on LHS
R(↖): $R_1 = \sqrt{3}g$, R(↗): $T = \frac{\sqrt{3}}{5}g + g$
For particle on RHS
R(↗): $R_2 = \frac{\sqrt{3}}{2}mg$, R(↘): $T = \frac{1}{2}mg - \frac{\sqrt{3}}{5}mg$
For maximum value of m, R(↘) is equal to R(↗):
$\frac{1}{2}mg - \frac{\sqrt{3}}{5}mg = \frac{\sqrt{3}}{5}g + g$
$\Rightarrow \frac{1}{2}m - \frac{\sqrt{3}}{5}m = \frac{\sqrt{3}}{5} + 1$
$\Rightarrow \left(\frac{1}{2} - \frac{\sqrt{3}}{5}\right)m = \frac{\sqrt{3}}{5} + 1$
$\Rightarrow m = \frac{10 + 2\sqrt{3}}{5 - 2\sqrt{3}}$
b 2.70 m s^{-2} (3 s.f.)
6 a 3 m s^{-2} b 10.88 N
c R(→): $T - 1.5\mu g = 4.5$
$\mu = \frac{10.88 - 4.5}{1.5g} = \frac{319}{735} = 0.434$ (3 s.f.)
d The string doesn't stretch so the tension in the string is constant.

Challenge

a For particle on LHS: R(↗): $T = \frac{1}{2}m_1 g$
For particle on RHS: R(↖): $T = \frac{\sqrt{3}}{2}m_2 g$
To prove, equate values of T.
b If (attempted) motion is down slope on RHS
Consider particle on LHS
$T = \frac{1}{2}m_1 g + \frac{\sqrt{3}}{2}\mu m_1 g$
Consider particle on RHS
$T = \frac{\sqrt{3}}{2}m_2 g - \frac{1}{2}\mu m_2 g$
Equate values of T to find $\frac{m_1}{m_2}$
If (attempted) motion is down slope on LHS
Consider particle on LHS
$T = \frac{1}{2}m_1 g - \frac{\sqrt{3}}{2}\mu m_1 g$
Consider particle on RHS
$T = \frac{1}{2}\mu m_2 g + \frac{\sqrt{3}}{2}m_2 g$
Equate values of T to find $\frac{m_1}{m_2}$

Mixed exercise 7

1 a 32.3° (3 s.f.) b 16.3 N (3 s.f.)
2 a 18.0° (3 s.f.) b 43.3 N (3 s.f.)
3 $T_1 = 1062$ N, $T_2 = 1013$ N
4 12 N (2 s.f.)
5 a 12.25 N b 46.6 N (3 s.f.)
c F will be smaller
6 a R(↑): $T\cos 20 = 2g + T\cos 70$
$T = \frac{2g}{\cos 20 - \cos 70}$
$= 33$ N (2 s.f.)
b 42 N (2 s.f.)
7 a 364 N (3 s.f.) b Hill unlikely to be smooth.
8 R(→): $F = N$
Taking moments about A
$Fa\sin\theta + \frac{5}{2}mga\cos\theta = 5aN\sin\theta$
$\frac{5}{2}mg\cos\theta = 4F\sin\theta$
$\frac{5}{8}mg = F\tan\theta$
9 $\frac{7}{24}$
10 a $\frac{8W}{9}$
b R(↑): $R + \mu N \geqslant W$, R(→): $N = \frac{W}{3}$
$\frac{\mu W}{3} \geqslant W - \frac{8W}{9}$

c The ladder has negligible thickness/the ladder does not bend.

11 a Taking moments about point where ladder touches the ground
R(↑): $R = W$, R(→): $N = 0.3R$
$1.5W = 1.2W$. This cannot be true so the ladder cannot rest in this position.
b R(→): $F = N$
Taking moments about point where ladder touches the ground $1.5W = 4N$, $F = N = \frac{3W}{8}$
c $\frac{W}{4g}$

12 18N (2 s.f.)

13 0.070 (2 s.f.)

14 $6.35\,\text{ms}^{-2}$ (3 s.f.)

15 R(→): $T - \mu m_1 g = m_1 a$, R(↑): $T = m_2 g - m_2 a$
$m_1 a + \mu m_1 g = m_2 g - m_2 a$
$g(m_2 - \mu m_1) = a(m_1 + m_2)$

16 R(↗): $T - m_1 g \sin 30 = \frac{1}{2}m_1$,
R(↘): $m_2 g \cos 45 - T = \frac{1}{2}m_2$
and $T = \frac{\sqrt{2}}{2}m_2 g - \frac{1}{2}m_2$, $T = \frac{1}{2}m_1 + \frac{1}{2}m_1 g$
$\sqrt{2}\,m_2 g - m_2 = m_1 + m_1 g$
$m_2(\sqrt{2}\,g - 1) = m_1(1 + g)$

Challenge

a 94.3N (3 s.f.)

b 80.6N (3 s.f.), 54.2° (3 s.f.) to the horizontal

Prior knowledge 8

1 a $19\mathbf{i} - 43\mathbf{j}$ b $-10\mathbf{i} + 11\mathbf{j}$ c $\frac{5}{13}\mathbf{i} - \frac{12}{13}\mathbf{j}$

2 a $23\,\text{m s}^{-1}$ b 52 m

3 a i $6e^{2x}$ ii $6\cos 3x$
b i $\frac{4}{3}e^{3x+1}$ ii $\frac{5}{2\pi}\sin 2\pi x$

Exercise 8A

1 a $6\mathbf{i} + 12\mathbf{j}$ b $-7\mathbf{i} + 4\mathbf{j}$ c $-2\mathbf{i} + 6\mathbf{j}$
d $10\mathbf{i} - 13\mathbf{j}$ e $2\mathbf{i} - 3\mathbf{j}$ f 4 s

2 $\frac{\sqrt{85}}{2}\,\text{m s}^{-1}$, 319°

3 2.5 s

4 a $\begin{pmatrix}120 - 30t\\ -10 + 40t\end{pmatrix}$ b 4 s

5 $2.03\,\text{m s}^{-1}$

6 a $7t\mathbf{i} + (400 + 7t)\mathbf{j}$, $(500 - 3t)\mathbf{i} + 15t\mathbf{j}$
b $350\mathbf{i} + 750\mathbf{j}$

7 a $\begin{pmatrix}\frac{3}{5}\\ \frac{4}{5}\end{pmatrix}\text{m s}^{-2}$ b $\begin{pmatrix}\frac{15}{2}\\ 10\end{pmatrix}\text{m}$

8 a $-\frac{5}{2}\mathbf{i} - \frac{7}{4}\mathbf{j}\,\text{m s}^{-2}$
b $\left(10 + 15t - \frac{5}{4}t^2\right)\mathbf{i} + \left(-8 + 4t - \frac{7}{8}t^2\right)\mathbf{j}\,\text{m}$

9 a $\begin{pmatrix}60\\ -15\end{pmatrix}\text{m s}^{-1}$ b 688 m

10 $\begin{pmatrix}40\\ -60\end{pmatrix}\text{m}$

11 a $12.1\,\text{m s}^{-1}$, $6.08\,\text{m s}^{-1}$ b $18\mathbf{i} - 3\mathbf{j}$
c $15\mathbf{i} - 12\mathbf{j}$

12 a $\mathbf{i} - 7\mathbf{j}\,\text{m s}^{-2}$ b $s = (-4t + 0.5t^2)\mathbf{i} + (8t - 3.5t^2)\mathbf{j}$
c 15:00 d $-160\mathbf{j}$

13 a North-east of O when **i** and **j** components are equal
$2t^2 - 3 = 7 - 4t \Rightarrow 2t^2 + 4t - 10 = 0 \Rightarrow t^2 + 2t - 5 = 0$
b 1.70 m c $7.83\,\text{m s}^{-1}$, 026.6° d 19.3 m

Challenge

24 s

Exercise 8B

1 a $(36\mathbf{i} + 27.9\mathbf{j})\,\text{m}$ b $13\,\text{m s}^{-1}$ (2 s.f.)

2 a $\mathbf{r} = (4t)\mathbf{i} + (5t - 5t^2)\mathbf{j}$ b 1.25 m

3 a Either answer with justification
e.g. The sea is likely to be horizontal and relatively flat, whereas the ball is subject to air resistance, so the assumption that sea is a horizontal plane is most reasonable.
Or e.g. Although the sea is horizontal it is unlikely to be flat because of waves, so the assumption that the ball is a particle is most reasonable.
b $\mathbf{v} = (6.9\mathbf{i} - 17\mathbf{j})\,\text{m s}^{-1}$ (both values to 2 s.f.)
c $5.5\,\text{m s}^{-2}$ (2 s.f.)

4 a R(↑): $0 = 4ut - \frac{g}{2}t^2 \Rightarrow t = \frac{8u}{g}$
R(→): $750 = 3ut = \frac{24u^2}{g} \Rightarrow u^2 = \frac{750g}{24} \Rightarrow u = 17.5$
b 250m c 22° (nearest degree)

5 a 48 m b 120 m (2 s.f.)
c $T = 2.5\,\text{s}$, $\mathbf{r} = (20\mathbf{i} - \frac{45}{8}\mathbf{j})\,\text{m}$

6 a $x = at \Rightarrow t = \frac{x}{a}$
$y = bt - 5t^2 \Rightarrow y = b\left(\frac{x}{a}\right) - 5\left(\frac{x}{a}\right)^2 \Rightarrow y = \frac{bx}{a} - \frac{5x^2}{a^2}$
b i $X = 1.6b$ ii $Y = 0.05b^2$

Exercise 8C

1 a $v = t + \frac{\cos \pi t}{\pi} - \frac{1}{\pi}$ b $s = \frac{t^2}{2} + \frac{\sin \pi t}{\pi^2} - \frac{t}{\pi}$

2 a $v = -\frac{\cos 3\pi t}{3\pi} + \frac{2}{3\pi}$ b $\frac{1}{\pi}$
c $s = -\frac{\sin 3\pi t}{9\pi^2} + \frac{2t}{3\pi} + 1$

3 a $v = -\frac{\sin 4\pi t}{4\pi}$ b $\frac{1}{4\pi}$
c $s = \frac{\cos 4\pi t}{16\pi^2} - \frac{1}{16\pi^2}$ d $\frac{1}{8\pi^2}$ e 16

4 a $1.18\,\text{m s}^{-1}$ b $-0.152\,\text{m s}^{-2}$
c $-0.759\,\text{N}$

5 a $0.5\,\text{m s}^{-1}$ b $0.1\,\text{m s}^{-1}$

6 a $12.9\,\text{m s}^{-1}$ in the direction of s increasing
b $24\,\text{m s}^{-1}$ in the direction of s decreasing
c 132 m
d 20.8 m and 118.5 m

7 3.31 s

8 a $k = 40$, $T = 25$ b $4\,\text{m s}^{-1}$
c $v = \frac{20}{\sqrt{t}}$, so for small t, the value of v is large
e.g. $t = 0.01$, $v = 200\,\text{m s}^{-1}$, so not realistic for small t.

9 a $k = \frac{1}{2}$ b $t = \pi, 3\pi$
c $a = 4\cos\left(\frac{t}{2}\right)$, $4a^2 = 64\cos^2\left(\frac{t}{2}\right)$
$v = 2 + 8\sin\left(\frac{t}{2}\right)$, $(v - 2)^2 = 64\sin^2\left(\frac{t}{2}\right)$
$4a^2 = 64 - (v - 2)^2 \Rightarrow 64\cos^2\left(\frac{t}{2}\right) = 64 - 64\sin^2\left(\frac{t}{2}\right)$
$\Rightarrow \cos^2\left(\frac{t}{2}\right) = 1 - \sin^2\left(\frac{t}{2}\right)$
$\Rightarrow \cos^2 T + \sin^2 T = 1$
d $10\,\text{m s}^{-1}$, $4\,\text{m s}^{-2}$

10 a $24\,\text{m s}^{-1}$ b 54.4 m c 8.43 s d 101.2 m

Exercise 8D

1 a $(3\mathbf{i} + 23\mathbf{j})\,\text{m s}^{-1}$ b $18\mathbf{j}\,\text{m s}^{-2}$

2 $(0.024\mathbf{i} + 0.006\mathbf{j})\,\text{N}$

3 a 0.305 s b $6\,\text{m s}^{-1}$
c **i**-component of velocity is negative, **j**-component of velocity = 0

4 a $20\,\text{m s}^{-1}$ b $10\,\text{m s}^{-2}$
b $a = 8\mathbf{i} - 6\mathbf{j}$, no dependency on t therefore constant.
$|a| = 10\,\text{m s}^{-2}$

5 a $6\sqrt{5}\,\text{m s}^{-1}$ b $t = 2$
c $(-16\mathbf{i} + 4\mathbf{j})\,\text{m}$ d 15.2 N (3 s.f.)

6 a $k = -0.5, -8.5$
b $10\,\text{m s}^{-2}$ for both values of k
7 a $52\,\text{m s}^{-1}$ b $(12\mathbf{i} + \frac{15}{2}\mathbf{j})\,\text{m s}^{-2}$
8 a 4 b $(-36\mathbf{i} + 8\mathbf{j})\,\text{m s}^{-2}$
9 $\mathbf{a} = 4\mathbf{i} + 2\mathbf{j}$, no t dependency so constant. $|\mathbf{a}| = 2\sqrt{5}\,\text{m s}^{-2}$
10 a $\sqrt{15}$ b $(-4\sqrt{30}\mathbf{i} + 2\sqrt{15}\mathbf{j})\,\text{m s}^{-2}$

Exercise 8E

1 a $16\mathbf{i}\,\text{m s}^{-1}$ b $128\mathbf{i} - 192\mathbf{j}\,\text{m}$
2 a $13\,\text{m}$ b $(4\mathbf{i} + 6\mathbf{j})\,\text{m s}^{-2}$
3 a $\mathbf{v} = \left(t^2 - 4t + 2\pi - \frac{\pi^2}{4}\right)\mathbf{i} - 6\cos t\mathbf{j}$ b $2\pi^2 - 4\pi$
4 a $\left(\left(\frac{5t^2}{2} - 3t + 2\right)\mathbf{i} + \left(8t - \frac{t^2}{2} - 5\right)\mathbf{j}\right)\text{m s}^{-1}$
b $t = \frac{1}{2}$ c $\frac{9\sqrt{2}}{8}\,\text{m s}^{-1}$
5 a $\left((t^2 + 6)\mathbf{i} + \left(2t - \frac{t^3}{3}\right)\mathbf{j}\right)\text{m}$ b $6\mathbf{j}\,\text{m}$
6 a $(-\mathbf{i} - \frac{1}{4}\mathbf{j})\,\text{m s}^{-1}$ b $(94\mathbf{i} - 654\mathbf{j})\,\text{m}$
7 a $((2t^4 - 3t^2 - 4)\mathbf{i} + (4t^2 - 3t - 7)\mathbf{j})\,\text{m s}^{-1}$
b $t = \frac{7}{4}$
8 a $\mathbf{r} = (2t^2 - 3t + 1)\mathbf{i} + (4t + 2)\mathbf{j}\,\text{m}$
b i 3.4 ii $\mathbf{r} = 36\mathbf{i} + 22\mathbf{j}$

Challenge

$\mathbf{r} = \left(\frac{3\pi}{2} + 1\right)\mathbf{i} + \left(\frac{5\pi^2}{8} + 1\right)\mathbf{j}$

Mixed exercise 8

1 $4.8\mathbf{i} - 6.4\mathbf{j}$
2 $10.1\,\text{m}$
3 a $2t\mathbf{i} + (-500 + 3t)\mathbf{j}$ b $721\,\text{m}$
4 a $(1 + 2t)\mathbf{i} + (3 - t)\mathbf{j}$, $(5 - t)\mathbf{i} + (-2 + 4t)\mathbf{j}$
b $\mathbf{r}_{BA} = \mathbf{r}_B - \mathbf{r}_A = (5 - t)\mathbf{i} + (3 - t)\mathbf{j} - ((1 + 2t)\mathbf{i} + (4t - 2)\mathbf{j})$
$= (4 - 3t)\mathbf{i} + (5 - 5t)\mathbf{j}$
c For A and B to collide $\mathbf{r}_A = \mathbf{r}_B$.
Equating $\mathbf{i} \rightarrow t = \frac{4}{3}$, equating $\mathbf{j} \rightarrow t = 1$. Times are not the same therefore the ships do not collide.
d $5.39\,\text{km}$
5 a $48\,\text{m}$ b $120\,\text{m}$ (2 s.f.)
6 a $p = 16$, $q = 19.4$ b $25.1\,\text{m s}^{-1}$
c $\frac{97}{80}$ d $3.50\,\text{s}$ (3 s.f.)
e e.g. weight of the ball, air resistance
7 a $76.6\,\text{m}$ (3 s.f.) b $110\,\text{m}$ (3 s.f.)
8 a $k = -4$ b $4\,\text{m}$ c 0.05
9 a $0.3\sqrt{3}\,\text{m}$ b $t = 3$ c $0.329\,\text{m s}^{-2}$ (3 s.f.)
10 a $(\ln 2 - 2)\,\text{m s}^{-2}$ in the direction of x increasing.
b $\frac{8}{e}\,\text{m}$
11 a V_P is $(6t\mathbf{i} + 2\mathbf{j})\,\text{m s}^{-1}$ and V_Q is $(\mathbf{i} + 3t\mathbf{j})\,\text{m s}^{-1}$
b $12.2\,\text{m s}^{-1}$ (3 s.f.)
c $t = \frac{1}{3}$
d Equate $\mathbf{i}$-components and solve to get $t = 1$.
Equate $\mathbf{j}$-components and solve to get $t = \frac{1}{3}$ or 1.
So $t = 1$ and $r = (7\mathbf{i} + \frac{3}{2}\mathbf{j})\,\text{m}$
12 a Differentiate: $\mathbf{v} = 6t\mathbf{i} - 8t\mathbf{j}$, $\mathbf{a} = 6\mathbf{i} - 8\mathbf{j}$ so constant
b $10\,\text{m s}^{-2}$
143.1° (nearest 0.1°)
13 a $-6\mathbf{i}\,\text{m s}^{-1}$
b Differentiate: $\mathbf{v} = -6\sin 3t\mathbf{i} - 6\cos 3t\mathbf{j}$,
$\mathbf{a} = -18\cos 3t\mathbf{i} + 18\sin 3t\mathbf{j}$
$|\mathbf{a}| = \sqrt{18^2(\cos^2 t + \sin^2 t)} = 18\,\text{m s}^{-2}$ so constant
14 a Differentiate: $\mathbf{a} = 4c\mathbf{i} + (14 - 2c)t\mathbf{j}$,
$\mathbf{F} = m\mathbf{a} = \frac{1}{2}(4c\mathbf{i} + (14 - 2c)t\mathbf{j})$
b $4, \frac{234}{29} \approx 8.07$
15 a $((2t^4 - 3t^2 - 4)\mathbf{i} + (4t^2 - 3t - 7)\mathbf{j})\,\text{m s}^{-1}$
b $t = \frac{7}{4}$
16 $10\sqrt{41}\,\text{m s}^{-1}$
17 a $\mathbf{v} = (2t^2 + 3)\mathbf{i} + (3t + 13)\mathbf{j}$ b $3.11\,\text{s}$ (3 s.f.)

Challenge

1 a 20
b $v = 0$ when $t = 1.64\,\text{s}$ (or $-2.44\,\text{s}$) so only changes direction once
c $-2\sqrt{20}(\sqrt{20} + 1)^{\frac{1}{2}}$
2 a Differentiate: $\dot{\mathbf{r}} = (6\omega\cos\omega t)\mathbf{i} - (4\omega\sin\omega t)\mathbf{j}$
$|\dot{\mathbf{r}}|^2 = (36\omega^2\cos^2\omega t)\mathbf{i} + (16\omega^2\sin^2\omega t)\mathbf{j}$
use $\sin^2\omega t + \cos^2\omega t = 1$ and $2\cos^2\omega t = \cos 2\omega t + 1$
b $v = \sqrt{26\omega^2 + 10\omega^2\cos 2\omega t}$ max when $\cos 2\omega t = 1$ and min when $\cos 2\omega t = -1$
c 109.8° (1 d.p.)

Review exercise 2

1 a 573 N (3 s.f.) b 427 N (3 s.f.)
c All his weight acts through a single point.
2 a Taking moments about C: $2.5mgl = 4R_B l$
So $R_B = \frac{5}{8}mg$ (as required)
b Taking moments about B: $1.5mgl = 4R_C l$
So $R_C = \frac{3}{8}mg$ (as required)
c i The weight of the plank acts through its midpoint.
ii By modelling a plank as a rod you can ignore its width.
3 a 125 N b 750 N
4 0.375 m
5 19.8 N (3 s.f.)
6 R(↖): $F\cos 30 - 2g\sin 45 = 4$
$\frac{\sqrt{3}}{2}F = 4 + \frac{2g\sqrt{2}}{2} \Rightarrow F = \frac{2}{\sqrt{3}}(4 + \sqrt{2}g)\,\text{N}$ (as required)
7 a 144 767 N (to the nearest whole number)
b 0.10 (2 s.f.)
c 0.74 seconds (2 s.f.)
d Will not remain at rest. $F_{MAX} = 15\,000\,\text{N}$ and component of weight down slope = 26 000 N which is greater.
8 a 0.404 s (3 s.f.) b 0.808 m (3 s.f.)
9 a $19.8\,\text{m s}^{-1}$ (3 s.f.)
b The ball as a projectile has negligible size and is subject to negligible air resistance.
Free fall acceleration remains constant during flight of ball.
10 a 2.7 s (2 s.f.) b 790 m (2 s.f.)
11 a $u = 3.6$ (2 s.f.) b $k = 86$ (2 s.f.)
c 43° (2 s.f.)
12 a $(36\mathbf{i} + 27.9\mathbf{j})\,\text{m}$ b $13\,\text{m s}^{-1}$ (2 s.f.)
13 a $u_y = u\sin\alpha$, $s_y = 0$, $a = -g$
Using $s = ut + \frac{1}{2}at^2$: $0 = ut\sin\alpha - \frac{1}{2}gt^2$
$\frac{1}{2}gt = u\sin\alpha$ so $t = \frac{2u\sin\alpha}{g}$
b $u_x = u\cos\alpha$, $s_x = R$, $a = 0$, $t = \frac{2u\sin\alpha}{g}$
Using $s = ut + \frac{1}{2}at^2$:
$R = ut\cos\alpha = \frac{2u^2\sin\alpha\cos\alpha}{g} = \frac{u^2\sin 2\alpha}{g}$
c $R = \frac{u^2\sin 2\alpha}{g}$ so $\frac{dR}{d\alpha} = \frac{2u^2\cos 2\alpha}{g}$
At a maximum $\frac{dR}{d\alpha} = 0$ so $\cos 2\alpha = 0$ and $\alpha = 45°$
d 12° and 78° (nearest degree)
14 a R(↑): $T\cos 30° = g + T\cos 60°$
$T = \frac{2g}{\sqrt{3} - 1}$ (as required)
b 36.6 N (3 s.f.)
c There are no frictional forces acting on the bead.
15 a Resolving perpendicular to the slope:
$R = 500g\cos\alpha$ and $\cos\alpha = \frac{24}{25}$
So $R = 480g$ (as required)
b $68g\,\text{N}$

16 2.46 m (3 s.f.)

17 $\mu = \frac{1}{2\tan\alpha}$

18 44.4 N (3 s.f.)

19 12.9 (3 s.f.)

20 a $\mathbf{r}_P = (400 + 300t)\mathbf{i} + (200 + 250t)\mathbf{j}$ and $\mathbf{r}_Q = (500 + 600t)\mathbf{i} - (100 + 200t)\mathbf{j}$

b $100\mathbf{i} - 300\mathbf{j}$

c 1390 km (3 s.f.)

21 a $v = 3 + \frac{4k}{(2t-1)^2}$

$t = 0 \Rightarrow v = 3 + 4k = 10 \Rightarrow k = \frac{7}{4}$

b $\frac{29}{6}$m

22 a $(t^2 + 2)\mathbf{i} + t\mathbf{j}$ b $27.5\,\text{m s}^{-1}$ (3 s.f.)

c $4.12\,\text{m s}^{-2}$ at an angle of 14° to the positive unit $\mathbf{i}$ vector

23 a $(24\mathbf{i} + 12\mathbf{j})\,\text{m s}^{-1}$ b $\frac{d^2r}{dt^2} = 8\mathbf{i} + 4\mathbf{j}$

24 24 m

25 a $\mathbf{v} = \left(t^2 - \frac{3}{4}t^4 + 3\right)\mathbf{i} - (4t^2 + 4t - 1)\mathbf{j}$

b 0.207 s (3 s.f.)

26 a $\mathbf{v} = (-2t^2 + 8)\mathbf{i} - 2t\mathbf{j}$ b $t = 2$ s

Challenge

1 Taking moments about B, in limiting equilibrium:
$0.4mgk = 0.4k \times 100 + 1.2Fk$
$12F = 4mg - 400$, so $F = \frac{1}{3}(mg - 100)$N
So in order for m to be lifted $F > \frac{1}{3}(mg - 100)$N

2 Max distance: 12.3 m, $t = 323.13$

3 R(→): $d\cos\theta = u\sin\theta \times t \Rightarrow t = \frac{d\cos\theta}{u\sin\theta}$
R(↑): $-d\sin\theta = u\cos\theta \times t - \frac{g}{2}t^2$
Substitute for t.
Rearrange to show that $d = \frac{2u^2}{g}\tan\theta\sec\theta$

Exam-style practice

1 a 0.4133

b n is large, p is close to 0.5

c 0.114 (3 s.f.)

d The probability is not significant at the 5% level so there is no evidence that the claim is incorrect.

2 a Venn diagram with three circles labelled A, B and C with circles A and B not intersecting. Values are 0.2 (just A), 0.2 (A and B), 0.207 (just B), 0.143 (B and C), 0.117 (just C) and 0.133 (outside).

b $0.4 \times 0.55 \neq 0.2$ c $\frac{4}{9}$, 0.444 to 3 s.f.

d $\frac{13}{40}$, 0.325

3 a Continuous – it is a measured variable

b 19.4 °C and 2.81 °C (both to 3 s.f.)

c Continuous data, unequal class widths

d 7.9 °C

e $H_0: \rho = 0$; $H_1: \rho > 0$. Critical value = 0.6215 so no evidence that the population correlation coefficient is greater than zero.

4 a Yes – the value of the PMCC is close to 1.

b $k = 1.19$, $n = 2.13$ (both to 3 s.f.)

c Value of t is a long way from the range of the experimental data so extrapolation – no it would not be sensible.

5 a 80.1% b 0.9402

c $H_0: \mu = 4.5$, $H_1: \mu \neq 4.5$ Probability = 0.059 so not significant – there is no evidence to suggest that the mean weight is not 4.5 kg.

6 0.731 (3 s.f.)

7 Position vector at $t = 3$ is $-43\mathbf{i} - 4\mathbf{j}$, distance = 43.2 m (3 s.f.)

8 a 10.2 s (3 s.f.) b 128 m (3 s.f.)

c $89.0\,\text{m s}^{-1}$ (3 s.f.)

9 a $30\mathbf{i} - 40\mathbf{j}$ b 50 m c 12 s d 102°

10 a 18.6 N

b For P, R(←): $T - \mu 3g = ma$ so $18.6 - 29.4\mu = 1.5$ and $m = 0.582$ (as required)

c 0.175 s

d Tension and acceleration are equal both sides of the pulley.

11 a Taking moments about B
$\frac{1}{2}mgl = Tx$ where $x = l\sin\alpha$ and $\sin\alpha = \frac{5}{13}$
$\frac{1}{2}mgl = \frac{5}{13}lT$ so $T = \frac{13}{10}mg$ (as required)

b $\frac{5}{12}$ or 0.42 (2 s.f.)

Online Full worked solutions are available in SolutionBank.

Index